Digital Innovations in Architecture, Engineering and Construction

Series Editors

Diogo Ribeiro⬤, Department of Civil Engineering, Polytechnic Institute of Porto, Porto, Portugal

M. Z. Naser, Glenn Department of Civil Engineering, Clemson University, Clemson, SC, USA

Rudi Stouffs, Department of Architecture, National University of Singapore, Singapore, Singapore

Marzia Bolpagni, Northumbria University, Newcastle-upon-Tyne, UK

The Architecture, Engineering and Construction (AEC) industry is experiencing an unprecedented transformation from conventional labor-intensive activities to automation using innovative digital technologies and processes. This new paradigm also requires systemic changes focused on social, economic and sustainability aspects. Within the scope of Industry 4.0, digital technologies are a key factor in interconnecting information between the physical built environment and the digital virtual ecosystem. The most advanced virtual ecosystems allow to simulate the built to enable a real-time data-driven decision-making. This Book Series promotes and expedites the dissemination of recent research, advances, and applications in the field of digital innovations in the AEC industry. Topics of interest include but are not limited to:

- Industrialization: digital fabrication, modularization, cobotics, lean.
- Material innovations: bio-inspired, nano and recycled materials.
- Reality capture: computer vision, photogrammetry, laser scanning, drones.
- Extended reality: augmented, virtual and mixed reality.
- Sustainability and circular building economy.
- Interoperability: building/city information modeling.
- Interactive and adaptive architecture.
- Computational design: data-driven, generative and performance-based design.
- Simulation and analysis: digital twins, virtual cities.
- Data analytics: artificial intelligence, machine/deep learning.
- Health and safety: mobile and wearable devices, QR codes, RFID.
- Big data: GIS, IoT, sensors, cloud computing.
- Smart transactions, cybersecurity, gamification, blockchain.
- Quality and project management, business models, legal prospective.
- Risk and disaster management.

Marcello Colledani · Stefano Turri
Editors

Systemic Circular Economy Solutions for Fiber Reinforced Composites

 Springer

Editors
Marcello Colledani
Department of Mechanical Engineering
Politecnico di Milano
Milan, Italy

Stefano Turri
Department of Chemistry, Materials
and Chemical Engineering
Politecnico di Milano
Milan, Italy

ISSN 2731-7269   ISSN 2731-7277   (electronic)
Digital Innovations in Architecture, Engineering and Construction
ISBN 978-3-031-22354-9   ISBN 978-3-031-22352-5   (eBook)
https://doi.org/10.1007/978-3-031-22352-5

This Springer imprint is published by the registered company Springer Nature Switzerland AG
The registered company address is: Gewerbestrasse 11, 6330 Cham, Switzerland

Contents

Introduction, Context, and Motivations of a Circular Economy for Composite Materials

Marcello Colledani⦿**, Stefano Turri**⦿**, Marco Diani**⦿**, and Volker Mathes**

Abstract Circular Economy is an emerging production-consumption paradigm showing the potential to recover and re-use functions and materials from post-use, end-of-life, products. Even if several barriers still exist at different levels, from legislation to customer acceptance, the transition to this sustainable industrial model has been demonstrated to potentially bring economic, environmental, and social benefits, at large scale. Composite materials, which usage is constantly increasing, are composed by a fiber reinforcement in a resin matrix. Among them, the most widely adopted are Glass Fiber Reinforced Plastics (GFRP) and Carbon Fiber Reinforced Plastics (CFRP). Their applications range from wind blades to automotive, construction, sporting equipment and furniture. The post-use treatment of composite-made products is still an open challenge. Today, they are either sent to landfill, where not banned, or incinerated. The application of Circular Economy principles may lead to the creation of new circular value-chains aiming at re-using functions and materials from post-use composite-made products in high value-added applications, thus increasing the sustainability of the composite industry as a whole.

Keywords Composites · Circular economy · GFRP · CFRP · Waste · Legislation · Circular value-chains

M. Colledani (✉) · M. Diani
Department of Mechanical Engineering, Politecnico Di Milano, Via La Masa 1, 20156 Milan, Italy
e-mail: marcello.colledani@polimi.it

S. Turri
Department of Chemistry, Materials and Chemical Engineering, Politecnico Di Milano, "Giulio Natta", Piazza Leonardo da Vinci 32, 20133 Milan, Italy

V. Mathes
AVK—Federation of Reinforced Plastics, Am Hauptbahnhof 10, 60329 Frankfurt Am Main, Germany

M. Colledani and S. Turri (eds.), *Systemic Circular Economy Solutions for Fiber Reinforced Composites*, Digital Innovations in Architecture, Engineering and Construction, https://doi.org/10.1007/978-3-031-22352-5_1

1 Introduction

Glass fiber (GF) and carbon fiber (CF) reinforced polymer composites have revolutionized important manufacturing sectors, such as transport (automotive, aircraft, boats) and construction (building and infrastructures, plants, wind turbines), due to their lighter weight and intrinsically better corrosion resistance with respect to metals. The composite market by volume is dominated by glass fiber reinforced plastics (GFRP), with around one third of the whole production volume manufactured for the transport sector (cars, commercial vehicles, boats and to a lesser extent, aircraft), and another third for the construction industry (buildings, infrastructures, and wind turbines) [1]. Carbon fiber reinforced plastics (CFRP) are mainly adopted within the aerospace industry, with 32% of the total CFRP global demand, followed by the automotive industry which currently makes up around 21.8% of the total demand [1]. In the light of the above, the relevance of composite materials in many strategic industrial sectors appears significant. Therefore, the development of sustainable Circular Economy solutions to the post-use management of composite materials represents one of the key challenges of the modern manufacturing industry.

This chapter introduces the issues and the possibilities of the implementation of Circular Economy principles in this sector, starting with a general discussion about circular economy, introducing composite material features, the current market situation and the current composite waste treatments, with reference to existing legislation.

2 Circular Economy

In 1989 Pearce and Turner, introduced the concept of Circular Economy (CE) [2]. In their book they underlined the limitations of the traditional "take, make, dispose" approach, in view of the exploitation raw materials and the generation of increasingly relevant quantities of waste. They proposed the idea of a new closed-loop concept in which fractions from End-of-Life products can be reinserted as raw materials for new productions. The final goals are waste reduction and improvement of environmentally friendly processes, creating new adaptable and resilient systems.

This idea has been further enforced in the last years by the Ellen MacArthur Foundation [3], outlining the opportunities of Circular Economy: "*A circular economy is restorative and regenerative by design, and aims to keep products, components, and materials at their highest utility and value at all times. The concept distinguishes between technical and biological cycles. As envisioned by the originators, a circular economy is a continuous positive development cycle that preserves and enhances natural capital, optimizes resource yields, and minimizes system risks by managing finite stocks and renewable flows. It works effectively at every scale*" [3]. Value-chains should ensure not only environmental but also economic sustainability. The three fundamental pillars of Circular Economy have been defined as

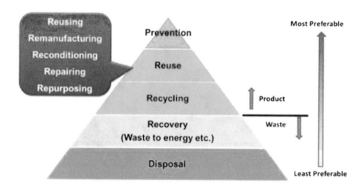

Fig. 1 Waste reduction strategies [4]

(i) preserve and enhance natural capital; (ii) optimize resource yields; (iii) foster systems effectiveness.

In practice, different Circular Economy options have been introduced to regain and re-use material fractions and component functions from post-use products, i.e. reaching the end of the first use cycles. The European Waste Framework Directive 2008/98/EC, introduced these options as the five-step hierarchy shown in Fig. 1 [4]. Solutions in the upper parts of the pyramid, which consider End-of-Life products as a resource, are preferable than those in the lower part. Prevention is considered as the most preferrable option, while disposal should be avoided whenever possible. The study identifies the role of legislation and policies in promoting and prioritizing the adoption of the most attractive Circular Economy options.

3 Current Circular Economy Solutions for Composite Materials

3.1 The Composite Material Market

Composite materials are a wide family of heterogeneous materials composed by two or more phases, with different physical properties. The resulting properties of composite materials are better than those of the constituting phases. The reinforcements could be in the form of short fibers, long fibers, filler, particulate and flakes, leading to different behaviors depending on the shape (see Fig. 2 [5]). Some examples of composite materials could be seen in nature. Wood is composed by cellulose fibers in a lignin matrix. The result is a flexible material with high elastic module. In the same way, bones are composed by collagen fibers in an apatite phase, leading to a material with low specific gravity and high mechanical properties.

Focusing on artificial composites, also called fiber-reinforced plastics, it is a wide group of materials mainly composed by fibers, resin matrix, additives and fillers.

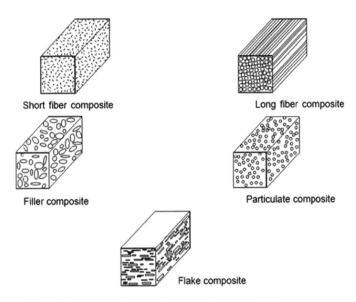

Fig. 2 Different fibers/matrix configuration in composite materials [5]

The two-phase composition results in high mechanical properties together with low weight, high corrosion resistance and, as a consequence, low maintenance needs. Different types of fibers are used in composite materials. More than 99% of the products currently on the market are composed by glass fibers, carbon fibers, natural fibers, aramid fibers and basalt fibers. Among them, glass fibers show the largest adoption rate, with 95% share of the worldwide production volume of composites [1]. This is due to the good mechanical properties that they bring, combined with their reasonable market price. For these reasons Glass Fibers Reinforced Plastics use is widespread in several sectors as wind blades, sports equipment, construction, marine and automotive (and transportation in general) and electronics (for E-glass fibers), as reported in Fig. 3 [1].

The second largest share of composite materials includes Carbon Fiber Reinforced Plastics. Carbon fibers present higher mechanical properties with respect to glass fibers, with higher cost (from 4 to 40 times the price of glass fibers depending on the specific application [6]). Due to their remarkable properties, CFRPs find application in high added-value sectors such as wind energy, sports equipment, sports car, construction, aircraft and aerospace [1].

In addition to the constituent material, fibers could be classified by their length. In particular, they could be short fibers, long fibers and endless fibers. Short fibers have a maximum length of 2 mm, long fibers are between 2 and 50 mm, while endless fibers are longer than 50 mm. The GFRPs market share shows a predominance of short fibers with respect to long and endless ones. Indeed, the European production volume of short fibers in 2021 has been of 1.51 million tons, while that of long and endless fibers has been of 1.1 million tons [1]. Even more, fibers are not only used

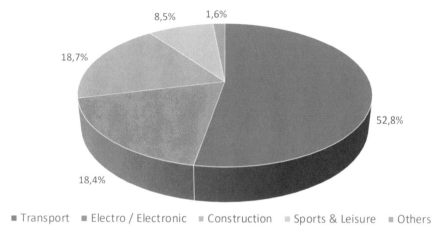

Fig. 3 GFRPs production in Europe by application industry [1]

as single chopped fiber or strand but also as semi-finished product as mats, textiles fabrics, knitted fabrics, non-woven and fiber pre-forms depending on the specific application.

Regarding the matrix, a spread variety of resins could be used. In the first analysis, they can be divided in thermosets, that could not be melted and reshaped once hardened, and thermoplastics, that can be reshaped several times through temperature. In particular, 86% of composite products made with long and endless fibers have a thermoset matrix [1]. Among thermosets plastics, the most used are unsaturated polyester, vinylester, epoxy and polyurethane, while regarding thermoplastics the most used in composites are polypropylene, polyamide, polystyrene, polyphenylene sulfide and polyetherether-ketone (also known as PEEK).

Finally, several fillers could be used to give specific characteristics to the final composite products. They are inert materials of mineral nature as calcium carbonate, frequently used as coating, aluminum trioxide and silica, as in piping and in polymer concrete.

In 2021 the production volume of CFRPs was approximately 52.000 tons, natural fibers reinforced plastics had a volume of more than 92.000 tons, while the volume of GFRPs was 2.910.000 tons. Aramid fibers and basalt fibers, with other types of fibers, have very low production volumes due to their very specialized applications. While the worldwide composite growth rate has been of about 8% in the last years, European composites production volume increased by 18.3%, returning to the pre-pandemic level [1]. In Fig. 4, the situation of composite production shares and volumes in Europe is reported.

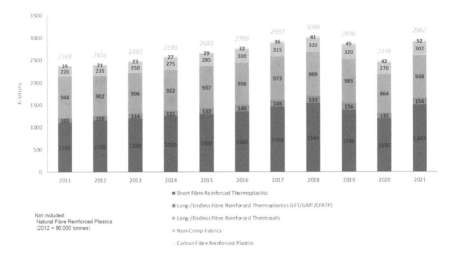

Fig. 4 Composites production volume in Europe in ktons excluding natural fibers [1]

3.2 Composite Waste Current Situation

Composite waste streams can be mainly divided into three groups. The first is composed by small components as consumables, sport and leisure equipment and design products, which are not systematically separately collected on a national or European level. Sorting of composite materials from mixed municipal waste streams is not performed due to technical and economic issues. This results in a high loss of composite materials and, as a consequence, of economic value.

The second group is composed by large infrastructures as wind blades, aircrafts, boats, construction structures and vehicles. The current situation and related opportunities of these groups will be better detailed in Chap. 2. As an example, the amount of End-of-Life wind blades, composed by GFRP, CFRP, steel and aluminum (in addition to other minor materials) is predicted to reach 30.000 tons per year in Europe by 2026, while it will increase up to 50.000 tons by 2030, as shown in Fig. 5 [7]. EoL leisure boats were also identified as significant GFRP waste stream with an estimated volume of up to 10.000 tons per year [8].

The third group of composite waste streams is formed by composite production waste. The collection and sorting of these materials is less problematic with respect to End-of-Life products. Moreover, in this case, the materials composition and the location are known. However, challenges are represented by the classification of these streams as "waste" and the related management challenges, as discussed in Sect. 4.

Fig. 5 Expected amount of End-of-Life wind blades in tons per year in different European Countries [7]

3.3 Recycling of Composite Materials

Recycling is the most widely investigated solution for the sustainable treatment of post-use composite products. Existing technologies and recycling methods are of three main types, namely mechanical, thermal or chemical [9]. In this section, a brief overview of the currently investigated solutions is reported.

Mechanical recycling: Mechanical recycling typically consists in one or more size-reduction stages (or shredding stages, also preceded by a coarse cutting step, if needed) to obtain particles, that remain composed of mixed fibers and resin, featuring desired dimensional distributions. The objective is to directly reuse the obtained fraction into new composite material formulations. To facilitate downstream reprocessing stages, the material flow in output from the size-reduction process can be sieved and divided into two or more fractions, with controlled characteristics [9]. The composite shredding powder residue instead can be used both as filler and as reinforcement. While the use of this powder as filler is possible in terms of physical and chemical properties, the costs are generally higher if compared to the costs of alternative virgin fractions, such as calcium carbonate or silica. In addition, in most applications, the weight fraction of composite powders that can be incorporated as a filler is limited, typically less than 10% [10]. This is mainly due to an increase in the viscosity of the compound, resulting in extended processing problems. For these reasons, powder fractions are usually incinerated to obtain energy due to their high content in resin. The possibility to use these materials for additive manufacturing has also been discussed and preliminarily demonstrated [11].

Concerning the reuse of granular fractions composed by fibers embedded within the resin, different studies have been performed both for thermosets and thermoplastics composite materials. As an example, up to 50% in weight of carbon fiber particles obtained through shredding have been successfully incorporated into new products with virgin PEEK resin through injection molding [12]. In addition, thermoplastic polymers could be subjected directly to reforming processes. Fiber-rich fractions of mechanically recycled thermoset composites are usually more complex to be reused as reinforcement, since the mechanical properties of the resulting materials are reduced due to poor bonding between recycled particles and the new resin [10].

Several works dealt with these issues finding innovative and promising solutions to use increasingly large fractions of recyclates, in particular focusing on Sheet Molding Compound (SMC) and Bulk Molding Compound (BMC) production techniques. As an example, it has been proven that longer mixing time of the paste with the recycled particles results in improved interface between recycled material and new resin. This leads to increased mechanical properties that are similar to those of the material obtained with virgin fibers [13].

In addition to the classical mechanical recycling processes, the possibility and potentiality to use an innovative process based on high-voltage fragmentation has been shown. The product is placed in a dielectric ambient as de-ionized water and a high-intensity and fast growing voltage (80–200 kV, <500 ns) is induced. As a consequence, a high electric field is generated, resulting in a dielectric breakdown and in the generation of a spark plasma channel with high pressure (10^{10} Pa) and temperature (10^4 K). After up to 1.000 discharge cycles the product is disintegrated. Working directly on the interfaces between different materials, this technology is able to clean the fibers from resin. Some works showed that applying this method to GFRPs it is possible to obtain lower amount of residual resins in comparison with comminution and a wider fibers length distribution [14]. The main disadvantages of this process remain the high cost and the low throughput [15].

Thermal recycling: Thermal processes used for composite materials recycling are pyrolysis, fluidized-bed pyrolysis and microwaves-assisted pyrolysis. These techniques work between 450 and 700 °C (depending on the resin), eliminating the matrix in controlled atmosphere. They are optimized to recover fibers and fillers but the resin is volatilized into lower-weight molecules, producing waste gases as carbon dioxide, hydrogen and methane and possible combustion residues on the fibers [9]. Exceeding the specific resin temperature, the process will concern combustion with energy recovery. This is typical in cement kilns in which the composite waste is converted into energy and fillers and fibers for the concrete.

Pyrolysis is a well-known process in which the temperature is higher than 450 °C in absence or in presence of oxygen, in a controlled atmosphere [16], to degrade the resin matrix. The output will be composed by solid products (in particular fibers, fillers and char), oil and gases. The char is stuck on the fibers, that need a post-treatment in a furnace at 450 °C, leading to more degraded fibers [17]. This effect is visible in particular for glass fibers. As an example, at the minimum temperature of 450 °C, mechanical properties of glass fibers are reduced by at least 50% of the original ones. In addition, as pyrolysis is an energy intensive and expensive process, the cost to recover glass fibers is higher than the market price of the virgin ones. For these reasons, pyrolysis process is mainly used to treat CFRPs. Furthermore, they are less sensitive to high temperatures, while char is still present on the surface of the fibers. To eliminate it in standard conditions, a post-treatment process with temperature up to 1300 °C is needed, resulting in significantly loss of strength of the fibers [18]. As a consequence, a compromise between resulting mechanical properties and the amount of remaining resin residues is needed. As an example, high tenacity carbon fibers after a two-steps process, the first one at 550 °C in nitrogen the second one at the same temperature in oxidant atmosphere, retained more than 95% of their tensile strength without resin residue on the surface [18]. A pyrolysis temperature between 500 and 550 °C seems to be the upper limit of the process in order to maintain acceptable strength of carbon fibers.

Fluidized-bed process pyrolysis uses a bed (e.g. composed by silica sand) made fluid by hot air, obtaining in this way oxidant conditions. In this way, it is possible to rapidly heat the materials and to release the fibers from the resin by friction. The organic fraction of the resin is then degraded through a second step at 1.000 °C for energy recovery [10]. Glass fibers treated with this process showed a reduction in tensile strength up to 50%, while carbon fibers of about 25% [19], probably due to damages from the fluidized sand. The added value of this process is the possibility to treat mixed and contaminated materials as products with painted surfaces or foam cores.

Microwave-assisted pyrolysis heats composite materials in an inert atmosphere, resulting in degradation of the matrix into gases and oil as for traditional pyrolysis. The main advantage is the energy savings due to the fast thermal transfer as the composites are heated directly in their core through microwaves [20]. This process has been used to treat both GFRPs and CFRPs. On the other hand, this process needs small particles in input. This results in an output composed by a wide length distribution of the fibers and in a reduction of resulting mechanical properties of the final product, with a limited percentage of recycled material that can be used (less than 25%) [21].

Chemical recycling: Chemical recycling of composite materials is performed through solvolysis, a chemical treatment to degrade a resin using a solvent. This technique has been widely used in the last 20 years, starting from the degradation of polyurethane into its monomers, carboxylic acids and glycols, and a styrene-fumaric acid copolymer [22]. Since this first positive attempt, several processes based on different solvents and techniques have been tried to recycle both thermoplastics and thermosets composite materials [9].

Solvolysis factors include the solvent, the working temperature, the pressure and the presence of a catalyst, depending on the resin to degrade. The reactive solvent diffuses into the composite and breaks specific bonds. In this way, it is possible to recover monomers from the resin, avoiding the formation of char residues.

To start the process, the activation energy of the polymer has to be reached. This could be achieved increasing the temperature or using a catalyst. The main advantage is that lower temperatures are required with respect to thermal treatment. On the other hand, reactors could be expensive, in particular for solvolysis in supercritical conditions (that is the process with best results), as they have to withstand high temperatures and pressures and corrosion due to modified properties of the solvents

[23]. As an example, polyester resins are easier to degrade through solvolysis than epoxy resins, leading to lower required temperatures.

The most used solvent is water, both alone and with co-solvent (e.g. alcoholic, phenolic, amine). Typical catalysts are alkaline catalysts as sodium hydroxide and potassium hydroxide or, less frequently, acidic catalysts (used in case of more resistant resins or lower temperatures). Other solvents are alcohols as methanol, ethanol, propanol and acetone or even glycols [9].

3.4 Existing Barriers

Despite the mature development phase of some of the reported technologies, key barriers exist bounding the systematic transition of the composite industry to a Circular Economy approach, including:

- *Governance*: significant fragmentation of stakeholders in different sectors, leading to poor-communication and inter-sectorial alignment of policies, thereby creating barriers to innovation; low priority of composite waste in the political agenda, thereby limiting stakeholder engagement; incomplete understanding of barriers to innovation, thereby limiting leadership and formation of an effective integrated innovation strategy;
- *End-users perception*: industry perception of composite waste as a cost to be minimized rather than a resource to be valorized;
- *Commercial uptake*: negative consumer and industry perception of products derived from recyclates and safety concerns;
- *Finance and regulation*: poor mobilization and alignment of finance across the sector; poor alignment of pricing methodologies and legislation, lack of suitable business models that support EoL composite valorization;
- *Technology development*: limited synergistic use of available inspection, repair and reprocessing technologies due to the different specialization areas.

4 Current Composite Waste Management Practices and Legislation

The following paragraphs provide a brief overview of the EU legislations concerning composite waste treatment and waste management.

End-of-Life GFRP and CFRP products are usually not systematically collected separately on a national or European level. Furthermore, sorting of composites from commercial or municipal mixed waste streams is not practiced due to technical challenges and economic considerations. Compared to other materials, the proportion of composites in residual waste as well as the commercial value of mixed GFRP and

CFRP waste are both relatively low, making manual or automated sorting unprofitable in practice.

As a result, once composites have entered a mixed or residual waste stream, they are not sent to recycling processes, regardless of the further treatment of the entire waste stream (as landfill, incineration or co-processing). This applies, in particular, to small scale consumption products, as sports equipment, consumer electronics or design objects, which are generally disposed by consumers via municipal residual waste.

Compared to small consumption products, EoL composite waste streams derived from large-scale consumption products, as leisure boats and vehicles, or large installations, as wind energy systems and construction waste, are significantly more attractive for centralized recycling approaches.

Despite the large amount of potential EoL-waste from these sectors, an estimation of the annually generated amount of waste is quite difficult and data about presence and treatment in different European countries are rare. Additionally, vehicles, as airplanes, cars, trucks and boats, as well as wind energy systems are assumed to have a relatively long lifecycle. Although the actual production capacities of these sectors are known quite well, the exact time of decommissioning is hard to predict. Furthermore, some products as vehicles underlie a massive exportation after their standard life-time, losing the possibility to be recycled and reused.

Contrary to composites EoL products, production waste appears as more reliable source of high-quality materials suitable for reuse and recycling processes. Their collection produces non-mixed waste fractions with well-known material properties and compositions. In addition, the amount of waste is constant or predictable and in a limited number of locations. Finally, databases elaborated for the estimation of current and future waste streams are easily accessible and precise.

Today, approximately 30.000 tons of post-consumer GFRP and CFRP-waste are produced annually in Europe. Additionally, 40–50.000 tons of commercial production waste are generated [24]. Due to an increasing use of composites in various applications, a steady increase of these amounts is expected.

Compared to other waste streams, the total amount of composite waste produced within Europe is relatively small. Therefore, this could lead to the assumption that an international collection and a centralized recycling could be economically beneficial. However, for the transportation of waste across borders—also within the European Union—some legal aspects must be considered. According to the European Waste Framework directive, establishments and undertakings which carry out waste treatment must obtain a permission from their corresponding competent authority. Within the EU, the import, export and transit of waste to, from or through EU Countries are subjected to approval procedures (called notifications), which are regulated in detail in the European Waste Shipment Regulation [25]. Transboundary movements of waste require not only a notification by the Country of waste origin, but also by the Country of destination and all others the transport will travelling through.

An exemption from notification can apply for non-mixed and non-hazardous waste. A final decision about an obligation for notification concerning a specific waste stream is up to each single national authority.

A single waste management company which intends to perform centralized GFRP and CFRP waste treatment using waste from different European countries, would have to arrange transport notifications with each single waste generator, independently from the type of the intended waste treatment (fiber/polymer recycling or thermal recovery in cement kilns). Since every notification procedure is time consuming and accompanied by considerable financial efforts, such a procedure for international waste collection does not seem practical. A more realistic approach could be a cooperation of waste management companies, which collect GFRP and CFRP waste on a national level and coordinate notifications for transboundary shipment to a centralized large-scale recycling plant.

In contrast to externally collected composite waste, an in-house reuse of residual materials from production processes is legally possible. Materials obtained from treatment of scraps (as recycled glass or carbon fibers), could be considered as product and not as waste, and would not be affected by the scope of waste management laws. However, also in this case a final decision about the classification of recovered materials as product, by-product or waste rests with the corresponding competent authority.

The already presented European Waste Framework Directive 2008/98/EC [4] sets the basic concepts and definitions related to waste management. It explains, through the so called "end-of-waste criteria", in which cases waste ceases to be waste and becomes a secondary raw material, and how to distinguish between waste and by-products. Although the Waste Framework Directive gives a lot of hints, explanations and definitions, a high degree of uncertainty is still present and different approaches concerning the handling of GFRP and CFRP waste in the EU Countries can be highlighted, mainly due to the absence of a clear classification of GFRP and CFRP.

Some examples from the different Countries are exemplified described in the following list.

- In the Netherlands LAP3 is the formal document of the Dutch Authorities regarding management of waste in general [26]. The document describes the waste stream plans for 85 sectors in Netherlands, including Plastics and Rubbers (considered in the Sector Plan 11). In this plan, for the first time, composites are mentioned, even if through a remark that specifies the difficulties to recycle fiber reinforced plastics. LAP 3 indicates for thermoplastics (including composites) that the minimum standard for waste management is reuse (or recycling). For thermosets plastics (including composites) the minimum standard for waste management is thermal recycling. Any further recycling technology "above" thermal recycling is acceptable, including the use in cement production. As in Netherlands most of the waste incinerators are qualified as thermal recycling installations, this means that landfill is not accepted anymore.
- In Belgium there is no specific rule for GFRP or CFRP. There are some companies collecting composite waste from the industry after waste sorting but there are no specific disposal rules.

- An interim solution concerning landfill can be found in Finland. There is a controversial situation in Finland regarding GFRP and CFRP. The landfill ban is in force as no more than 10% of organic content is allowed to be disposed. In the meanwhile, few landfills maintained a special exemption for GFRP to support the transition to a complete landfill ban. On the other hand, as there are not many real recycling processes for GFRP and CFRP, waste is processed by incineration (with energy recovery).
- Considering Europe (and not only EU), the situation is even more complex. As an example, in the UK landfill is still an opportunity. Landfill tax now stands at £98.60/ton (2022 rate), making the cost of landfill, including gate fees and transport, typically £140 to £150 per ton [27]. On the other hand, while sharp increases in landfill taxes are not expected, Germany and several other European countries have already largely banned landfill.

This small list shows the heterogeneous European situation that can affect the implementation of robust recycling and reuse of composites and the related circular value-chains. A common legislation at EU level is needed to coordinate actions and to promote best practices all around Europe.

5 Conclusions

Circular economy is showing its full potentiality only in the last years. The possibility to reduce environmental impacts while creating new circular value-chains acts as a game changer, not only at local but at global level. On the other hand, composite materials are more and more used in a widespread number of applications thanks to their peculiar characteristics of lightweight, high mechanical properties and corrosion resistance. Despite this, a systematic reuse of materials and components based on recycling and remanufacturing has not yet been implemented, even if different technologies are available. This is due for several reasons, from customer perception to the absence of a unique legislation at European level for EoL composites treatment. To solve these problems, and to transform them from an issue to an opportunity, a change at systemic level is needed, based on Circular Economy principles.

References

1. Witten, E., Mathes V.: The European market for fibre reinforced plastics/composites in 2021, market developments, trends, challenges and outlook. AVK Annual Report (2021)
2. Pearce, D.W., Turner R.K.: Economics of natural resourcing and the environment. Johns Hopkins University Press (1989)
3. MacArthur, E., et al.: Towards the circular economy. J. Ind. Ecol. **2**, 23–44 (2013)
4. European Commission: European Union Directive 2008/98/EC of the European parliament and of the Council of 19 November 2008 on waste and repealing certain directives, 51 (2008)

5. Sikarwar, S., Yadaw, S.B., Yadav, A.K., Yadav, B.: Nanocomposite material for packaging of electronic goods. International Journal of Scientific and Innovative Research **1**(02), 93–108 (2014)
6. Business Wire, Carbon fiber market by raw material (pan, pitch, rayon), fiber type (virgin, recycled), product type, modulus, application (composite, non-composite), end-use industry (a and d, automotive, wind energy), and region—global forecast to 2029. https://www.busine sswire.com/news/home/20191004005339/en/Carbon-Fiber-Market-by-Raw-Material-PAN-Pitch-Rayon-Fiber-Type-Virgin-Recycled-Product-Type-Modulus-Application-Composite-Non-composite---Global-Forecast-to-2029---ResearchAndMarkets.com. Last access 15 June 2022
7. DecomBlades Project. https://decomblades.dk/. Last access 15 June 2022
8. Final Report Summary—EURECOMP (Recycling Thermoset Composites of the SST). https://cordis.europa.eu/project/id/218609/reporting. Last access 15 Sept 2022
9. Oliveux, G., Dandy, L.O., Leeke, G.A.: Current status of recycling of fibre reinforced polymers: review of technologies, reuse and resulting properties. Prog. Mater Sci. **72**, 61–99 (2015)
10. Pickering, S.J.: Recycling technologies for thermoset composite materials—current status. Compos. A Appl. Sci. Manuf. **37**(8), 1206–1215 (2006)
11. Mantelli, A., Romani, A., Suriano, R., Diani, M., Colledani, M., Sarlin, E., Turri, S., Levi, M.: UV-assisted 3D printing of polymer composites from thermally and mechanically recycled carbon fibers. Polymers 13.5:726 (2021)
12. Schinner, G., Brandt, J., Richter, H.: Recycling carbon-fiber-reinforced thermoplastic composites. J. Thermoplast. Compos. Mater. **9**(3), 239–245 (1996)
13. Palmer, J., Ghita, O.R., Savage, L., Evans, K.E.: Successful closed-loop recycling of thermoset composites. Compos. A Appl. Sci. Manuf. **40**(4), 490–498 (2009)
14. Rouholamin, D., Shyng, T., Savage, L., Ghita, O.: A comparative study into mechanical performance of glass fibres recovered through mechanical grinding and high voltage pulse power fragmentation. In: 16th European conference on composite materials, ECCM 2014, 01 (2014)
15. Leisner, T., Hamann, D., Wuschke, L., Jackel, H.-G., Peuker, U.A.: High voltage fragmentation of composites from secondary raw materials-potential and limitations. Waste Manage. **74**, 123–134 (2018)
16. Torres, A., de Marco, I., Caballero, B.M., Laresgoiti, M.F., Legarreta, J.A., Cabrero, M.A., Gonzalez, A., Chomon, M.J., Gondra, K.: Recycling by pyrolysis of thermoset composites: characteristics of the liquid and gaseous fuels obtained. Fuel **79**(8), 897–902 (2000)
17. Cunliffe, A.M., Williams, P.T.: Characterisation of products from the recycling of glass fibre reinforced polyester waste by pyrolysis. Fuel **82**(18), 2223–2230 (2003)
18. Meyer, L.O., Schulte, K., Grove-Nielsen, E.: CFRP-recycling following a pyrolysis route: process optimization and potentials. J. Compos. Mater. **43**(9), 1121–1132 (2009)
19. Hyde, J.R., Lester, E., Kingman, S., Pickering, S., Wong, K.H.: Supercritical propanol, a possible route to composite carbon fibre recovery: a viability study. Compos. A Appl. Sci. Manuf. **37**(11), 2171–2175 (2006)
20. Lester, E., Kingman, S., Wong, K., Rudd, C., Pickering, S., Hilal, N.: Microwave heating as a means for carbon fibre recovery from polymer composites: a technical feasibility study. Mater. Res. Bull. **39**(08), 1549–1556 (2004)
21. Akesson, D., Foltynowicz, Z., Christeen, J., Skrifvars, M.: Microwave pyrolysis as a method of recycling glass fibre from used blades of wind turbines. J. Reinf. Plast. Compos. **17**(01), 1136–1142 (2013)
22. Yoon, K.H., DiBenedetto, A.T., Huang, S.J.: Recycling of unsaturated polyester resin using propylene glycol. Polymer **38**(9), 2281–2285 (1997)
23. Kritzer, P.: Corrosion in high-temperature and supercritical water and aqueous solutions: a review. The Journal of Supercritical Fluids **29**(1), 1–29 (2004)
24. Woidasky, J.: Weiterentwicklung des Recyclings von faserverstärkten Verbunden. Proceedings of Berliner Recycling und Rohstoffkonferenz, 241–260 (2013)
25. European Parliament and Council: Regulation No 1013/2006 of the European Parliament and the Council of 14 June 2006 on shipments of waste. Official Journal of the European Union 1–98 (2006)

26. Dutch National Waste Management Plan. https://lap3.nl/service/english/. Last access 15 Sept 2022
27. UK Landfill Taxes rates. https://www.gov.uk/government/publications/rates-and-allowances-landfill-tax/landfill-tax-rates-from-1-april-2013. Last access 15 June 2022

The FiberEUse Demand-Driven, Cross-Sectorial, Circular Economy Approach

Marcello Colledani⬤, Stefano Turri⬤, and Marco Diani⬤

Abstract Composite materials are widely used in several industrial sectors such as wind energy, aeronautics, automotive, construction, boating, sports equipment, furniture and design. The ongoing increase in composites market size will result in relevant waste flows with related environmental issues and value losses if sustainable solutions for their post-use recovery and reuse are not developed and upscaled. The H2020 FiberEUse project aimed at the large-scale demonstration of new circular economy value-chains based on the reuse of End-of-Life fiber reinforced composites. The project showed the opportunities enabled by the creation of robust circular value-chains based on the implementation of a demand-driven, cross-sectorial circular economy approach, in which a material recovered from a sector is reused within high-added value products in different sectors. A holistic approach based on the synergic use of different hardware and digital enabling technologies, compounded by non-technological innovations, have been implemented to develop eight demonstrators grouped in three use cases, fostering different strategies. In particular, Use Case 1 focused on the mechanical recycling of short glass fibers, Use Case 2 on the thermal recycling of long fibers, while Use Case 3 focused on the inspection, repair and remanufacturing of carbon fiber reinforced plastics products and parts.

Keywords Composite materials · FiberEUse project · Cross-sectorial approach · Demand-driven approach · Recycling technologies

M. Colledani (✉) · M. Diani
Department of Mechanical Engineering, Politecnico Di Milano, Via La Masa 1, 20156 Milan, Italy
e-mail: marcello.colledani@polimi.it

S. Turri
Department of Chemistry, Materials and Chemical Engineering, Politecnico Di Milano, "Giulio Natta", Piazza Leonardo da Vinci 32, 20133 Milan, Italy

M. Colledani and S. Turri (eds.), *Systemic Circular Economy Solutions for Fiber Reinforced Composites*, Digital Innovations in Architecture, Engineering and Construction, https://doi.org/10.1007/978-3-031-22352-5_2

1   Introduction

Composites are relatively young and intrinsically durable materials, but composite-based components or products have a definite lifetime, normally shorter than 20–30 years. For example, wind turbines are predicted to have a lifecycle of 20–25 years, with related dismantling and recycling issues emerging after the use phase. Similarly, many of composite-made boats are still operating, but the average lifetime for recreational boats is around 10 years, and up to 30 years for sailboats [1]. Again, the average lifetime of composite components in car bodies does not exceed 10 years. As the demand for composites in these sectors is continuously growing it is evident that a correct waste management will become an important issue for industrial stakeholders operating in these business sectors. A systemic and systematical waste management strategy has to be defined, supporting the transition from a linear to circular economy. Indeed, the currently limited recycling of composites is seen as an industry challenge that bounds market growth in strategic sectors, such as e-mobility. For example, according to the EU ELV (End-of-life Vehicles) legislation [2], the 95% in weight of vehicles disposed after 2015 must be recyclable. The possibility to reach this strategic goal in the automotive sector, undergoing a profound transformation towards E-mobility, may be bounded if proper solutions for composite materials, very attractive for e-vehicles due to the lightweight properties compensating the additional weight of battery packs, are not found.

Innovative technological approaches and economical models have to be jointly developed considering not only the compliance with environmental regulations, but also the generation of profit. Current waste management practices in composites are dominated by landfilling which is still a relatively cheap option, however it is the least preferred by the legislation and doesn't comply with the European Waste Framework Directive [3]. It is recognized that landfilling will become unviable mainly due to the legislation-driven cost increases. Landfilling of composite waste is actually already forbidden in Germany and Austria, and other EU countries are expected to follow this approach.

The preferred strategy for EoL composite components is certainly passing through recycling and reuse. Mechanical, thermal and chemical recycling methods are well known at state of the art [4–6]. An alternative option is incineration with energy recovery and incorporation of fiber residues in ashes or cement.

Not only research institutions but also industries and sectorial associations, such as WindEurope, are sensitive to the problem of a correct management of EoL composite products and components. The official position of the European Composite Industry Association is to promote co-processing of glass fiber waste in the cement kiln route [7]. According to this view, Life Cycle Assessment studies showed that significant reduction of CO_2 emission of the clinker manufacturing process can be obtained (up to 16%). However, it has been also shown that no more than 10% of the fuel input to a cement kiln route could be substituted by Glass Fiber Reinforced Plastics (GFRP) waste [6]. Indeed, the presence of boron in E-glass fibers negatively impacts the performance of the cement. Even if co-processing of composite waste in the cement kiln route is compliant with the current EU legislation, technical added-value of fibers and polymer matrices are lost in this approach. Co-processing offers environmental benefits but low, if any, profitability.

Among stakeholder associations, the European Boating Association (EBA) presented a position paper on the problem of EoL boats, underlining that there is a compelling need for a specific legislation considering the concept of extended producer responsibility as the best route to take [8], and highlighting the difficult transition from linear to circular economy in the sector.

In particular, the material composition and the cross-linked nature of fiber-reinforced plastics, especially thermosets, make recyclability complex. The existing limitations for a transition to a sustainable circular economy approach for composite-made parts are discussed in the following.

- Lack of a systemic value-chain integration for re-use: Although research effort has been devoted to the development of isolated technologies and processes for recycling composites, the opportunity for a circular economy model for post-use fiber reinforced plastics parts in Europe is still unexploited, mainly because of the lack of a systematic approach for integrating different stakeholders in the value-chain in a stable, trustful, and efficient way.
- Low volume, unstable and unpredictable supply of post-use products for material and components for re-use: At industrial level, it is usually very complex to guarantee a constant, in terms of volume, and replicable supply of input post-use products if a single sector is considered. This can be overcome through the implementation of a *cross-sectorial approach*, exploiting the innovative idea of *demand-driven circular economy*, consisting in adapting the recycling and re-use strategy and processes in order to meet, in certified and traceable way, the requirements of high added-value products in other sectors.
- Poor consumers acceptance of products embedding recovered materials: A general belief that products embedding recycled materials present lower performance than products made of virgin materials still exists. Awareness raising and citizens involvement are essential to foster circular economy implementation.
- Lack of circular business models for boosting profitability: Even in the cases when recycling and re-use is technically feasible, it is usually very hard for companies to prove the profitability of the business through sustainable business cases. This is mainly due to the lack of service-oriented business models, targeted to boost virtuous circular economy practices by enabling a direct accessibility to post-use parts and a certain level of control on the reverse logistics flows, to properly bound costs.

Many of these barriers are the results of fragmentation, poor communication, limited knowledge on the market demand, lack of common vision and strategy both across vertical and horizontal value chains, as well as low revenue and a paucity of supporting legislation and incentives to encourage composites reuse.

The EU H2020 FiberEUse project (GA No. H2020-730,323-1, www.fibereuse.eu) was set-up with the target objective to overcome these limitations by developing and demonstrating industry-mature circular economy practices for composite-made products, inspired by the aforementioned principles of cross-sectorial and demand-driven circular economy. FiberEUse involves a consortium of 21 partners from 7 EU Countries (as shown in Fig. 1) representative of reference stakeholders of the targeted circular value-chain. The project aimed at integrating different innovation actions through a holistic approach, enhancing the profitability of composite recycling and reuse in new value-added products. The project developed and demonstrated an integrated technically and economically viable system for the valorization of composite recyclates through innovation in recycling, reprocessing and remanufacturing technologies, compounded by innovation at non-technological, value-chain integration oriented, level. Through these achievements, FiberEUse established a cross-sectorial, cross-industry, technological and innovation platform able to promote an efficient and cost-effective recycling and reuse of post-use glass and carbon fiber composites, at the same time bringing benefit to the environment, end-users, and the sector as a whole.

This book presents the results obtained during the project, emphasizing the achieved maturity level of the developed solutions in view of their future up-scale at wider industrial value-chain level within different composite-relevant sectors and regional areas at worldwide scale.

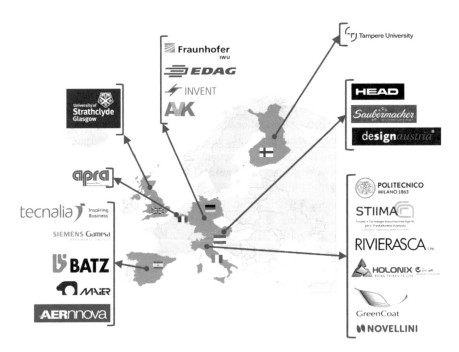

Fig. 1 The FiberEUse project consortium

2 The FiberEUse Demand-Driven, Cross-Sectorial Circular Economy Approach and Vision

The FiberEUse vision is based on the observation that circular value-chains for composite materials can be made economically viable for the involved stakeholders only if the value of the obtained output fractions is maintained high throughout multiple use-cycles, i.e. without significant value losses caused by excessive material property down-cycling. In line with this vision, the target objective of the FiberEUse circular strategies is to obtain output fractions featuring sufficient physical properties to match the input requirements and specifications imposed by the target output applications. This is made possible by a novel demand-driven, cross-sectorial circular economy approach. The rationale of this approach is explained in the next section.

2.1 The Concept of Cross-Sectorial Circular Economy Approach

The essence of the proposed cross-sectorial circular economy approach consists in the recycling, reprocessing and re-use of materials from products in sectors characterized by higher requirements and more demanding technical specifications in new products and applications featuring less demanding, more liberal, specifications. As a matter of fact, this approach exploits the technical degrees of freedom provided by the different input material specification sets characterizing different applications and sectors. The practical implication of this approach is explained in the following through concrete examples.

In Figs. 2 and 3 [9] the characteristics industry requirements and specifications for composite materials (both glass and carbon fiber based) in different sectors are shown, in terms of tensile strength and tensile module. As it can be noticed, for both composite materials, the specification areas characterizing different application scenarios are not overlapping. For example, for both CFRP and GFRP, the typical tensile strength required by the aeronautics sector is almost twice as large as that required by automotive applications. Similarly, considering GFRPs, the typical tensile strength required by wind energy applications is approximately 30% higher than that required by the boating industry. This means that, although recycling processes may contribute to a deterioration of the mechanical properties of materials extracted from post-use products, if these characteristics are properly controlled and managed then the output material fractions may still maintain sufficient residual properties for other applications, across two or more use-cycles.

This technical observation highlights the opportunity for establishing cross-sectorial value-chains for the implementation of new circular economy business cases, provided that innovation in the recycling, reprocessing and re-use process-chain can make it flexible and adaptable to varying target material properties

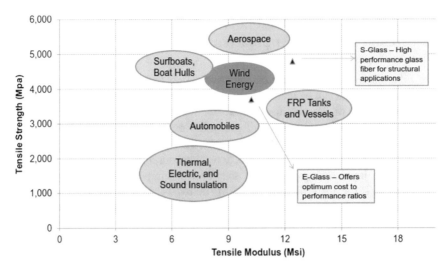

Fig. 2 Industry requirements for GFRP made parts in different sectors [9]

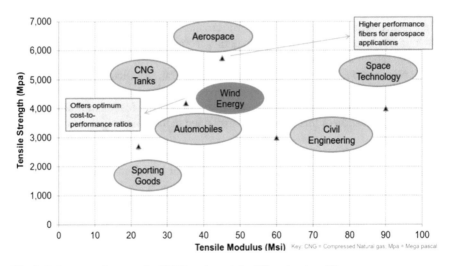

Fig. 3 Industry requirements for CFRP made parts in different sectors [9]

required by different applications. This is the essence of the demand-driven approach explained in detail in the next paragraph.

2.2 The Concept of Demand-Driven Circular Economy Approach

To enable this cross-sectorial approach, a recycling, reprocessing and re-use process-chain have to be implemented, that is able to provide to the output fractions containing fibers and resin the input characteristics needed by the sector in which they will be used as secondary raw materials, in partial or complete substitution or virgin raw materials. The FiberEUse demand-driven approach is formalized to support this vision.

Current fiber reinforced plastics industrial value-chains, for example in the wind energy and boat industries, are essentially linear as represented in Fig. 4, where the main product flow is captured by horizontal directed arrows, while the secondary resource flows contributing to the transformation in each value-chain phase are represented through vertical arrows crossing each transformation box. The only exception is the existence of few industrial players with capability to process composite materials limiting to production waste and not accepting post-use products in input. The main destination of these materials and components from composite post-use products today is the disposal in landfills, or the non-profitable incineration in cement kiln factories. For example, wind energy parks operators and producers are struggling finding a solution to their post use products as no industrial stakeholders accepts their post-use wind blades for recycling and re-use.

As reported in Chap. 1, the scientific literature has developed and demonstrated potentially applicable technical solutions for composite recycling. However, these solutions are grounding on the recent but traditional concept of circular economy. As a matter of fact, the current approach is essentially "push" in the sense that the goal is to maximize the quality and quantity of the materials recovered by recycling processes and, only later, to scout potential re-use applications based on the obtained material properties (Fig. 5). In this way, only few high added-value re-use applications in very specific sectors can be achieved, covering a minor fraction of the EoL composite potential.

In order to unlock high-value circular value-chains for fiber reinforced plastics, a new concept of demand-driven circular economy has to be adopted that will revolutionize the current circular economy oriented industrial practices in the composites field. The key FiberEUse concept of demand-driven circular economy solution is

Fig. 4 Current linear value-chain for composite post-use products (100% landfilled)

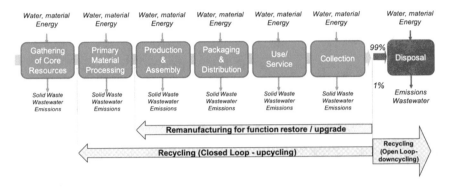

Fig. 5 "Push" traditional circular value-chains

represented in Fig. 6. The core idea is that the circular value-chain is transformed into a "pull" system, where the requirements and specifications on the materials and components to be re-used are transferred directly from the demand side in terms of characteristics and functionalities of the high added-value products reusing such materials and components, and, in turn, propagated upstream to recycling, reprocessing or remanufacturing, in case of component function recovery, stages. This approach will guarantee that the materials and components obtained through circular economy practices are actually re-usable into new products that are demanded by the market, thus bringing circular value-chains at the same level of maturity as linear value-chains and fostering the implementation of the aforementioned cross-sectorial approach. This will also support the creation of sufficiently high volume of waste streams across composite use sectors, crucial to build economies of scale and industrialization of recycling. This approach has been preliminarily exploited in the FiberEUse project for the development of the demonstrators. Considering its potential, the implementation of the demand-driven approach has to be considered as a future innovation priority in circular economy.

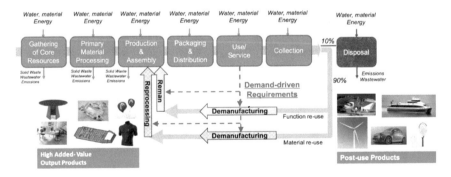

Fig. 6 "Pull" demand-driven circular value-chains

3   The FiberEUse Use Cases and Demonstrators

The project developed and demonstrated at a large scale a set of environmentally and simultaneously economically profitable solutions for the treatment and valorization of EoL composite waste deriving from different manufacturing sectors. A holistic approach based on the synergistic use of different enabling technologies has been implemented in the realization of three large scale use cases shown as yellow, blue and green lines in Fig. 7. Each of these large use case generated several other demo-cases (8 in total) to close the loop of composites lifecycle in different industrial sectors from a circular economy viewpoint.

More in detail, Use Case 1 (represented in Fig. 8) was dedicated to mechanical recycling of short GFRP from post-use products from wind energy, construction and sanitary sectors and their reuse in high added-value customized applications, including furniture (Demo-Case 1), design and creative products (Demo-Case 2) and sports equipment (Demo-Case 3) through additive remanufacturing, compounding, extrusion, molding and finishing. The process-chain involved in Use Case 1 is represented in Fig. 9.

Use Case 2 (represented in Fig. 10) aimed at thermal recycling of long fibers (glass and carbon) and re-use in high-tech, high-resistance applications. Input products was End-of-Life wind turbines and aerospace components and the reuse of obtained composites in automotive (Demo-Cases 4 and 5) and construction sector (Demo-Case 6) has been demonstrated by applying controlled pyrolysis and compounding, molding and extrusion. The process-chain involved in Use Case 2 is represented in Fig. 11.

Use Case 3 (represented in Fig. 12) focused on the inspection, repair and remanufacturing of End-of-Life CFRP products and parts in high-tech applications (in particular automotive). Several technologies have been implemented as non-destructive

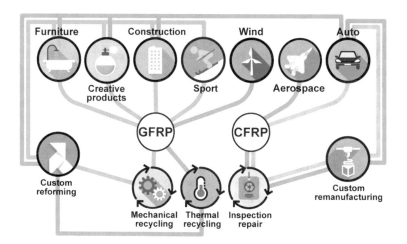

Fig. 7 Visualization of use cases of the FiberEUse project (credits to S. Ridolfi)

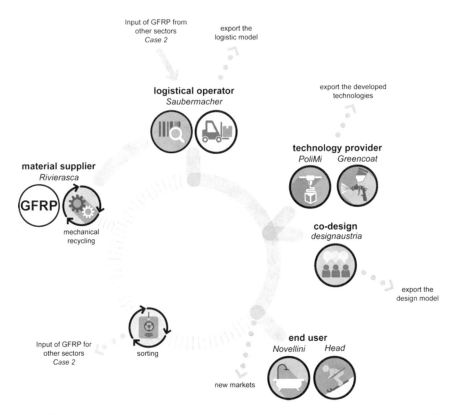

Fig. 8 Use Case 1 value-chain key stakeholders, dedicated to mechanical recycling of short GFRP (credits to S. Ridolfi)

inspection techniques, laser cutting and repair, and innovative design to allow a complete circular economy demonstration in the automotive sector (Demo-Cases 7 and 8). The process-chain involved in Use Case 2 is represented in Fig. 13.

The FiberEUse technological and non-technological pillars have supported the realization of each demo-case and will be amenable to industry upscale to support the generation of further demo-cases for opening the demand of pos-use composite fractions, thus leading to a wider impact in the future. This has been achieved by demonstrating a comprehensive, modular solution that combines: (a) optimized mechanical and thermal composite recycling, inspection, repair and reprocessing technologies; (b) specific value-chain integration actions aimed at improving the market of the recyclates and the profitability of the target reuse options through the demonstration of three large use cases. The FiberEUse technological and non-technological innovations are further detailed in the following sections.

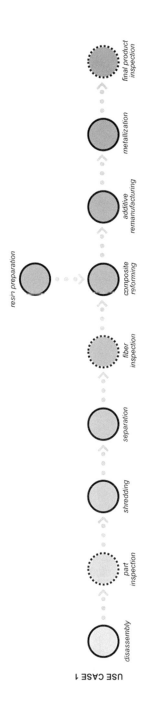

Fig. 9 Use Case 1 process-chain (credits to S. Ridolfi)

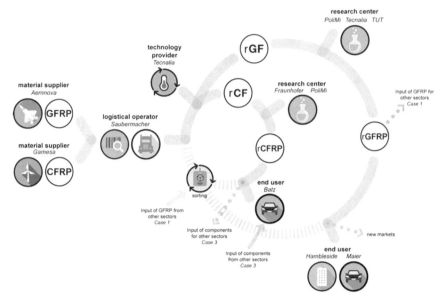

Fig. 10 Use Case 2 value-chain key stakeholders, dedicated to thermal recycling of long fibers (credits to S. Ridolfi)

4 The FiberEUse Technologies

The realization of the aforementioned eight FiberEUse demo-cases and the validation of the achieved performance has been supported by the development and validation of a set of technologies supporting the FiberEUse demand-driven, cross-sectorial circular economy vision. A brief discussion of each technical pillar follow.

Disassembly of large infrastructures. The first stage of the process chain to treat EoL products is disassembly. It is a fundamental step, in particular when dealing with composite-rich large infrastructures, as wind blades. It can influence both the quality of the following processes (as recycling) and the economic sustainability of the entire process. Different route can be followed, with tasks that can be performed both in situ or directly in the recycling plant, and several jobs have to be optimized (as cutting) to facilitate recycling. Chap. 3 of this book focuses on the optimization of the disassembly procedure of large infrastructures, presenting a model that is able to minimize the costs considering all the different aspects, from machining to logistics.

Low-cost and highly-controlled mechanical grinding of GFRP waste. As already presented before, mechanical recycling is one of the most used solutions to treat GFRP. This is due in particular for the low cost of these materials that requires non-expensive processes. On the other hand, non-optimized processes hinder the possibility to reuse this material in new high-added value products. To overcome this issue, the design of optimized suitable solutions, with high throughput and low costs, following a demand-driven approach, is fundamental. Chapter 4 of this book

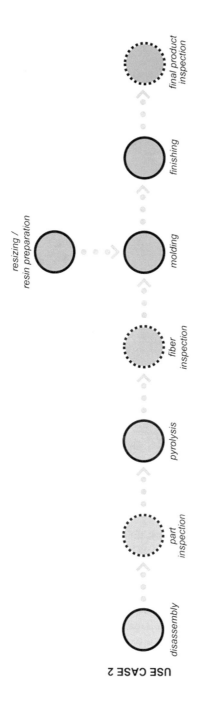

Fig. 11 Use Case 2 process-chain (credits to S. Ridolfi)

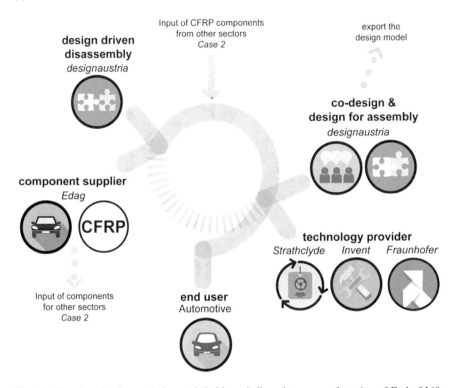

Fig. 12 Use Case 3 value-chain key stakeholders, dedicated to remanufacturing of End-of-Life CFRP products and parts (credits to S. Ridolfi)

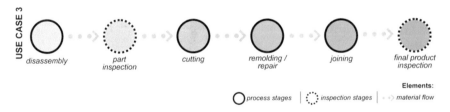

Fig. 13 Use Case 3 process-chain (credits to S. Ridolfi)

is dedicated to the presentation of the results obtained in smart mechanical recycling processes of composite materials with the related developed models.

Innovative pyrolysis process. Thermal recycling is the most suitable solution to treat composites to obtain long fibers. Traditional processes show good results in terms of fibers cleaning but suffer from two aspects. The first one is the possible degradation of the fibers that can affect their reuse in high-added value products. The second one is the cost, mainly due to the high energy consumption, limiting their application to CFRP (that have higher selling price). In addition, even if the fibers are recovered, the resin cannot be recycled to be reused in new applications.

Chapter 5 of this book will show the results obtained during FiberEUse project on thermal demanufacturing for long fibers recovery, presenting different approaches and an innovative thermal depolymerization process.

Reformulation. The resins currently used in manufacturing of composite materials are based on styrene. This is due mainly on the characteristics of this monomer, in particular the possibility to finely tune the final properties of the cured material through the reduction of the viscosity and improving the processability. This results in thermoset resins with high mechanical properties and good heat and chemical resistance. On the other hand, styrene shows low environmental sustainability, together with health and safety issues as it is hazardous, flammable and volatile. Chapter 6 of this book is dedicated to the work held during FiberEUse project on the development of new styrene-free liquid resins for composite reformulation.

Resizing and reprocessing. In the production of composite products, a fundamental role is played by the interface between the matrix and the reinforcement. If its quality is too low, this results in bad adhesion between the fibers and the matrix, and, as a consequence, in low mechanical properties. To create a good interface, a surface treatment is performed on the fibers, called sizing. During recycling processes (in particular thermal and chemical ones), the original sizing is removed. For this reason, a new surface treatment is needed, called resizing. Chapter 7 of this book presents the developed resizing processes and the results obtained substituting virgin fibers in compounding of automotive products.

Additive manufacturing with customized 3D-printing machinery and software. In the last years, additive manufacturing technologies, or 3D-printing, show a rapidly growth for a variety of reasons, in particular for the easiness in personalization of production with active involvement of end-users in the co-design of the final product and the possibility to realize complex geometries, unattainable through conventional manufacturing techniques. Many of those appealing features are typical of open, low cost 3D-printers which are fed by thermoplastic filaments like in the case of FDM machines. These printers must be modified in order to allow a correct composite reprocessing. In particular, both discontinuous fibers and grinded recyclates have to be correctly reprocessed, including thermosets composites. Chapter 8 is dedicated to the solutions developed during FiberEUse project on the additive remanufacturing of recycled composites.

Environmental friendly surface finishing for aesthetic and functional performance enhancement. Enhancing the performances of remanufactured or repaired goods through suitable surface treatments and coatings is a mandatory step to increase the perceived quality of the product. Surface enhancement can occur through the application of both organic and inorganic environmental friendly coatings. Composites suffer from poor erosion and temperature resistance, which partially limit their use in metal part replacement. Metallization can minimize wear or degradation. Surface properties can be improved by metallization through environmental friendly magnetron sputtering physical vapor deposition (PVD) and thermal spraying. Chapter 9 is dedicated to the presentation of the results obtained during FiberEUse project in composite finishing for reuse.

Development and automation of laser assisted cutting and repair technologies for CFRP component reuse. Currently, the repair of fiber reinforced structures is a difficult and time consuming process. In many cases the damaged part is ground manually in order to prepare the resulting cavity to be refilled with a patch of new plies. This repair preparation procedure requires experienced and highly qualified workers and it is not easy to generalize. Furthermore, the manual grinding process results in an accelerated tool wear and a costly replacement. Replacing the manual grinding process with an automated laser cutting process has the potential advantage to precisely cut the composite structure by locally focusing a small laser spot on the area to be removed. In addition, non-destructive techniques can be used to inspect parts and components before and after repair. Chapter 10 is dedicated to the work done in FiberEUse project on composite repair and remanufacturing, based on both traditional and innovative solutions.

Design guidelines. Design plays a fundamental role in circular economy. As an example, it can hinder the disassembly of some components. As a result, a destructive disassembly has to be performed, preventing the reuse of a part or component. This is relevant in particular in sectors as automotive, in which the End-of-Life parts still have a high added value. For this reason, it is important to design new products avoiding fixing systems as glue or rivets, in favour of solutions that can facilitate their second life. On the other hand, to increase customer acceptance of products manufactured with recycled material, a solution can be the involvement of final users since the design phase following a co-design approach. Chapters 11 and 12 of this book will concentrate on these two design opportunities, both for co-design of creative products embedding recycled fibers and on an innovative modular car design concept for reuse. In addition, Chapter 13 will give hints on product redesign through specific guidelines.

ICT solutions. Even if all the previous technologies will be implemented and adopted by different companies, one of the most critical issue can be the communication between them. Currently, there is not a robust value-chain to rely on to implement circular economy in composites sector. To overcome this issue, an ICT tool has to be developed to connect all the actors along the value-chain, from the wind blades producers to the final manufacturers embedding recycled material in their products. Chapter 14 presents the digital cloud-based platform developed during FiberEUse project for the value-chain integration.

5 The FiberEUse Non-technological Pillars

The FiberEUse cross-cutting innovation has been achieved through the scalable and replicable development of the aforementioned technologies, compounded by the following set of non-technological innovation actions.

Scouting new markets for recycled material. The research included the identification and documentation of the properties and market value of the GF and CF recyclates in an accessible user-friendly manner, as well as current legislative barriers and propositions on potential supporting legislative incentives to enable (i) enhanced decision making on EoL scenarios of recyclates (ii) symbiosis between sectors such as energy, building, automotive, and aircraft. A series of physical and virtual tools for a further stimulation of potential end-user have been developed, namely through the realization of a physical and virtual product and material library, and through the management of data in a user-friendly, cloud-based ICT platform.

New service-oriented business model development. New business models supporting the FiberEUse close-loop manufacturing vision have been designed, with the active involvement of end-users. The main building blocks have been identified for each target application (value proposition, revenue model, collection strategy, supply chain) and then hierarchically organized in a coherent business model.

Optimization of logistical infrastructures. Efforts have been dedicated to explore appropriate collection systems and infrastructures. Alternative scenarios for reverse logistics network design have been modelled and evaluated and optimized for each demo-case for both cost and environmental impact minimization objectives, to match their individual requirements.

Sustainable co-design with end-users. Sustainable co-design aimed at the realization of products with better usability, easier disassembly and user friendliness has been studied involving end-users and stakeholders associations. Co-design of remanufactured products involves the realization of an array of different product concepts, their prototyping and dissemination among end-users. The customer feedback collection, performed exploiting IT tools, allows a direct contact with end-users, and addresses an iterative process of product revision leading to a final wider social and market acceptance for goods derived from composite recyclates.

Dissemination, communication and training. A key barrier to overcoming the challenges resulting from fragmentation is dissemination, education and communication. Such communication offers truly disruptive potential. FiberEUse implemented robust tools to ensure effective education and training (training modules, white book, workshops) and dissemination and communication (working groups, website, social networks, workshops and seminars) across the value chain.

Vision, strategy and roadmap. Based on the outcome of the innovation system assessment and working closely with the key stakeholder platform, the FiberEUse consortium defined and agreed a common vision, strategy and roadmap for overcoming the identified cross cutting challenges, creating a framework to facilitate the take-up of the proposed solutions.

Legislation and incentives. The research efforts here included the identification of the range of legislation and instruments that would forward the composite recycling, remanufacturing and reuse industry. Examination of the current EU legislation and a consultation among the stakeholders identified the gaps, and guidelines for the alignment of regional legislations has been individuated, while a prioritization of needed actions have been outlined using robust methods.

Stakeholders engagement and eco-system building. By mobilizing key knowledge holders, facilitators, clusters, decision makers, providers, beneficiaries, etc. within a single platform, FiberEUse involved the necessary critical mass of capabilities, resources and competences to be able to support a real transition to its innovative circular value-chain.

6 Conclusions and Structure of This Book

Composite materials are widely used in several sectors, from wind energy to design. The market size of both GFRP and CFRP products, and the related amount of waste are constantly increasing. The recycling and reuse of these materials constitutes a relevant opportunity at economic, environmental and social level. FiberEUse worked in the implementation of a demand-driven, cross-sectorial approach, in which the material recycled from a sector will be re-used in another sector requiring less demanding characteristics, avoiding value losses. This approach will be the basis of the innovative concept of demand-driven circular economy, in which the End-of-Life product is treated through processes that are optimized based on specific requirements that depends on the properties of the final product embedding that recycled material, maximizing in this way the recovery of the residual value.

This book reports the results obtained during the entire FiberEUse project. Chapters 1 and 2 introduced the context, the motivations and the opportunities of implementing circular economy in composites, with a specific focus dedicated to the demand-driven cross-sectorial approach and its potentiality in GFRP and CFRP sector. Following this introduction, several topics of scientific and industrial relevance, able to allow a robust circular economy in composites, are described in depth.

The first macro group of chapters focuses on EoL products treatment as the disassembly of large composite-rich installations (Chap. 3), smart composite mechanical demanufacturing processes to obtain particles with specific dimensions that could be reused in new products (Chap. 4), and thermal demanufacturing processes for long fibers recovery (Chap. 5).

The second part is dedicated to the reuse of the recovered particles. Chapters present composite reformulation form granulates, with a focus on innovative resins (Chap. 6), fiber re-sizing, compounding and inspection (Chap. 7), additive manufacturing of recycled composites (Chap. 8), surface finishing for reuse (Chap. 9) and composite repair and remanufacturing (Chap. 10).

Third group of chapters underlines the role of design for de- and remanufacturing in composites sector. Innovative approaches as the co-design of creative products embedding recycled fibers are explored (Chap. 11) and a new modular car design concept for reuse is shown (Chap. 12), together with products re-design guidelines (Chap. 13).

Chapter 14 is dedicated to the cloud-based platform for value-chain integration developed in FiberEUse, to connect all the different stakeholders and to facilitate the exchange of materials among them.

The objective of the next group of chapters is to present the results obtained in the three use cases of FiberEUse. As already stated, the focus is on mechanical recycling of short fibers for Use Case 1 (Chap. 15), thermal recycling of long fibers for Use Case 2 (Chap. 16) and modular car parts disassembly and remanufacturing for Use Case 3 (Chap. 17).

Chapter 18 is dedicated to the presentation of the material library systems that will facilitate the adoption of the developed solutions through a tangible-intangible interaction approach.

Next chapters show the opportunities enabled by the implementation of circular economy solutions in composite sectors. New circular business models are presented (Chap. 19), together with their economic and risk assessment (Chap. 20).

Finally, the book ends with a specific focus on the analysis policy actions to be considered and implemented in the next years to favor the industrial implementation of the FiberEUse results (Chap. 21).

References

1. European Boating Association. http://www.eba.eu.com/positions. Last access 15 Sept 2022
2. European Commission. European Union directive 2000/53/EC of the European parliament and of the Council of 2000 on wind turbine and automotive EoL
3. European Commission. European Union directive 2008/98/ec of the European parliament and of the council of 19 November 2008 on waste and repealing certain directives, 51 (2008)
4. Pickering, S.J.: Recycling technologies for thermoset composite materials—current status. Comp. A **37**, 1206–1215 (2006)
5. Rubicka, J., et al.: Technology readiness level assessment of composites recycling technologies. J. Cleaner Prod. **112**, 1001–1012 (2016)
6. Oliveux, G., Dandy, L.O., Leeke, G.A.: Current status of recycling of fibre reinforced polymers: review of technologies, reuse and resulting properties. Prog. Mater Sci. **72**, 61–99 (2015)
7. European Composites Industry Association. http://www.eucia.eu/about-composites/sustainability. Last access 15 Sept 2022
8. BoatDIGEST project. http://www.boatdigest.eu/inde.asp. Last access 15 Sept 2022
9. Frost&Sullivan. Trends and opportunities for composite materials in wind blade manufacturing. https://www.frost.com/. Last access 16 Oct 2017

Disassembly of Large Composite-Rich Installations

Marco Diani ⓘ**, Nicoletta Picone** ⓘ**, Luca Gentilini** ⓘ**, Jonas Pagh Jensen** ⓘ**, Alessio Angius** ⓘ**, and Marcello Colledani** ⓘ

Abstract Considering the demanufacturing of large infrastructures (as wind blades and aircrafts) rich in composite materials, the most impacting step in terms of costs is disassembly. Different routes could be followed for dismantling and transportation and several factors influence the final result (as the technology used, the logistic and the administrative issues). For this reason, it is fundamental to understand which solution has to be followed to reduce the impact of decommissioning on the overall recycling and reusing cost. This work, after the formalization of the different possible disassembly scenarios, proposes a Decision Support System (DSS) for disassembly of large composite-rich installations, that has been designed and implemented for the identification of the most promising disassembly strategy, according to the process costs minimization. The mathematical models constituting the core of this tool are detailed and the DSS is applied to disassembly of onshore wind blades, underling the importance of similar systems to optimize demanufacturing costs.

Keywords Disassembly · Decommissioning · DSS · Cutting · Mechanical treatments · Wind blades · Aircraft

1 Introduction

Disassembly is the most impacting step in demanufacturing, in particular for large infrastructures as wind turbines and aircrafts.

Wind turbines are composed of materials both easily recyclable through well-established practices, as iron and steel, and materials that require innovative solutions,

M. Diani (✉) · L. Gentilini · A. Angius · M. Colledani
Department of Mechanical Engineering, Politecnico Milano, Via La Masa 1, 20156 Milan, Italy
e-mail: marco.diani@polimi.it

N. Picone
STIIMA-CNR, Via Alfonso Corti 12, 20133 Milan, Italy

J. P. Jensen
Siemens Gamesa Renewable Energy A/S, Fiskergade 1-9, 7100 Vejle, Denmark

© The Author(s) 2022
M. Colledani and S. Turri (eds.), *Systemic Circular Economy Solutions for Fiber Reinforced Composites*, Digital Innovations in Architecture, Engineering and Construction, https://doi.org/10.1007/978-3-031-22352-5_3

37

as composites. The protection of the environment and the notion of circular economy is gaining increased momentum, which contradicts with the difficulty of recycling the composite parts (mainly blades and nacelle housing) [1].

The composite parts typically consist of fiberglass and a cured resin matrix, namely Glass Fibers Reinforced Plastics (GFRP). The strength of the composites comes from the glass fibers, and therefore, the manufacturers aim to have as much glass in the composite as possible. The normal measurement of the amount of glass in composites is the fiber volume fraction (standard: 55%–60% v/v). The matrix is important for keeping the glass in shape and as a 'carrier' of forces to the glass fibers. These are very strong forces, so the matrix is chemically (i.e. covalently) bonded to the glass fiber, and at the same time the matrix is cross-linked through formation of chemical bonds. This results in an incomparable material when in use, but difficult to recycle.

Due to the fact that recycling of these materials is complex, a wide range of recycling processes has been proposed and tried out—some in commercial settings, others in lab scale attempts [2]. This ranges from architectural reuse of the blade for bike shed, play ground or even walking bridge to more traditional recycling (material recycling), where six main recycling routes, with varying technology readiness levels, are commonly referred to (although the majority of the blades are still reported to end up in landfill) as co-processing in cement kiln, mechanical grinding, pyrolysis, solvolysis, High Voltage Pulse Fragmentation and fluidized bed or gasification. In any case, all these steps are subsequent to the disassembly (or decommissioning) phase.

The decommissioning of a wind farm constitutes the final stage of a project when service life extension or repowering is not a financially feasible practice. It represents the least desirable End-of-Life (EoL) scenario. The main objective of this stage is returning the farm to its original conditions prior to initial deployment. In the decommissioning procedure of a wind farm, all wind tower elements are dismantled (Fig. 1): firstly all blades (GFRP based), nacelle (GFRP based) and the tower (steel based) will be disassembled and hoisted down by crane; then the posterior elements will be disjointed and reduced into smaller pieces suitable for scrap. It is important to underline that the technologies, as well as the qualification and crew for the decommissioning activities used, are comparable to those of the commissioning stage.

Concerning the geographical location, the ideal wind farm installation would have a near constant flow of non-turbulent wind throughout the year, with a minimum likelihood of sudden powerful bursts of wind. An important factor of turbine siting is also access to local demand or transmission capacity. According to these requirements, both onshore and offshore installations would be possible, with limited accessibility in the latter case (only the onshore installations have been considered in FiberEUse).

On the other hand, the handling of EoL aircraft is a relatively young research topic and little knowledge about the aircraft EoL process is available [3]. There is a lack in the norms, since the handling of EoL aircraft has not been legally regulated yet. The common practice for the final disposal of aircraft was to store them besides airports or in deserts around the globe until a few years ago. For decades, thousands of retired aircraft have been stored in so-called aircraft "hot spot" or "graveyards".

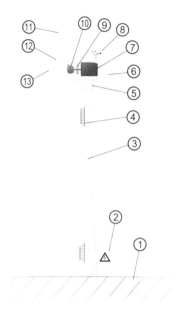

1	Foundation
2	Connection on the electric grid
3	Tower
4	Access ladder
5	Wind orientation control (Yaw control)
6	Nacelle
7	Generator
8	Anemometer
9	Electric or mechanical brake
10	Gearbox
11	Rotor blade
12	Blade pitch control
13	Rotor hub

Fig. 1 Wind turbine components

Recently, two largest aircraft manufacturers (Airbus and Boeing) began to develop alternative approaches proposing a three steps process approach of handling EoL aircraft, the so called "3D Approach", based on decommissioning, disassembly and smart dismantling [4]. According to this innovative approach, during the decommissioning process, the aircraft is taken out of service to be inspected, cleaned and decontaminated. Furthermore, all operating liquids are removed and either re-sold for direct re-use or disposed in specific recovery channels. The second step includes the disassembly procedure, defined as a systematic physical separation of a product into its constituent parts, components or other groupings. During this step, knowledge about the specific aircraft type, such as structure, material and part composition needs to be gained in order to define an efficient disassembly planning. The third step takes into consideration recycling and valorization channels, including best practice recommendations and full compliance to applicable regulation. The developed 3D Approach showed the possibility to increase ratio of value creation up to 80–85% (instead of 50–60%), demonstrated a reuse/recycling ratio >70%, showed strong reduction of landfilled waste (<15% instead of 40–45%).

As can be noticed in Fig. 2, composite materials constitute almost 50% of new aircraft design (in this figure a specific example of Boeing 787 is reported), with an average weight saving of 20%. Selecting the optimum material for a specific application meant analyzing every area of the airframe to determine the best material, given the operating environment and loads that a component experiences over the life of the airframe. For example, aluminum is sensitive to tension loads but handles compression very well. On the other hand, composites are not as efficient in dealing with compression loads but are excellent at handling tension. The expanded use of composites, especially in the highly tension-loaded environment of the fuselage, greatly reduces maintenance due to fatigue when compared with an aluminum structure. This type of analysis has resulted in an increased use of titanium as well. Titanium can withstand comparable loads better than aluminum, has minimal fatigue concerns, and is highly resistant to corrosion. According to the aircraft material composition, almost all the parts could be potentially dismantled for fibers recovery (both CFs and GFs).

From this analysis it is evident the importance of decommissioning and transportation of these parts which are the most impacting factors on the demanufacturing costs. As a consequence, a deep understanding of decommissioning process and an optimization of the demanufacturing process chain is fundamental to achieve a robust circular economy for these products, with a special focus on the cross-sectorial approach. In addition, the large scale dimension of the products requires specific in site disassembly and handling procedures, also enabling possible preliminary on-site treatment (e.g. cutting and mechanical preparation). For this reason a Decision Support System (DSS) for demanufacturing of large infrastructures has been developed and it will be presented in the next Sections.

2 State of the Art

De- and remanufacturing includes the set of technologies and systems, tools and knowledge-based methods to recover and re-use functions and materials from industrial waste and post-consumer high-tech products, to support a sustainable implementation of a new producer-centric Circular Economy paradigm. The goal of de- and remanufacturing systems is inherently different from the goal of a manufacturing system. While manufacturing transforms raw materials into products meeting the customer requirements, de- and remanufacturing transform post-consumer products into valuable materials/new products meeting the customer requirements, for secondary use. However, manufacturing and de- and remanufacturing objectives are clearly not independent. The products that are manufactured and sold to the market today are the products that will be collected and processed in input at the de- and remanufacturing system after the use phase at the customer side. As a consequence, the rapid introduction of new products and the increasing product variety experienced by manufacturers in the last decade is being reflected in the continuous evolution of post-consumer products received in input by de- and remanufacturing systems. This trend represents the main challenge for the development of demanufacturing technological solutions embedding the reconfigurability and adaptability features, efficiently facing the continuous evolution (i.e. complexity) of pre-use products, and the lack of availability and traceability of post-use products information (e.g. design, structure, materials, variability and uncertainty of EoL quality conditions), useful for adapting de- and remanufacturing decisions and operations accordingly.

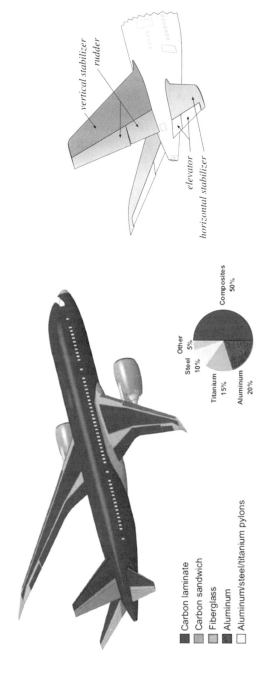

Fig. 2 Composite materials used in Boeing 787 (left); tail composite parts (right)

Therefore, for the full implications of any reuse of recycled Carbon Fibers (rCF) and recycled Glass Fibers (rGF) to be considered, processes, performance, product quality, and markets demand need to be combined to quantify cost benefits of EoL products demanufacturing operations. To this purpose, an in-depth literature review of demanufacturing technologies and processes has been performed.

Concerning the disassembly of EoL products, many studies have been conducted for relatively small products from automotive, electric and electronic sectors (e.g. mechatronic products) as well as for large infrastructure in aeronautic and wind energy sectors [4–6]. In both cases, it has been proved that the disassembly process cannot be simply considered as the reverse of assembly. This is largely due to additional sources of uncertainties, mainly related to the unpredictable characteristics of the returned products (cores) both in terms of quality and quantity [7]. These result from: (i) component defects, (ii) upgrading or downgrading during usage and (iii) damage during the disassembly operation [8]. Therefore, disassembly task planning results to be more complex combinatorial problem than assembly planning. The main aim of disassembly task planning is to find the optimum disassembly path, which is cost effective, improves the value of recovered component and returned material, and respects fixed constraints. Theoretically, the number of possible disassembly sequences increases exponentially with the number of product components and disassembly operations. As a result, finding the optimal solution is an NP-complete optimization problem [8]. In [9] the most effective methodologies and techniques for disassembly task planning are summarized. They include mathematical programming (MP), heuristic methods, to find near-optimal solutions to the disassembly sequencing problem [9], artificial intelligence (AI) methods, e.g. simulated annealing, genetic algorithms (GA), fuzzy sets, neural networks, multi-agent systems, and Bayesian networks, and adaptive planning, for example by Petri Nets, which is used to generate a disassembly sequence with respect to the uncertainties and unexpected circumstances encountered during the disassembly operations.

Concerning the recycling processes, many different technologies have been studied for the last two decades: mechanical processes [10–12], pyrolysis and other thermal processes [13–18], and solvolysis [19, 20]. Some of them, particularly pyrolysis, have even reached an industrial scale, and are commercially exploited: for example, ELG Carbon Fibre Ltd. in United Kingdom use pyrolysis, Adherent Technologies Inc. in USA use a wet chemical breakdown of composite matrix resins to recover fibrous reinforcements and, in France, Innoveox proposes a technology based on supercritical hydrolysis.

According to this review, it is quite clear that currently solutions do exist for the demanufacturing of composite materials. It can be seen in the literature that many different processes and methods have been applied and have shown the feasibility of recovery such materials, some of them being more commercially mature than others. However, industrial applications using recycled fibers or resins are still rare, partly because of a lack of confidence in performance of rGF and rCF, which are considered as of lower quality than virgin ones, but also because rGF and rCF are not completely controlled in terms of length, length distribution, surface quality (adhesion to a new matrix) or origin (often different grades of fibers are found in a batch of recycled composites coming from different manufacturers) [2].

3 Rationale of the Work

Figure 3 represents an overview of the possible demanufacturing routes for the identified FiberEUse target GFRP/CFRP EoL products/parts, including disassembly, cutting, mechanical and thermal treatments. Literature review on key processes and technologies highlights that many researchers have attempted to study and thus optimize the process parameters in both traditional and non-traditional machining of GFRP and CFRP, to reduce or eliminate the problem of matrix cracking, fiber pull-out, swelling and delamination, thus increasing the surface quality [21]. Failure behaviors do not only arise from the heterogeneous and anisotropic structure, but also from the machining methods and their interactions. These problems are not always significant in the case of demanufacturing of such products: if they should not be remanufactured or reused, the process could be destructive and the potential damages caused during the processing are not relevant. On the other hand, many different techniques for mechanical and thermal treatment, have been studied for the last two decades also reaching high Technology Readiness Level (TRL) and industrial scale applications.

From Fig. 3, the most important processes to be modelled could be derived. In particular, they are: disassembly, cutting, mechanical treatments and thermal treatments. In addition, to develop a robust DSS, it is fundamental to consider also the process boundaries and the macro-constraints as logistic, administrative procedure, operator safety, processing capacity and material requirements. Starting from the analysis of these factors, two different macro-scenarios on the basis of the output material requirements and four different sub-scenarios, according to the possibility to perform a *full in plant treatment*, including exceptional load transport, or a *mixed treatment*, i.e. partially on-site and partially in plant, have been outlined in the next sub-sections and summarized in Fig. 4.

3.1 Selective Cut Macro-Scenario

The application of a selective cut scenario is feasible when the product is characterized by regions with a higher content of fibers. In that case, it could be possible to isolate the fiber rich areas from the rest of the product in order to obtain two output fractions: a high fibers content fraction, to be reused for high level applications, and a low fibers content fraction, to be reused for less demanding applications.

One example is represented by the wind turbine blades. By analyzing their composition, it could be observed that the central longitudinal part (highlighted in red in Fig. 5) shows a higher percentage of GF and/or CF, to withstand the mechanical

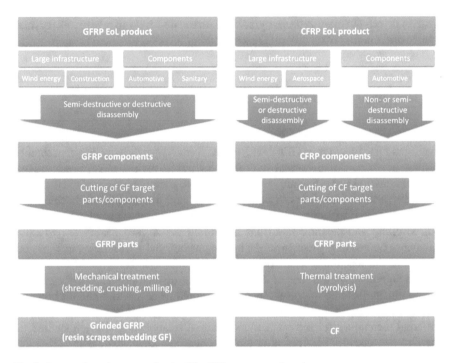

Fig. 3 Demanufacturing routes for the FiberEUse target products/parts

stresses to which they are likely to be subjected. A selective cut following the red dot lines could be a feasible and optimal strategy to enabling two different recycling routes. Following this approach, the material recovered by the central part could be reused for the production of high value products. The two lateral parts, poorer in fiber content, could enter in lower value reuse route.

Another possible approach, is to optimize the disassembly and cutting procedure according to an average fiber content (material requirement), in order to homogenize the output product, both in terms of dimension and fiber concentration. Considering the wind blade structure, the optimal solution (Fig. 6), consists in a longitudinal cut (red dot line) and different transversal cut (green dot lines). In this way, the cut sections have the same percentage of fibers, leading to a homogeneous output.

This selective cut approach is not interesting in the case of parts coming from aeronautic, construction and automotive sectors, which do not present heterogeneity in fibers concentration.

Fig. 4 Considered sub-scenarios. From top to bottom: full in plant treatment; on-site cutting and in plant shredding; on-site cutting, on-site shredding and in plant shredding; on-site cutting, in plant cutting and in plant shredding

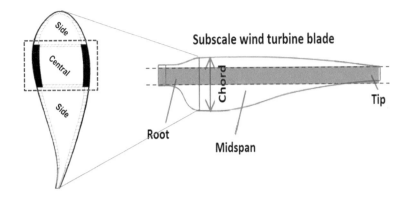

Fig. 5 Top view (right) and cross section (left) of a wind blade. Selective cut scenario for a wind blade providing two different fiber content output fractions

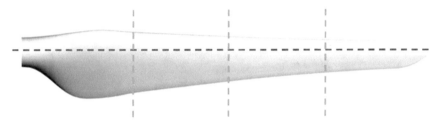

Fig. 6 Selective cut scenario for a wind blade providing average fiber content in output fractions

3.2 Non-selective Cut Macro-Scenario

The application of non-selective cut scenario could be performed for parts characterized by homogeneous fibers content. This is the case of parts coming from aeronautic and construction sectors.

3.3 Mixed Treatment Sub-Scenario

In this sub-scenario different routes are possible: after the preliminary product disassembly, one or more processes could be performed in situ (i.e. cutting and shredding) and one or more in plant (i.e. cutting, shredding, and pyrolysis). In particular, there are five different routes depending on the material under treatment (i.e. GFRP or CFRP). Three of them are related to GFRP products (i.e. GF wind blades and construction components). In this case the identified routes are the following:

- EoL products are cut in situ to increase truck saturation and reach acceptable dimension for further shredding treatment, transported through regular transport and shredded in plant to reach the target output;
- EoL products are cut and shredded in situ to increase truck saturation, transported through regular transport and shredded again in plant to reach the target output dimension;
- EoL products are cut in situ to increase truck saturation, transported through regular transport, cut again in plant to reach acceptable dimension for further shredding treatment and shredded in plant to reach the target output dimension.

The other two routes, related to CFRP and G&CFRP products (i.e. CF wind blades and aircraft parts), are:

- EoL products are cut in situ to increase truck saturation and reach acceptable dimensions for further thermal processing, transported through regular transport and pyrolyzed in plant;
- EoL products are cut in situ to increase truck saturation, transported through regular transport, cut in plant reaching acceptable dimensions for thermal processing and pyrolyzed in plant.

3.4   Full in Plant Treatment Sub-Scenario

In this sub-scenario the disassembled product will be transported as it is to the recycling plant. The most relevant advantages of this solution, in terms of costs, are related to the save of: (i) the administrative costs for in-situ treatment permission (high variable depending on the region of the installation and on the period of the year in which the treatment is performed), (ii) the travel costs for the operators, (iii) the transportation and set-up costs for in-situ treatment machines and (iv) the higher in-situ energy costs (highly variable depending on the geographic area of the installation). On the contrary, this scenario undergoes to higher logistic costs due to the need to perform an exceptional load transport (non-saturated), consequently resulting in a non-optimized transport cost.

The full in plant treatment sub-scenario includes two routes. The one related to GFRP products (i.e. GF wind blades and construction components) is:

- EoL products are dismantled in-situ and transported by using an exceptional load transport to the recycling plant. Here the products are cut to reach acceptable dimension for further shredding treatment.

The route related to CFRP and G&CFRP products (i.e. CF wind blades and aircraft parts) is:

- EoL products are dismantled in-situ and transported by using an exceptional load transport to the recycling plant. Here the products are cut to reach acceptable dimension for further thermal treatment.

4 Methodology

The mathematical model here presented has the scope to create a DSS software tool able to assess the overall treatment costs related to the different scenarios described in Sect. 3. The model is governed at the first hierarchical level by the arbitrary decisions between selective or non-selective cut, and between thermal and mechanical recycling (depending on the type of fibers to be recovered from EoL product). Mathematically:

Variables name	Description	Unit of measure	Variable type
C	Cost of treatment	€	Cumulated
b_{SC}	Boolean: selective cut	Boolean	Input
C_{SC}	Cost of selective cut approach	€	Cumulated
$C_{non\text{-}SC}$	Cost of non-selective cut approach	€	Cumulated

$$C = b_{SC} * C_{SC} + (1 - b_{SC}) * C_{non-SC} \tag{1}$$

Variables name	Description	Unit of measure	Variable type
C_{SC}	Cost of selective cut approach	€	Cumulated
b_{MEC}	Boolean: mechanical recycling	Boolean	Input
$C_{SC\text{-}MEC}$	Cost of mechanical recycling approach	€	Cumulated
$C_{SC\text{-}PYRO}$	Cost of thermal recycling approach	€	Cumulated

$$C_{SC} = b_{MEC} * C_{SC-MEC} + (1 - b_{MEC}) * C_{SC-PYRO} \tag{2}$$

Variables name	Description	Unit of measure	Variable type
$C_{non\text{-}SC}$	Cost of non-selective cut approach	€	Cumulated
b_{MEC}	Boolean: mechanical recycling	Boolean	Input
$C_{non\text{-}SC\text{-}MEC}$	Cost of mechanical recycling approach	€	Cumulated
$C_{non\text{-}SC\text{-}PYRO}$	Cost of thermal recycling approach	€	Cumulated

$$C_{non-SC} = b_{MEC} * C_{non-SC-MEC} + (1 - b_{MEC}) * C_{non-SC-PYRO} \tag{3}$$

Resulting in a cost for transportation C_{TR} equal to:

$$C_{TR} = b_{SC} * \left[b_{MEC} * C_{SC-MEC} + (1 - b_{MEC}) * C_{SC-PYRO} \right]$$
$$+ (1 - b_{SC}) * \left[b_{MEC} * C_{non-SC-MEC} + (1 - b_{MEC}) * C_{non-SC-PYRO} \right]$$
$$(4)$$

An application example about the non-selective cut scenario for GFRP product is modelled in the following section. In the case of different input flows, the core of the mathematical model remains the same and it's scalable to the other macro- and sub-scenarios, for each of which specific objective function should be derived. The main differences to be implemented for the other cases are here reported:

- In case of selective cut strategy, a first stage selective cut has to be added both for GFRP and CFRP products. The cut can be performed either on-site or in plant. Please notice that selective cut on-site enables all the other on-site treatments. For the selective cut contour, resulting in the cutting length, the cutting optimization algorithm can be exploited. Once performed the selective cut, the main optimization model can be applied separately to the N shapes resulting from the first cut.
- In case of CFRP products, shredding costs have to be properly substituted by pyrolysis costs. On-site coarse shredding to enable transport saturation through size reduction is not a viable option in case of thermal recycling.

4.1 Application Example: Non-Selective Cut Scenario for GFRP Products

According to the previous analysis, the model has the objective to select the cheapest option among the following possible routes:

- Route 1: EoL products are cut in situ to increase truck saturation and reach acceptable dimension for further shredding treatment, transported through regular transport and shredded in plant to reach the target output dimension;
- Route 2: EoL products are cut and shredded in situ to increase truck saturation, transported through regular transport and shredded again in plant to reach the target output dimension;
- Route 3: EoL products are cut in situ to increase truck saturation, transported through regular transport, cut again in plant to reach acceptable dimension for further shredding treatment and shredded in plant to reach the target output dimension.
- Route 4: EoL products are dismantled in-situ and transported by using an exceptional load transport to the recycling plant. Here the products are cut to reach acceptable dimension for further shredding treatment.

The objective function behind this model is reported in the following:

$$C_{non-SC-MEC} = min \sum_{i=1}^{4} b_{MEC-i} * C_{non-SC-MEC-i} \qquad (5)$$

Variables name	Description	Unit of measure	Variable type
$C_{non\text{-}SC\text{-}MEC}$	Cost of mechanical recycling approach	€	Cumulated
b_{MEC-i}	Boolean: route i	Boolean	Decision
$C_{non\text{-}SC\text{-}MEC-i}$	Cost of route i	€	Cumulated

Constrained by:

$$\sum_{i=1}^{4} b_{MEC-i} = 1 \qquad (6)$$

The objective function has been derived for the four possible routes. For space reasons, the procedure will be shown only for one route.

Route 1: full on-site cutting and full in plant shredding

Variables name	Description	Unit of measure	Variable type
$C_{non\text{-}SC\text{-}MEC-1}$	Cost of route 1	€	Cumulated
$C_{OS\text{-}cutS-1}$	Cost of on-site cutting	€	Cumulated
C_{TR-1}	Ordinary transport cost	€	Cumulated
$C_{IP\text{-}shred-1}$	Cost of in plant shredding	€	Cumulated

$$C_{non-SC-MEC-1} = C_{OS-cut-1} + C_{IP-shred-1} + C_{TR-1} \qquad (7)$$

Variables name	Description	Unit of measure	Variable type
$C_{OS\text{-}cut-1}$	Cost of on-site cutting	€	Cumulated
$C_{OS\text{-}cut\text{-}admin}$	On-site cutting administrative cost	€	Fixed
$C_{OS\text{-}cut\text{-}setup}$	On-site cutting set-up cost	€	Fixed
N	Number of blades	Pure, int	Input
$C_{OS\text{-}op}$	On-site cost of operator	€/h	Fixed
$C_{OS\text{-}cut}$	Energy and wear cutting-related costs	€/h	Fixed
L_1	Total cutting length	m	Algorithm

(continued)

(continued)

Variables name	Description	Unit of measure	Variable type
Th	Average thickness of the blade	m	Input
s_c	Cutting speed	m^2/h	Fixed

$$C_{OS-cut-1} = C_{OS-cut-admin} + C_{OS-cut-steup}$$
$$+ N * (C_{OS-op} + C_{cut}) * \frac{L_1 * 2 * th}{s_c} \tag{8}$$

The total cutting length is calculated by a dedicated developed algorithm presented in Sect. 4.2. The optimization algorithm works with a one-sided 2D image. For this reason the real cutting length is the double of the one received from the algorithm.

Variables name	Description	Unit of measure	Variable type
$C_{IP-shred-1}$	Cost of in plant shredding	€	Cumulated
$C_{IP-shred-setup}$	In plant shredding set-up cost	€	Fixed
N	Number of blades	Pure, int	Input
C_{shred}	Energy and wear shredding-related costs	€/h	Fixed
A	Surface of the blade (one side only)	m^2	Input
Th	Average thickness of the blade	m	Input
TH_{sh-1}	Shredder throughput	m^3/h	Algorithm

$$C_{IP-shred-1} = C_{IP-shred-setup} + N * C_{shred} * \frac{2 * A * th}{TH_{sh-1}} \tag{9}$$

The shredding throughput depends on the target output particle size, which determines the shredder grate size. Shredding throughput modelling breakdown is modeled in a dedicated algorithm and here is just summarized.

Variables name	Description	Unit of measure	Variable type
TH_{sh-1}	Shredder throughput	m^3/h	Algorithm
K_{sh}	Technology dependent constant	m^2/h	Fixed
d_{grate}	Grate size	mm	Input

$$TH_{sh-1} = K_{sh} * \frac{d_{grate}}{1000 \, mm/m} \tag{10}$$

Variables name	Description	Unit of measure	Variable type
C_{TR-1}	Ordinary transport cost	€	Cumulated
N_{TR-1}	Number of transports needed	Pure, int	Cumulated
$C_{TR\text{-fixed}}$	Transportation fixed cost	€	Fixed
$C_{TR\text{-km}}$	Transportation cost-per-kilometer	€ / km	Fixed
$D_{plant\text{-farm}}$	Wind farm—recycling plant distance	km	Input
V_{real-1}	Gross volume of blade pieces	m^3	Algorithm
$V_{max\text{-truck}}$	Maximum volume of a container	m^3	Fixed

$$C_{TR-1} = N_{TR-1} * \left[C_{TR-fixed} + C_{TR-km} * D_{plant-farm} \right] \qquad (11)$$

$$N_{TR-1} = int^+ \left\{ \frac{V_{real-1}}{V_{max-truck}} \right\} \qquad (12)$$

The gross volume to be transported is calculated by a dedicated algorithm and here, only the final mathematical formula is reported.

Variables name	Description	Unit of measure	Variable type
V_{real-1}	Real volume to be transported	m^3	Algorithm
V_{net}	Net GFRP volume	m^3	Cumulated
n_{cuts}	Number of cuts to be performed on-site	Pure number	Decision
A	Surface of the blade (one side only)	m^2	Input
th	Average thickness of the blade	m	Input

$$V_{real-1} = \left[\frac{n_{cuts} + 4}{n_{cuts} + 1} + \frac{1}{\sqrt[3]{n_{cuts} + 3}} \right] * V_{net} \qquad (13)$$

where:

$$V_{net} = 2 * A * th \qquad (14)$$

4.2 Cutting Process Modelling

Cutting length and number of cuts. A dedicated optimization algorithm (*Cutting Optimizator CO*) has been developed to evaluate a case-by-case optimal cutting strategy. The algorithm is able to treat a general digital image, here representing the wind blade, extrapolate the contour, and investigate an optimal cutting strategy to satisfy the model limitations (as pieces final size and concentration) maximizing or minimizing a set of pre-defined variables, as total cutting length, homogeneity within pieces, etc.

The *Cutting Optimizator* (CO) has been implemented in Python 3.0 and consists in a set of classes and methods that allows the user to find a sub-optimal cutting strategy for a surface having general shape. The CO offers methods to retrieve the contour of a surface and define lines (cuts) on it in order to split the surface and analyze the properties of the resulting parts. The CO provides also the possibility to couple the surface with a function that specify the concentration of material for each point contained in the surface.

The core of CO is a method, called "*ComputeCutting*" that takes in input:

- the set of points describing the contour of the surface to be cut,
- the maximal length and width of the parts that arise from the cutting,
- the maximal concentration of material of the parts that arise from the cutting.
- a general user-defined function $\mathbf{f}(-)$

The output is a set of cuts that optimize the splitting of the surface according to $\mathbf{f}$. The function $\mathbf{f}$ has been left general in such a way that the user can specify his own criteria. However, the input parameters of $\mathbf{f}$ are fixed. In particular, the function receives:

- A surface to be split,
- Two points, namely $\mathbf{a}$ and $\mathbf{b}$, that define the line of the cut,
- The description of the current best cut (two points, score value and the two parts that are generated by the cut),
- Three possible weights that can be used for general purposes.

Function $\mathbf{f}$ returns the best cut between the cut defined by $\mathbf{a}$ and $\mathbf{b}$, and the best cut.

The method "*ComputeCutting*" works on any kind of shape as long as its contour is specified as an array of x–y coordinates. Despite this, the CO offers a method that retrieves the set of coordinates from a digital image. This functionality has been included to avoid the manual generation of the shape of the surface under investigation.

The shape recognition is implemented by using the OpenCV library [22].

The optimization method. The optimization method implements a "*divide et impera*" strategy. The starting point is the original surface. The method finds the best cut to split the surface in two by using $\mathbf{f}$ to compare all the possible cuts. The best cut is used to generate two smaller parts. These two parts are stored and split in two parts each by using the same strategy that has been used for the original surface. The method cycles until all the parts satisfy the dimension and concentration criteria. Each cycle works with parts that are smaller than the predecessor.

Function f. At this moment, the function **f** is defined in such a way that it considers the minimal difference between:

- the length and the width of the projection of each part on a rectangle (denoted with **d**)
- the (approximated) concentration of each part (denoted with **c**)

and the length of the cut (denoted with **l**). The objective function is the weighted sum of these three measures. The best cut is the one able to minimize the sum. The measure **d**, **c** and **l** are normalized in such a way that no measure can dominate the others. Furthermore, each measure is associated with a weight that allows the user to increase the impact of one on the others.

On-site optimal cutting length and number of cuts. The cutting optimization algorithm provides a set of cuts with an inner hierarchy. There is a first cut, a second cut, a third and so on. Each of these cuts is associated with its own length. In the table below an example of a set of cuts provided by the algorithm, each one with its cutting length. The total cutting length exploited in all the other scenarios (full on-site cutting or full in plant cutting) is the sum of the single lengths.

Cut (#)	Length (m)
1	1.3
2	0.9
3	1
4	0.8
5	0.5
6	0.6
7	0.4
L, total cutting length (m)	5.5

The overall optimization model takes this cutting hierarchy and analyzed for each cutting level the associated on-site treatment costs and transportation cost, under the main assumptions that the more on-site cuts the more transport saturation.

For example:

Cut (#)	Length (m)	Number of on-site cuts	L*	L − L*
1	1.3	1	1.3	4.2
2	0.9	2	2.2	3.3
3	1	3	3.2	2.3
4	0.8	4	4	1.5
5	0.5	5	4.5	1
6	0.6	6	5.1	0.4
7	0.4	7	5.5	0

(continued)

(continued)

Cut (#)	Length (m)	Number of on-site cuts	L*	L − L*
L, total cutting length (m)	5.5			

For each of the seven hierarchy levels:

– The number of on-site cuts is exploited by the saturation algorithm to estimate the transport saturation and so the transportation costs.
– The on-site cutting length L* is exploited to calculate the costs of on-site cutting.
– The in plant cutting length L − L* is exploited to calculate the costs of in plant cutting.

Having these data, the main model is able to calculate the overall treatment costs associated to the on-site cutting for each hierarchy level and highlight the best solution.

5 Numerical Results

The development of the DSS as an ICT solution to drive the operator in the selection of the best demanufacturing route for each specific product under treatment has been based on an approach including the following main phases:

- identification of target composite products/parts in the four FiberEUse industrial sectors (i.e. wind energy, construction, aerospace and automotive);
- definition of a set of potentially significant quality characteristics to be evaluated before disassembly (e.g. type of connection, the critical issues related to the current disassembly procedures, disassembly technologies, as well as the logistics requirements);
- collection and analysis of requirements for each input product;
- clustering of the collected information in two macro-categories: products integrated in large infrastructures (i.e. wind blades, aircraft parts and construction components) and components from automotive sector (i.e. seat structure, rear panel, front-end, gear tunnel, leaf spring, monocoque, roof stiffener, roof bow, resonator, seat shell);
- definition of the recycling oriented integrated disassembly and transport problem;
- definition of the disassembly planning problem following two approaches, one for large infrastructures (innovative mathematical model) and one for components (common disassembly planning algorithms);
- definition of the DSS features for the identification of the most promising disassembly strategy for large infrastructures, according to the process costs minimization;
- implementation and validation of the DSS.

A model, implemented in Python 3.0 and MatLab, was developed to identify the most promising disassembly strategy in terms of costs minimization. The basic equations behind this model are addressed to analyze the identified scenarios and solve the related recycling oriented integrated disassembly and transport problem, in order to optimize the disassembly planning problem.

The main objective function is reported in the following:

$$minC = C_{OS-TREATMENTS} + C_{TRANSPORT} + C_{IP-TREATMENT} \quad (15)$$

$$minC = C_{OS-cut} + C_{OS-sh} + C_{TRANSPORT} + C_{IP-cut} + C_{IP-sh} + C_{IP-th} \quad (16)$$

Each term of the objective function is then related to specific variables described in the following:

- C_{OS-cut}. The cost for on-site cutting treatment depends on the: (i) cutting technology (feed rate, speed), (ii) energy consumption and tool wear, (iii) personnel costs, and (iv) set-up costs,
- C_{OS-sh}. The cost for on-site shredding treatment depends on the: (i) shredding technology (feeder dimension, feed rate, speed), (ii) energy consumption and tool wear, (iii) personnel costs, (iv) set-up costs and (v) output particles size (grate dimension),
- C_{IP-cut}. The cost for in plant cutting treatment depends on the: (i) cutting technology (feed rate, speed), (ii) energy consumption and tool wear, and (iii) personnel costs,
- C_{IP-sh}. The cost for in plant shredding treatment depends on the: (i) shredding technology (feeder dimension, feed rate, speed), (ii) energy consumption and tool wear, (iii) personnel costs, and (iv) output particles size (grate dimension),
- C_{IP-th}. The cost for thermal treatment depends on the: (i) technology (capacity), (ii) energy consumption, (iii) personnel costs, and (iv) residence time.

In order to provide a software tool that could be used in different disassembly problem, a specific software module has been designed and implemented. This software takes in input information about the product, technologies and logistic and administrative aspects (Table 1). Starting from these data, the tool runs the optimization and provides in output to the user the optimized disassembly strategy to isolate the target composite-made parts/components.

Preliminary validation tests have been performed for the demonstration of the efficiency of the developed DSS software tool. In the demonstration phase, the values of the input model variables have been provided by the FiberEUse industrial partners (i.e. material providers, material processors or end-users, logistic operators) directly involved at different levels of the value-chain.

Table 1 Input model variables

Category	Name	Description	Unit of measure
Product	N	Number	–
	D_1	Lenght	m
	D_2	Width	m
	D_3	Height	m
	th	Thickness	m
Selective disassembly	SD	Y/N	–
Personnel cost	C_{OS-OP}	On-site operator costs	€/h
	C_{IP-OP}	In plant operator costs	€/h
On-site cutting techonology	$C_{OS-CUT-E}$	Energy consumption and tool wear	€/h
	s_{OS-CUT}	Cutting speed	m min^{-1}
	fr_{OS-CUT}	Feed rate	m min^{-1}
	$C_{OS-CUT-SET}$	Set-up costs	€
In plant cutting techonology	$C_{IP-CUT-E}$	Energy consumption and tool wear	€/h
	s_{IP-CUT}	Cutting speed	m min^{-1}
	fr_{IP-CUT}	Feed rate	m min^{-1}
	$C_{IP-CUT-SET}$	Set-up costs	€
On-site shredding technologies	sh_{OS-l}	Feeder lenght	m
	sh_{OS-w}	Feeder width	m
	C_{OS-SH}	Energy consumption and tool wear	€/h
	s_{OS-SH}	Feed rate (model)	m^3/h
	d_{OS-SH}	Output particles dimension	mm
	$C_{OS-SH-SET}$	Set-up costs	€
	sat_{OS-SH}	Saturation (model)	–
	su_{OS-SH}	Downstream shred speed-up (model)	–
In plant shredding technologies	sh_{IP-l}	Feeder lenght	m
	sh_{IP-w}	Feeder width	m
	C_{IP-SH}	Energy consumption and tool wear	€/h
	s_{IP-SH}	Feed rate (model)	m^3/h
	d_{IP-SH}	Output particles dimension	mm
	$C_{IP-SH-SET}$	Set-up costs	€
In plant thermal technologies	th_{IP-l}	Feeder lenght	m
	th_{IP-w}	Feeder width	m

(continued)

Table 1 (continued)

Category	Name	Description	Unit of measure
	C_{IP-TH}	Energy consumption	kw/h
	s_{IP-TH}	Capacity	m^3/h
Logistic	C_{TR-OVS}	Exceptional load transport (included personnel)	€/km
	C_{TR-REG}	Non-exceptional load transport (included personnel)	€/km
	C_{TR-FEE}	Fee exceptional load	€
	C_{TR-ML}	Max load	m^3
	d_{TR}	Distance between installation and plant	km
Administrative	$C_{PER-OS-CUT}$	Permission on site cut	€
	$C_{PER-OS-SH}$	Permission on site shredding	€

The software tool has been validated according to the non-selective cut scenario for GFRP wind blades. Some of the input data of the demonstration test are reported in the following:

- product: GF wind blade;
- number of products: 18 wind blades (wind turbines: 6);
- strategy: non-selective cutting;
- site-plant distance: 500 km;
- target output particles dimension: 6 mm.

The model has the objective to select the cheapest option among the following possible routes:

- Route 1: GF wind blades are cut in situ to increase truck saturation and reach acceptable dimension for further shredding treatment, transported through regular transport and shredded in plant to reach the target output dimension;
- Route 2: GF wind blades are cut and shredded in situ to increase truck saturation, transported through regular transport and shredded again in plant to reach the target output dimension;
- Route 3: GF wind blades are cut in situ to increase truck saturation, transported through regular transport, cut again in plant to reach acceptable dimension for further shredding treatment and shredded in plant to reach the target output dimension.
- Route 4: GF wind blades are dismantled in-situ and transported by using an exceptional load transport to the recycling plant. Here the products are cut to reach acceptable dimension for further shredding treatment.

The results of validation are reported in Table 2.
According to the input values, the optimal disassembly scenario is Route 3.

Table 2 Demonstration test results: non-selective cut scenario for GFRP wind blades	Cost for Route 1 (€)	Cost for Route 2 (€)	Cost for Route 3 (€)	Cost for Route 4 (€)
	79.28992	81.90088	78.67105	91.94732

6 Conclusions

In this Chapter, effective and efficient solutions and approaches for the disassembly of EoL composite products coming from different sectors have been analyzed in depth, in order to derive the most promising disassembly strategy to isolate the target composite-made parts/components.

Different macro- and sub-scenarios have been analyzed and specific mathematical models for large infrastructures have been developed and proposed.

Finally, the developed models have been validated and a specific software tool, based on a DSS approach, has been designed and implemented for the automatic identification of the most promising disassembly strategy. This software takes in input information about the characteristics of EoL product, the key processes and technologies, logistic aspects, the required target output product (both in terms of dimension and fibers content) and provides to the user the cutting path (if needed) and the optimal disassembly scenario, according to the process costs minimization. Similar conclusions can be outlined for the disassembly of EoL airplanes, considering the different technologies to treat CFRP, for which the mixed treatment is the best solution.

References

1. MacArthur, E., et al.: Towards the circular economy. J. Ind. Ecol. **2**, 23–44 (2013)
2. Oliveaux, G., Dandy, L., Leeke, G.: Current status of recycling of fibre reinforced polymers: review of technologies, reuse and resulting properties. Progress in Materials Science 61–99 (2015)
3. Perry, J.: Sky-high potential for aircraft recycling. Aircraft Maintenance, pp. 2–5 (March 2012)
4. Ribeiro, J.S., de Oliveira Gomes, J.: Proposed framework for end-of-life aircraft recycling. Procedia CIRP **26**, 311–316 (2015)
5. Dayi, O., Afshar, A.: A lean based process planning for aircraft disassembly. IFAC-PapersOnLine **49**(2), 054–059 (2016)
6. Martinez Luengo, M., Kolios, A.: Failure mode identification and end of life scenarios of offshore wind turbines: a review. Energies **8**(8), 8339–8354 (2015)
7. Vongbunyong, S., Chen, H.W.: Disassembly automation, pp. 25–54. Springer, Cham (2015)
8. Gungor, A., Gupta, S.M.: An evaluation methodology for disassembly processes. Comput Ind Eng **33**(1–2), 329–332 (1997)
9. Lambert, A.J.D.: Disassembly sequencing: a survey. Int. J. Prod. Res. **41**(16), 3721–3759 (2003)
10. Pickering, S.J.: Recycling technologies for thermoset composite materials—current status. Compos. A **37**, 1206–1215 (2006)

11. Schinner, G., Brandt, J., Richter, H.: Recycling carbon-fiber-reinforced thermoplastic composites. J Thermoplast Compos Mater **9**, 239–245 (1996)
12. Kouparitsas, C.E., Kartali, C.N., Varelidis, P.C., Tsenoglou, C.J., Papaspyrides, C.D.: Recycling of the fibrous fraction of reinforced thermoset composites. Polym Compos **23**, 682–689 (2002)
13. Torres, A., de Marco, I., Caballero, B.M., Laresgoiti, M.F., Legarreta, J.A., Cabrero, M.A.: Recycling by pyrolysis of thermoset composites: characteristics of the liquid and gases fuels obtained. Fuel **79**, 897–902 (2000)
14. Cunliffe, A.M., Williams, P.T.: Characterisation of products from the recycling of glass fibre reinforced polyester waste by pyrolysis. Fuel **82**, 2223–2230 (2003)
15. Feih, S., Boiocchi, E., Mathys, G., Mathys, Z., Gibson, A.G., Mouritz, A.P.: Mechanical properties of thermally-treated and recycled glass fibres. Compos. B **42**, 350–358 (2011)
16. Gosau, J.M., Tyler, F.W., Allred, R.E.: Carbon fiber reclamation from state-of-art 2nd generation aircraft composites. In: Proceedings of the international SAMPE symposium and exhibition. Baltimore, MD, USA (2009)
17. Åkesson, D., Foltynowicz, Z., Christéen, J., Skrifvars, M.: Microwave pyrolysis as a method of recycling glass fibre from used blades of wind turbines. J Reinf Plast Compos **31**, 1136–1142 (2012)
18. López, F.A., Rodríguez, O., Alguacil, F.J., García-Díaz, I., Centeno, T.A., García-Fierro, J.: Recovery of carbon fibres by the thermolysis and gasification of waste prepeg. J. Anal. Appl. Pyrol. **104**, 675–683 (2013)
19. Iwaya, T., Tokuno, S., Sasaki, M., Goto, M., Shibata, K.: Recycling of fiber reinforced plastics using depolymerization by solvothermal reaction with catalyst. J. Mater. Sci. **43**, 2452–2456 (2008)
20. Pinero-Hernanz, R., Dodds, C., Hyde, J., Garcia-Serna, J., Poliakoff, M., Lester, E.: Chemical recycling of carbon fibre reinforced composites in nearcritical and supercritical water. Compos. A **39**, 454–461 (2008)
21. Chandramohan, D., Murali, B.: Machining of composites—a review. Academic Journal of Manufacturing Engineering **12**(3), 67–71 (2014)
22. Open CV Homepage. https://opencv.org/. Last accessed 15 March 2021

Smart Composite Mechanical Demanufacturing Processes

Marco Diani⬡, Nicoletta Picone⬡, and Marcello Colledani⬡

Abstract Recycling of Glass Fibers Reinforced Plastics (GFRP) can be preferentially performed through mechanical processes due to the low cost of virgin fibers. Because of the poorer mechanical properties after comminution, the most interesting solution to reuse this material is a cross-sectorial approach, in which particles obtained through shredding of products from one sector are used in another sector. To allow this, a fine control on the particles dimension is fundamental, together with the minimization of operational costs. In this chapter, after a deep analysis on the available size reduction technologies and a preliminary feasibility analysis on the products involved in Use-Case 1 of the FiberEUse project, a 2-step architecture to optimize these two characteristics is presented. The models for both steps are shown and the developed solutions is applied to the End-of-Life products, demonstrating the potential of this approach, leading to optimal dimension of the particle with operational costs lower than both virgin fibers and disposal costs.

Keywords GFRP · Mechanical recycling · Optimization · Dimensional distribution · Operational costs

1 Introduction

Due to their applications and, in particular, to lower price of virgin fibers, Glass Fibers Reinforced Plastics (GFRP) are currently inserted in co-processing of cement [1] or recycled through mechanical processes, in particular using shredding technologies. In addition, comminution reduces length of the fibers, decreasing related mechanical properties, hindering the possibility of a closed-loop recycling. Moreover, a complete liberation of the fibers from the resin matrix residues is impossible.

M. Diani (✉) · M. Colledani
Department of Mechanical Engineering, Politecnico Di Milano, Via La Masa 1, 20156 Milan, Italy
e-mail: marco.diani@polimi.it

N. Picone
STIIMA-CNR, Via Alfonso Corti 12, 20133 Milan, Italy

M. Colledani and S. Turri (eds.), *Systemic Circular Economy Solutions for Fiber Reinforced Composites*, Digital Innovations in Architecture, Engineering and Construction, https://doi.org/10.1007/978-3-031-22352-5_4

To overcome these limitations, a cross-sectorial approach has to be followed, in which shredded GFRP from sectors with higher requirements on mechanical properties could be reused in new products with lower requirements. Several possible applications for GFRP in function of the required tensile strength and tensile modulus are possible. As an example, particles in output from a mechanical recycling process of End-of-Life wind blades could be used as fillers for thermal, electric and sound insulation but also in higher-added value applications as automotive. Furthermore, they could be used in other sectors as sanitary, sports and leisure equipment or design products, which form the Use-Case 1 of FiberEUse project. Different reuse options need different dimensional and morphological properties.

Leveraging on a cross-sectorial approach can open new potentials for composite made parts recycling, remanufacturing and re-use under a systemic circular economy perspective. To enable this cross-sectorial approach, a recycling driven by the characteristics needed by the sector in which the fibers will be used as secondary raw material (that are driven especially by particles dimensions) has to be implemented. In addition, due to the low price of virgin glass fibers, a minimization of the process cost is needed to make the recycled GFRP competitive as secondary raw materials.

2 State of the Art

As shown in [2], recycling of composite materials is performed following three different procedures: mechanical, thermal and chemical methods. Focusing on mechanical recycling, it has been applied in particular on GFRP, in particular to Sheet Moulding Compounds (SMC) and Bulk Moulding Compounds (BMC) [3]. The shredding process aims to create particles that can be included as reinforcement in new products. Palmer et al. [4] have shown also the possibility to have a real closed-loop recycling of thermoset composites reintegrating the recovered fibers as reinforcement in SMC and BMC automotive.

Size reduction processes aim to obtain high liberation of target materials and they are able to create homogenous (both in shape and in size) particles mixtures at a desired dimension [5]. Palmer et al. [4] underlines that higher degrees of liberation are achieved by a particle made of a small number of materials and the highest possible liberation is obtained for particles made of only one material.

Different attempts to develop mathematical models of a shredding process are available in literature, in particular focused on the mining field. Gaudin [6] proposed the first model based on mineral texture (called by the author "Mineral Dressing"), simplifying the mineralogical texture of an ore and predicting the particles distribution as a function of the size. King [7] improved this approach, proposing an equation to predict linear liberation distribution as a function of the particle size. Instead of transforming the original ore texture, the author used a linear probe across the image of a polished section of an ore to characterize it. Meloy [8] developed a texture transformation for the original ore texture to a simple geometry as spheres or cubes. This allows to consider the shredding process as the broken of the regular geometric model into smaller particles with the same shape and to calculate the liberation distribution through geometrical formulas. Barbery [9] improved the approach using a Boolean model. In this way, under some assumptions (in particular the non-preferential breakage), he was able to calculate the fraction of liberated particles. King [10] derived an analytical solution to the multidimensional integrodifferential equation applied to a shredding batch process for multi-component mineral system. The obtained solution was compared to experimental data, showing that the model is reliable.

More recently, several approaches tried to model the evolution along time of the particle size distribution through the determination of the probability that particles of a specific liberation class generate particles of another liberation class. One of the most studied approach is the "Textural Modelling". It used the mineralogical information acquired during the process to predict the dimensional distribution evolution. Gay [11] developed a mechanistic method in which ore properties are direct consequence of the changes in composition distribution.

One the other hand, literature studies are focused on dimensional distribution. Gay [12] takes the "kernel estimation" approach and applied it to this problem. The kernel represents the volumetric frequency of the event that one particle with a specific composition and dimension (*parent particles*) will generate another particle with different characteristics (*progeny particles*). Once the kernel has been determined, it could be applied to a new set of input particles to predict composition and distribution of progeny particles. The author used a probabilistic approach to infer the kernel properties without deriving mineral texture and mechanistic properties but using only effective experimental data. The considered method is the maximum entropy principle. The author exploited it to explain and predict multi-sized progeny particles from single-sized parent particles.

Despite the several examples in mineral field, shredding process in recycling has been addressed by few works. In particular, [13] modelled the relationship between product mineralogy and size reduction and liberation during the EoL vehicles comminution to maximize the efficiency of following separation processes. They also underlined the differences in shredding between mineral and recycling fields. In particular, recycling is a continuous process with multi-material products, with strongly dependency on product design. This concept has been improved and demonstrated through simulations by Castro [14]. This approach requires a relevant number of experiments to train the model.

For this reason, the approach used to model the shredding process in FiberEUse project is based on *Population Balance Models (PBMs)*. Population Balance Models are mathematical models able to represent the evolution of particles characteristics through three different quantities, namely the percentage of each particle in a specific pre-determined size class per each time unit, the evolution of every single size classes per time unit and the proportion of particles that are able to exit from the comminution chamber per time unit.

PBMs have been largely used in literature, dividing the models depending on time, which could be considered discrete or continuous. Due to the relatively short average residence time of the particles, the discretization of size classes and the simplicity in describing the process as a series of short elementary breakage events, discrete time PBMs have been used, both in literature and in Task 2.2.

Different examples of PBMs for shredding are present in literature. A recent work by Bilgili [15] analyzes the non-linear effects of particles in comminution dividing them in three different types and underling the consequent deviations from the linear behavior.

Bilgili [16] developed a model for long time size reduction process including all these deviations. In particular, they decompose the breakage rate in one linear and one depending on the population. This second breakage rate is described by a functional. This last one explicitly considers the three types of deviations and the interactions between particles with different dimensions. This model is a time continuous non-linear model valid for dense-phase comminution process with extended shredding time.

These models require several experiments for the training. For this reason, a revision of the assumptions has to be done to adapt them to recycling field and allowing the control and optimization of recycling processes reducing the number and the cost of experiments to estimate the parameters.

3 Rationale of the Work

To shred GFRP EoL products, a size reduction procedure as in Fig. 1 has been developed. A preliminary coarse shredding is performed to reduce the dimension of the products, achieving suitable size for the fine comminution. This process-chain allows a fine control on the final particles dimensions together with a maximization of the throughput.

Different comminution technologies are commercially available. A deep analysis on most of them has been carried on and it is summarized in Table 1.

Following this analysis, the first shredding step has been performed using a single shaft shear shredder by Erdwich (model M600M-400), reaching a particles dimension lower than 10 mm. The obtained shredded material has been treated with a cutting mill by Retsch (model SM-300) using several different grids.

While the coarse shredding aims to create a particles mixture suitable for the second stage, the fine comminution step is responsible for the dimensional distribution of the particles and it mostly impacts on the operational costs. As stated before, different reuse options need different particles characteristics (in particular dimensions). As a consequence, the control of the shredding process is fundamental. A first formalization has been developed as in Fig. 2.

The input material, composed by entire products or coarse shredded particles, needs to be characterized. In particular, the most interesting information are the dimensional distribution of particles and, if needed for following steps, all the geometrical and morphological information on the mixture. These data are gathered and stored in a dedicated module, that could be useful both for model training or for future purposes. A simulation module receives the information, predicting

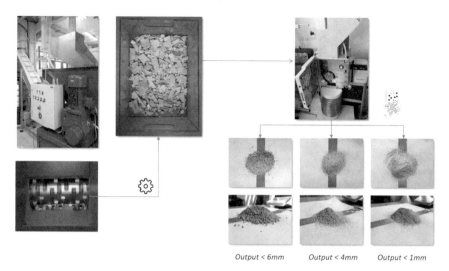

Output < 6mm   Output < 4mm   Output < 1mm

Fig. 1 Two stage size reduction process-chain

Table 1 Comparison of the different shredding technologies

Technology	Throughput	Efficiency	Suitability for composites
Two shafts shear shredder	High	Risk of fibers wrapping	Good for coarse shredding
Single shaft shear shredder	High	Good	Good for coarse shredding
Hammermill	High	Good for thermoset matrixes, not suitable for thermoplastic matrixes	Good for coarse shredding of thermoset composites
Impact crusher	High	Not suitable for thermoplastic matrixes	Low
Jaw crusher	High	Not suitable for thermoplastic matrixes	Low
Cutting mill	Mid	Good	Best for fine shredding
Disc mill	Low	Not suitable for thermoplastic matrixes	Low
Ultracentrifugal mill	Low	Good	Good for fine pulverization (preparation for 3D printing

the evolution over time of dimensional distribution and of all the other interested characteristics. Then an optimization module, able to exchange information with the simulation one, processes these data together with information about target distribution and characteristics. The optimized actions suggested by this module are able to change parameters of the physical shredding process, leading to optimal output. In addition, information on the machine status (as energy consumption) could be

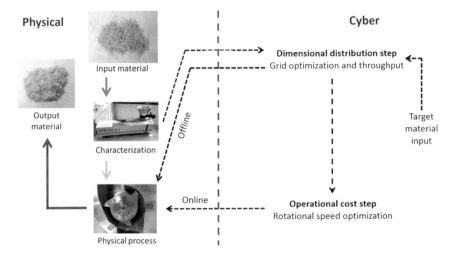

Fig. 2 First procedure for shredding control

acquired and processed in the same way, to obtain a process optimized not only in terms of material characteristics but also in cost, fundamental in recycling. From this first formalization, it is evident the needs to have good quality and fast data gathering systems in the physical world and robust and validated models in the cyber one. In addition, it is fundamental to understand which parameters could be controlled in a shredding process and how they could be changed.

To design the best control procedure, a deep analysis on common characteristics among the different technologies has been done and the most important parameters have been studied, dividing it in controllable and non-controllable ones. These parameters are described below and summarized in Table 2.

- *Volume of the chamber.* This design parameter controls the maximum amount of material that can be processed simultaneously. This mass is function not only of the volume of the chamber but also of the material density (high density results in higher quantity of material that can be treated and vice versa).
- *Number of breaking/cutting elements.* Every technology is based on different breaking principle but all the machines have number of tools as design parameter. As an example, in cutting mill this is the number of inserts, in hammermill the number of hammers and in impact crusher the number of breaking plates.

Table 2 Design and controllable parameters of a shredding process

Design parameters	Controllable parameters
Volume of the chamber	Grate size (offline)
Number of cutting elements	Rotational speed (online)
Number of cutters per cutting elements	Feed rate (online)

- *Number of breakers/cutters per cutting element.* The breaking or cutting tools could be designed with one or multiple elements that are able to perform size reduction. As an example, in cutting mills there are typically four cutters per insert but only one working at a time with different principles and under different stresses and strains. On the other hand, impact crusher has usually only one face of the breaking plate which undergoes the impact force.
- *Grate size or free falling aperture dimension.* The mechanism of particles exit in case of continuous flow processes (the most interesting for recycling) is a controllable parameter. Typically, this could be done through two different approaches. The first one is based on a grate, that could be easily changed, with a specific hole size and shape. Particles that have a dimension lower than the hole size are able to exit, while the others remain in the chamber to be reduced in dimension again. This is used for example in cutting mill, shear shredder or hammermill. The second one is the so called free falling aperture, in which there is a space between two elements of the shredding machine, allowing particles that can pass through that space to exit from the process, as in case of impact crusher and jaw crusher. In both cases, grate size influences both throughput and dimensional distribution of the particles in output.
- *Speed of the breaking/cutting mechanism.* Speed of the breaking or cutting mechanism is typically a controllable parameter that strongly influences both throughput and costs of the process. As an example, the moving part in cutting mills and in shear shredders is the rotor on which the cutting tools are mounted while in hammermills, chain shredders and ultra centrifugal mills the central shaft moves, in impact crushers the drum with the hangling systems, in jaw crushers the moving jaw and in disc mills the rotating disc itself.
- *Feed rate.* Feed rate is common to all the processes and could be controlled regulating the material entering in the chamber every time unit. There is an upper limit to this value that depends on several factors as grate size (or free falling aperture) and, in part, speed of the size reduction mechanism.

From this table it is evident that grate size on one side and rotational speed and feed rate on the other need two different control approaches. A machine stop is needed to change the grid and this task is performed manually. This results in loss of time both for the change and for transition to stationary process. On the other hand speed of the rotor and throughput (which is equivalent to the feed rate in stationary process) could be controlled online, avoiding stops. As a consequence, a 2-step approach has been developed as in Table 3.

The first step is dedicated to the optimization and control of the offline parameters, in particular the grate size. The objective is to optimize the dimensional and morphological distribution of particles in output, increasing the liberation of target materials and obtaining particles suitable for following processes (as recycling or direct reuse). This step takes as inputs the dimensional distribution of the particles to process and the target output distribution, typically suggested by an operator. The output will be the best grate to use in the size reduction process and the throughput expressed in mass per time interval.

Table 3 2-step approach for control of size reduction processes

Step	Objective	Input	Output	How
1	Optimize dimensional distribution of output particles	– Dimensional distribution of input particles – Target output distribution	– Optimal grid size – Related throughput	Population Balance Model (PBM) and least squares method (optimization)—offline
2	Minimize operational costs (energy consumption and tool wear)	– Throughput (from step 1)	– Rotational speed	Cyber-Physical System—online

The second step has as objective the minimization of operational costs, in particular due to energy consumption and tool wear. It takes as input the unitary throughput calculated in Step 1 and it gives as results the rotational speed and the throughput expressed in mass per time unit (as seconds or hours).

In the next section, the models developed for both steps will be presented, with an emphasis on the first one.

4   Methodology

4.1   Feasibility Analysis

A preliminary analysis has been performed on different EoL samples (e.g. sports equipment, sanitary, wind blades and construction). After the process-chain explained in Fig. 1, the obtained particles have been analyzed using two different technologies as shown in Fig. 3. The first one is an analytical vibrating sieve by Retsch which divides the sample in 9 different size classes, from 63 μm up to 10 mm. The second technology used is an optical Computerized Particles Analyser CPA 2-1 by Haver & Boecker. It is able to perform a real time dimensional and morphological analysis giving different distributions and the particle list with all the information available to be used (examples of typical output are in Figs. 4 and 5). Figure 6 shows a typical output result of a dimensional and morphological analysis (obtained on particles from shredding of sanitary products).

Three repetitions have been performed for every EoL product at three different rotational speeds. These analyses underline the importance of the grid on the density distribution. Changes in the grate size do not result in morphological characteristics variation while show relevant differences in density distribution. In addition, from the analysis on the results at different rotational speeds could be inferred that this parameter does not affect the final dimensional distribution.

Fig. 3 Analytical vibrating sieve on the left (Retsch) and optical computerized particles analyzer on the right (CPA 2-1 of Haver & Boecker)

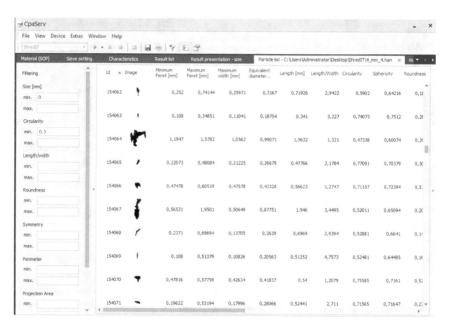

Fig. 4 Example of output particles list of CPA 2-1

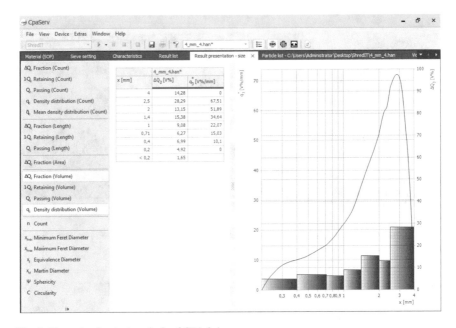

Fig. 5 Example of output analysis of CPA 2-1

4.2 Criticalities Analysis

As EoL products from different sectors has been shredded, several criticalities depending on material and shape have been found during the performed preliminary experiments. Issues and adopted solutions are summarized in Table 4 only for products that raised problems.

4.3 Step 1: Dimensional Distribution Model

The objective of the developed model is to predict the continuous time evolution of mass distribution of the particles inside and outside a size reduction machine. Due to the nature of the process, a discrete-time Population Balance Model (PBM) has been developed. The time interval, also called breakage interval, has been denoted with Δ, which represents the smallest time interval in which a breakage of a particle may occur (or, equivalently, it represents the time between two consecutive breakages). It is a function of rotational speed and number of cutting elements and it could be defined as

$$\Delta = \frac{1}{n_{cutters} \cdot \omega} \tag{1}$$

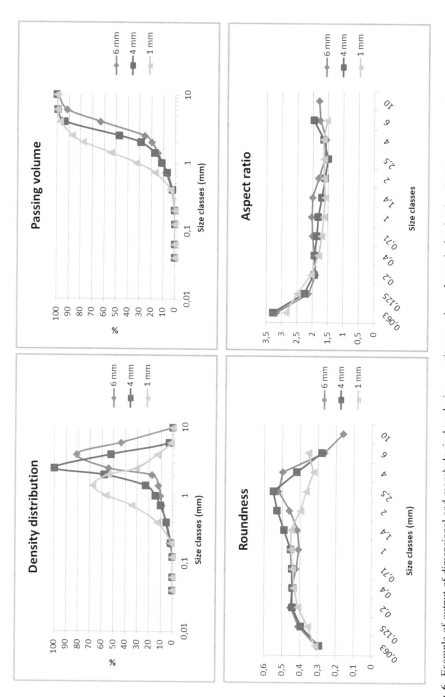

Fig. 6 Example of output of dimensional and morphological analysis on sanitary products. In particular density distribution (top left), passing volume (top right), roundness (bottom left), aspect ratio (bottom right)

Table 4 Criticalities analysis and adopted solutions

Sector	Products	Criticalities (coarse)	Criticalities (fine)	Adopted solution	Impact on the process
Sports equipment	Pre-preg	Samples entwined around the shaft	Melting of uncured resin	• Manual cut of the samples • Increase in speed (shorter residence time)	Medium
Sanitary	Bath tube scraps	Presence of PE film	Melting of resin	• PE film manual removal • Increase in speed (shorter residence time) • Intermediate step with larger grate size	High
Wind blades	EoL blades—Central part	Longer parts get stuck in the chamber		Properly cut of the EoL samples	Low

where ω is the rotational speed of the rotor expressed in round per minute (rpm) and $n_{cutters}$ is the number of cutting elements. The product at the denominator ($\omega \bullet n_{cutters}$) represents the number of breakages per minute.

In addition, also a discrete particle division into n size classes with each class indexed by $i = 1;...;n$ has been considered. Classes with larger dimensions have higher index number, while the lower ones are for smaller particles. The total number n of classes and the related sizes should be decided on the basis of both interesting information to derive from the model (as the target dimensions of the particles) and the resolution of the measurement technique and instruments used for mixture characterization (as the particle analyzer presented in Sect. 4.1).

Defining the breakage and selection matrix **P**, describing the probability of a particle to pass from one class to a lower one, as

$$P = \begin{bmatrix} p_{11} & \cdots & p_{1n} \\ \vdots & \ddots & \vdots \\ 0 & \cdots & p_{nn} \end{bmatrix} \tag{2}$$

where the element p_{ij} represents the probability of a particle in class I to move in class j after a time interval and the discharge matrix **D** as a diagonal matrix, describing the probability of a particle in one class to exit from the chamber, as

$$D = \begin{bmatrix} d_{11} & \cdots & 0 \\ \vdots & \ddots & \vdots \\ 0 & \cdots & d_{nn} \end{bmatrix} \tag{3}$$

where the element d_{ii} represents the probability of a particle in class i to leave the size reduction chamber, the mass evolution of the particles during the shredding process could be written as

$$M(k) = \mathbf{P} \cdot M^{CH}(k-1) + \mathbf{P} \cdot M^{IN,ADJ}(k-1) \tag{4}$$

$$M^{CH}(k) = (I - \mathbf{D}) \cdot M(k) \tag{5}$$

$$M^{OUT}(k) = \mathbf{D} \cdot M(k) \tag{6}$$

where $\mathbf{M}(k)$ is the distribution of the mass under process at time k, $\mathbf{M}^{IN,ADJ}(k)$ is the distribution of the mass entering the chamber at time k considering the available space, $\mathbf{M}^{CH}(k)$ is the distribution of the mass in the chamber at time k and $\mathbf{M}^{OUT}(k)$ is the distribution of the mass which exits from the chamber at each time step.

The distribution of the mass entering the chamber at time k considers the saturation of the chamber and is defined as

$$M^{IN,ADJ}(k) = \begin{cases} M^{IN}(k) & if \ \sum_{i=1}^{n} M_i^{CH}(k) + \sum_{i=1}^{n} M_i^{IN}(k) \leq m^{max} \\ \alpha(k)M^{IN}(k) & otherwise \end{cases} \tag{7}$$

where

$$\alpha(k) = \frac{m^{max} - \sum_{i=1}^{n} M_i^{CH}(k)}{\sum_{i=1}^{n} M_i^{IN}(k)} \tag{8}$$

The explained PBM is Markovian, as $\mathbf{P}$ and $\mathbf{D}$ are two transition matrices and the state $\mathbf{M}(k+1)$ at time $k+1$ only depends on the state $\mathbf{M}(k) = \mathbf{M}^{CH}(k) + \mathbf{M}^{IN,ADJ}(k)$ at time k.

Two different hypotheses have been introduced in this model (and validated in a previous work [17]) to simplify it, making it possible to use for real process model and control. These two assumptions have been named *multiplication* and *homogeneity*.

The multiplication assumption affirms that the size distribution of particles in output only depends on the number of breakage intervals, independently from the rotor speed ω and shredding time τ that generates that given number of breakage intervals. As a consequence, the output size distribution only depends on k and it is the same for all pairs $(\omega;\tau)$ such that the product $\omega \bullet \tau$ is constant.

The homogeneity assumption states that the transition matrix P does not depend on time (either calendar time t or the number of breakage events k). Thus, the comminution process does not depend on how long the rotor was previously running.

These two hypotheses considerably reduce the number of experiments to train the model and enable the usage of this model to directly control the process.

As could be noticed the distribution of the mass which exits from the chamber at each time step $\mathbf{M}^{OUT}(k)$ is a function of discharge matrix $\mathbf{D}$ and, as a consequence, it strongly depends on the mounted grid. It is possible to use the developed PBM model to predict the output distribution and to find the optimal grate size for the comminution and to achieve target requirements on particles dimensions for following processes.

4.4 Step 2: Operational Cost Model

Operational costs are fundamental to implement processes that are not only environmentally but also economically sustainable. This is more important when working with low cost virgin materials, as glass fibers composite plastics. In addition, to incentivize recycling it is important, if possible, to obtain materials with a cost that is lower than disposal costs. In shredding processes, the operational costs could be divided in costs due to energy consumption and to tool wear. While shredding at higher speeds results in lower residence times, with reduced energy consumption, tool wear increases. For this reason, an optimization of total operational costs is fundamental. In particular

$$C = C_{ec} + C_{tw} \tag{9}$$

where C is the total cost of the process, C_{ec} is the cost due to energy consumption and C_{tw} is the cost due to tool wear. These two quantities will be analyzed in this section.

Energy consumption in a shredding process depends on several factors as follows.

- *Absorbed power.* The absorbed power is technology and machine dependent.
- *Throughput.* The throughput influences the particles residence time and, as a consequence, the energy consumption.
- *Residence time.* The longer the particles remains in the chamber, the higher is the energy consumption to comminute them.
- *Rotational speed.* Higher rotational speeds typically means higher energy consumption but in a shorter time and vice versa.
- *Material.* Harder materials requires higher power while softer materials have to be shredded for longer time.
- *Saturation of the chamber.* The saturation of the chamber influences the throughput. In addition, higher saturation level results in more particles that could be broken at each time step.
- *Dimensional gap between input and output particles.* Higher differences in the dimensions between input and output particles result in higher energy consumption.

In addition, the process cost due to energy consumption depends also on the cost of electric energy, that is Country and Region dependent. It is also function of the availability of renewable energy and of the accessibility of supply.

The cost for energy consumption depends on the absorbed power as (from classical physics)

$$C_{ec} = C_{ee} \cdot P \cdot t \tag{10}$$

where C_{ec} is the cost for the electric consumption, C_{ee} is the cost for electric energy in €/kWh (Country and Region dependent), P is the total power absorbed and t is the time of the process. Dividing the absorbed power into two different factors [18], total absorbed power P could be written as

$$P = P_0 + P_{shredding} \tag{11}$$

where P_0 is the power at zero load absorbed to run both mechanical and electrical parts of the machine that depends on the rotational speed and $P_{shredding}$ is the power absorbed for the physical shredding process (considered as the act of breaking particles), function of material to treat, size reduction technology, tools type and shape and rotational speed. These two terms have been experimentally studied, finding for both a linear behavior as

$$P_0 = k_0 \cdot \omega + c_0 \tag{12}$$

$$P_{shredding} = (k_{sh} \cdot \omega + c_{sh}) \cdot S \tag{13}$$

where k_0, c_0, k_{sh} and c_{sh} are experimental parameters describing the power needed for the functioning of electrical and electronic equipment and the linear behavior of the absorbed power while increasing the rotational speed, and S is the saturation of the chamber. Taking into consideration the definition of the discrete time presented in Sect. 4.3 and defining P_{sh} as

$$P_{sh} = k_{sh} \cdot \omega + c_{sh} \tag{14}$$

the cost due to energy consumption could be written as

$$C_{ec} = C_{ee} \cdot (P_0 + P_{sh} \cdot S) \cdot \frac{m_{tot}}{Th_{tu}} \cdot \frac{1}{n_{cutters} \cdot \omega} \tag{15}$$

where m_{tot} is the total mass to treat and Th_{tu} is the throughput expressed in kg per time interval.

On the other hand, the cost due to tool wear depends on different factors as follows.

- *Cost of the tool/insert.* Tools and inserts cost depends on several characteristics as material, shape and geometry, number of cutting elements and surface finishing. More performing tools (with higher mechanical properties) cost more but the life of the tool is usually longer (if the material of the tool is well coupled with the material to treat).
- *Cost for tool/insert change.* The substitution of worn tools and inserts has a cost due both to the time needed to perform this job and to the stop of the shredding

machine. Easily replaceable tools and inserts have to be preferred in order to reduce this term together with the possibility to mount inserts with multiple cutters.

- *Number of tools/inserts.* A higher number of tools and inserts increases the efficiency of the shredding machine, in particular in terms of throughput. At the same time, more cutting elements results in higher costs both for the elements themselves and for substitution.
- *Number of cutters.* Several shredding machines are designed including multiple cutters inserts, with only one cutter working at a time. This configuration gives the possibility to increase the lifetime of the inserts and decrease the time need for inserts change (avoiding to actually substitute the insert every time), leading to a significant cost reduction.
- *Residence time.* The longer the particles remain in the chamber, the higher will be the tool wear per kg of material treated.
- *Rotational speed.* Higher rotational speeds result in higher tool wear.
- *Throughput.* The throughput influences the particles residence time and, as a consequence, the tool wear.
- *Material to treat.* Comminution of hard materials results in higher tool wear while shredding of soft and deformable materials could lead to longer residence times. The coupling of both material to treat and material of the tool has been investigated in literature leading to the two coefficients n and C of the Taylor's law in machining (i.e. $v_c T^n = c$ with v_c cutting speed and T lifetime of the tool) [19].

In addition to the described factors, the inhomogeneity in tool wear has the capability to relevantly affect the total cost for shredding. Considering the definition of the discrete time presented in Sect. 4.3, it could be noticed from this equation that Δ (and as a consequence the throughput expressed in kg/s) heavily depends on the number of cutting elements. If an insert fails, the others could make up for its lack. This results in a longer breakage interval Δ and, as a consequence, in an increase in operational costs (both due to tool wear and energy consumption). If another tool fails, the other ones could continue the shredding process but increasing the breakage interval and, as a consequence, the cost and so on, following an exponential curve. The process could continue until the cost is lower than the maximum acceptable cost (or, in an equivalent way, if the revenues are higher than the target revenues). To overcome this issue, an average value for failure time of the tools has been considered. At this time, the shredding process is stopped and the inserts are changed.

Assuming that the time for tool change is equal with respect to the time for cutter change (this hypothesis is valid for long time processes as in the case of shredding) the cost due to the tool wear could be written as

$$C_{tw} = n_{tools} \cdot \left(C_{tc} + \frac{C_{tool}}{n_{cutters}} \right) \cdot \frac{t}{T} \tag{16}$$

where C_{tw} is the cost due to tool wear, n_{tools} is the number of tools, C_{tc} is the cost for the tool change (or, equivalently, the cost for cutter change), C_{tool} is the market cost of a single tool, $n_{cutters}$ is the number of cutters for every tool, t is the time of the process and T is the average lifetime of a single tool.

Taking into consideration the definition of the discrete time presented in Sect. 4.3 and adapting the Taylor's law for tool wear in machining [19], the cost due to tool wear in a shredding process could be written as

$$C_{tw} = \left(C_{tc} + \frac{C_{tool}}{n_{cutters}} \right) \cdot \frac{m_{tot}}{Th_{tu}\left(\frac{c}{r}\right)^{\frac{1}{n}}} \cdot \frac{1}{\omega^{\frac{n-1}{n}}} \tag{17}$$

where m_{tot} is the total mass to treat, Th_{tu} is the throughput expressed in kg per time interval, r is the radius of the shredding rotor, ω is the rotational speed and c and n are the experimental parameters of the Taylor's law.

Finally, the final model for operational costs of a shredding process is

$$C = C_{ee} \cdot (P_0 + P_{sh} \cdot S) \cdot \frac{m_{tot}}{Th_{tu}} \cdot \frac{1}{n_{cutters} \cdot \omega} + \left(C_{tc} + \frac{C_{tool}}{n_{cutters}} \right) \cdot \frac{m_{tot}}{Th_{tu}\left(\frac{c}{r}\right)^{\frac{1}{n}}} \cdot \frac{1}{\omega^{\frac{n-1}{n}}} \tag{18}$$

As a consequence, the minimization of the cost could be done optimizing the rotational speed using this equation. In particular, depending on the material to treat, the experimental parameters and, as a consequence, the behaviour of cost due to electric energy consumption and cost due to tool wear could be opposite, leading to a situation as in Fig. 7, in which the optimal rotational speed is in between the minimum and the maximum acceptable rotational speeds.

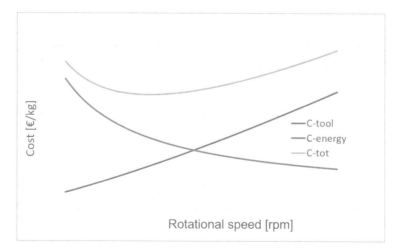

Fig. 7 Cost due to tool wear with respect to rotational speed: red line is the cost due to energy consumption, blue line is the cost due to tool wear and grey line is the total process cost

In addition, the effect of tools failure, explained in Sect. 4.4, could be exploited to control the overall status of the machine. If a tool is completely worn, the throughput in kg per time unit and, as a consequence, the throughput in kg per seconds, decreases. Constantly measuring the throughput of the process, it is possible to derive the number of failed tools and, based on the target revenues, to evaluate if the process should be stopped to change the tools or not. In addition, considering this effect, it is possible to continuously optimize the process cost, finding the rotational speeds that minimizes it considering the current throughput and changing it automatically, avoiding loss of time and money.

5 Application of the Solution and Obtained Results

The models presented in Sects. 4.3 and 4.4 have been implemented in a dedicated tool to achieve a complete optimization of the shredding process, both in terms of dimensional distribution and operational costs.

Through a dedicated Graphical User Interface (GUI) the operator is able to insert the target output distribution. As a result, after the acquisition and the elaboration of the information on the dimensional distribution of the material in input, the tool gives in output a representation of the predicted output distribution and it suggests the best grate size to use.

After this first step, the operator is able to insert all the information needed about the machine for the operational costs model. In particular, the operator has to insert the number of tools, the number of cutters, the cost for one tool, the cost for the tool change, the total mass to treat and the energy fixed cost. From a database it is possible to select the material to treat and the material of the tool, while the throughput in time unit is acquired directly from the first step. As a result, the tool gives in output a representation of the costs due to the energy consumption and to the tool wear and the total cost for the shredding process, suggesting the optimal rotational speed to use.

As an example, considering GFRP sanitary products composed by 30% of GF in a polyester resin, the software gives in output an optimal total cost of the process equal to about 75 €/ton with a rotational speed equal 2300 rpm to obtain particles with a final dimension of 1 mm. This is a good result as it is lower than the cost of new material and, in particular, also than the cost for landfill of this materials. As an example, the actual cost for landfill of EoL GFRP material in Italy is equal to 235 €/ton. The result obtained for EoL wind blades composed by 65% of glass fibers in epoxy resin is equal to 30 €/ton with a rotational speed equal 700 rpm to obtain particles with a final dimension of 1 mm.

6 Conclusions

In this Chapter an efficient comminution process-chain for the size reduction of GFRP End-of-Life products from different sectors (construction, sanitary, sports equipment and wind energy) has been presented.

A deep analysis of the different available size reduction technologies has been presented. The advantages and disadvantages of each shredding machine have been detailed.

Preliminary experiments have been done on the EoL products in a feasibility analysis. Criticalities have been detailed and a solution has been proposed. The acquired data have been used for the process modelling.

A 2-step architecture has been developed. The first step describe the evolution of the dimensional distribution along the comminution and suggests the optimal grate size for the process. The second step aims to minimize the operational costs.

This approach has been applied to EoL products in Use-Case 1 of FiberEUse project, showing the importance to control the process, optimizing the dimensional distribution of the particles and minimizing the operational costs (which result lower than both virgin material and disposal costs).

References

1. https://eucia.eu/about-composites/sustainability/
2. Oliveaux, G., Dandy, L., Leeke, G.: Current status of recycling of fibre reinforced polymers: Review of technologies, reuse and resulting properties. Progress in Materials Science 61–99 (2015)
3. Pickering, S.: Recycling technologies for thermoset composites—current status. Composites Part A 1206–1215 (2006)
4. Palmer, J., Ghita, O., Savage, L., Evans, K.: Successful closed-loop recycling of thermoset composites. Composites Part A 490–498 (2009)
5. Colledani, M., Tolio, T.: Integrated process and system modelling for the design of material recycling. CIRP Ann. Manuf. Technol. **62**, 447–452 (2013)
6. Gaudin, A.: Principles of mineral dressing, vol. 351. McGraw-Hill, New York (1939)
7. King, R.: A model for the quantitative estimation of mineral liberation by grinding. Int. J. Miner. Process. **6**, 207–220 (1979)
8. Meloy, T.: Liberation theory—eight modern usable theorems. Int. J. Min. Process **22**, 41–58 (1984)
9. Barbery, G.: Liberation 1, 2, 3: theoretical analysis of the effect of space dimension on mineral liberation by size reduction. Miner. Eng. **5**(2), 123–141 (1992)
10. King, R.S.: Mineral liberation and the batch comminution equation. Miner. Eng. **11**, 1143–1160 (1998)
11. Gay, S.W.: Liberation modelling using a dispersion equation. Miner. Eng. **12**, 219–227 (1999)
12. Gay, S.: A liberation model for comminution based on probability theory. Min. Eng. **17**, 525–534 (2004)
13. Schaik, A.R.: The influence of particle size reduction and liberation on the recycling rate of end-of-life vehicles. Miner. Eng. **17**, 331–347 (2004)
14. Castro, M.R.: A simulation model comminution-liberation of recycling streams. Relationships between product design and the liberation of materials during recycling. International Journal of Mineral Processing **75**, 251–281 (2005)

15. Bilgili, E.S.: Population balance modeling of non-linear effects in milling processes. Powder Technol. **153**, 59–71 (2005)
16. Bilgili, E.C.: Quantitative analysis of multi-particle interactions during particle breakage: a discrete non-linear population balance framework. Powder Technol. **213**, 162–173 (2011)
17. Diani, M., Pievatolo, A., Colledani, M., Lanzarone, E.: A comminution model with homogeneity and multiplication assumptions for the waste electrical and electronic equipment recycling industry. J. Clean. Prod. **211**, 665–678 (2019)
18. Shuaib, N.M.: Energy demand in mechanical recycling of glass fibre reinforced thermoset plastic composites. J. Clean. Prod. **120**, 198–206 (2016)
19. Taylor, F.: On the art of cutting metals. American Society of Mechanical Engineers, New York (1907)

Thermal Demanufacturing Processes for Long Fibers Recovery

Sonia García-Arrieta⬥, Essi Sarlin⬥, Amaia De La Calle⬥,
Antonello Dimiccoli⬥, Laura Saviano⬥, and Cristina Elizetxea⬥

Abstract The possibility of recycling glass (GF) and carbon fibers (CF) from fiber-reinforced composites by using pyrolysis was studied. Different fibers from composite waste were recovered with thermal treatment. The recycled fibers were evaluated as a reinforcement for new materials or applications. The main objective was to evaluate the fibers obtained from the different types of industrial composite waste considering the format obtained, the cleanliness and the amount of inorganic fillers and finally, the fibers quality. These characteristics defined the processes, sectors and applications in which recycled fibers can replace virgin fibers. These fibers were also evaluated and validated with tensile testing and compared to the tensile strength of virgin GF and CF.

Keywords Pyrolysis · Recycled fibers · Recycled carbon fiber · Recycled glass fiber

1 Introduction

GF and CF composites are interesting materials, they have many attractive properties such as high strength, good corrosion resistance, and light weight. They are widely used for applications such as automobiles, building insulation, leisure boats, construction materials, and leisure products [1]. The increase in the use of GF and CF composite materials due to their interesting properties and their light weight has generated an increment in waste of component at the end of life. Recycling of this

S. García-Arrieta (✉) · A. D. L. Calle · C. Elizetxea
TECNALIA, Basque Research and Technology Alliance (BRTA), Mikeletegi Pasealekua, 2, 20.009, Donostia-San Sebastián, Spain
e-mail: sonia.garcia@tecnalia.com

E. Sarlin
Materials Science and Environmental Engineering, Tampere University, Korkeakoulunkatu 6, 33720 Tampere, Finland

A. Dimiccoli · L. Saviano
Korec SpA, Via M. Polo 81, 56031 Bientina, PI, Italy

© The Author(s) 2022
M. Colledani and S. Turri (eds.), *Systemic Circular Economy Solutions for Fiber Reinforced Composites*, Digital Innovations in Architecture, Engineering and Construction, https://doi.org/10.1007/978-3-031-22352-5_5

81

waste is complex due to its inherent non-homogeneous nature. However, today there are different methods to recycle composite materials through mechanical, thermal, chemical processes and combinations of them [2].

Carbon fiber composites are currently being thermally recycled by companies such as ELG, with good results and with a potential market yet to be explored. Universities, Research Centers and companies have researched on composites recycling in the last 10 years [3]. The industrial thermal recycling of GF composites is currently challenging, mainly due to the low cost of the virgin raw material, that currently hinder obtaining an economically viable fiber material. This is the reason why the GF is recycled mainly by mechanical processes. Alternative processes are the subject of many research, particularly chemical processes which can recover value from the resin chemicals. Also, variants of pyrolysis, e.g. with a fluidised bed, or using microwave energy, have potential to provide cleaner fibers or use less energy. Environmental impact of different recycling processes shows that energy demand for chemical processes is typically higher than others. Thermal recycling is in the intermediate range, but only around 10% of the energy input required to produce virgin fiber. Mechanical grinding uses very little energy in comparison but produces a lower value product.

For CF there is scope to introduce variants of pyrolysis processing to optimize fiber surface properties and reduce energy consumption. The thermal process conditions for recover composites affects directly to the mechanical properties of the recycled fibers. In the case of recycled carbon fiber (rCF) the tensile strength could be reduced from 40 to 90% and in the case of recycled glass fiber (rGF) from 52 to 64%. Moreover, the thermal process removes the sizing from the fiber that must be replaced for a new sizing if the recycled fiber wants to be used in a technical application. The University of Strathclyde studied a suitable post-treatment for rGF that would permit their use in compounding application for automotive, but it would be necessary to scale-up this treatment with an economic feasibility able to compete with the virgin GF [4].

The present study has focused on developing and validating first, a preliminary laboratory-scale pyrolysis process for analysis the recovery of long GF and CF and after that, a pilot plant pyrolysis technology to validate the fiber in industrial application. In some of the cases it has been started with complete components and in another case the waste has been previously chopped. Thermally recovered GF and CF have been reused in high-strength, high-tech applications for the automotive and construction industries (Chap. 16). This study has focused on analysing the requirements and technical aspects necessary to obtain good quality rGF and rCF that can be used later in typical processes for plastics and composites sector.

1.1 Existing Pyrolysis Processes for Long Fiber Composites

Carbon fiber reinforced plastics (CFRP) pyrolysis is exploited by companies as ELG Carbon Fibrein UK, CFK Valley Stade Recycling and Hadeg Recycling Ltd both in

Germany, Carbon Conversions and Material Innovation Technologies RDF in USA or Karborek in Italy [5]. CFs are recovered in a controlled temperature and atmosphere furnace. They obtain fiber with 90% of the virgin properties [4].

Other companies, as Reciclalia or Formoso Technologies Group in Spain, Rymyc in Italy or Carbon Fiber Recycle Industry Co Ltd in Japan have pilot plants for fiber recovery [2]. Reciclalia for example, is working in the feasibility of recycling GF and CF Eolic blades to obtain long fibers to manufacture new hybrid thermoplastic fabric with high quality for automotive sector.

Figure 1 shows a CF recycling plant [6]. In this plant different stages are shown from the CF waste to the final product:

- Presorting: Crushing and sorting of materials according to type of fiber and state of processing: dry CF scraps, prepreg materials, end-of-life parts.
- Pyrolysis: Thermal treatment excluding oxygen in order to recover pure CF completely by means of thermal oxidation of pyrolysis gases.
- Refinement: Customized conditioning of fiber surface
- Cutting: Processing of CF into models "chopped" and "milled" accurately cut to desired fiber length
- Final product: At the end of the production process the client receives a customized product which absolutely fulfills the requirements

This means that not only the cost of the pyrolysis process must be considered in the final price of the rCF, but also of the rest of the pre-processes and post-processes.

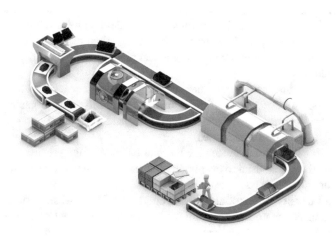

Fig. 1 CPK—valley pyrolysis plant [6]

1.2   Basic Principles of Thermal Demanufacturing Processes

Thermal processes include pyrolysis, fluidised-bed pyrolysis and pyrolysis assisted with microwaves [2]. Pyrolysis process consists of the chemical decomposition of the resin in an atmosphere without oxygen (e.g. nitrogen) and at high temperatures between 450 and 700 °C [7]. In these conditions, the resin (matrix) does not burn, it decomposes into lower-weight molecules and different sub-products are produced such us carbon dioxide, hydrogen and methane for example, and an oil fraction [2] that sometimes can be used by the petrochemical industry. Also, these molecules evaporated from the material can be used as an energy source because of their high heat capacity. Some studies suggest that the heat energy of the resin can be recovered, making the pyrolysis process self-sustaining [5]. Under these conditions, the fibers do not decompose, which makes this process interesting for recycling the fibers. However, several factors must be considered. GF suffer from the high temperatures and their mechanical properties can decrease. CF are less sensitive to temperature but alongside the CF, pyrolytic carbon coexists, which influences the mechanical properties of the regenerated fibers and depends to a large extent on the process parameters (furnace atmosphere, temperature, heating ramps and others). It is therefore possible to influence the properties of the fiber by acting on these parameters. This means that process optimization can and should be adapted to the mechanical requirements of the parts in which the recycled fibers are to be incorporated [8]. The parameters to consider are:

Temperature: Temperature is an important parameter in all stages and, therefore, in the final yield of the process. In particular, the proportions between solids, liquids and gas in the pyrolysis product are closely dependent on heating rate and the final temperature reached. At high heating rates and high final temperatures, most gas is produced, while at lower end temperatures and heating rates, most liquids or solids are produced.

Pressure: An increase in pressure makes gasification reactions more difficult, increasing the proportions of hydrocarbons and tars. Moving bed gasifiers usually operate at atmospheric pressure and fluidized bed gasifiers usually operate at pressure, reaching up to 30 bar in some cases.

Humidity: Moisture influences the thermal balance of the process as part of the heat produced must be used to evaporate this amount of water. It also influences the composition of the flue gases, even displacing some reactions.

Apparent density of material loads: Properly cut CFRP waste should be placed on trays, hooks, or charges. This implies a loss of volume to be treated in each batch with respect to the theoretical capacity of the kiln. The free spaces facilitate the gasification of organic matter and the movement of these gases towards the upper part of the installation for their elimination by burning.

Pyrolizer opening temperature: Cycle time on an industrial scale depends on the heating and cooling phases. The heating time depends mainly on the power of the oven. The cooling time depends exclusively on the temperature at which the system is opened. A quick opening will mean a higher emission of unburned gases into the

Table 1 Description of the studied material

Ref.	Quantity	Sector	Fiber/matrix and others	Manufacturing process	Part detail
A1 A2	200 g 9 kg	Construction	GF/polyester	Lamination	End of life composite roof
B	25 kg	Automotive	30% GF/polyester, inorganic fillers	SMC	Defective parts
C1 C2 C3	7.4 kg 130 kg 250 kg	Wind sector	GF/epoxy, foam, cores and gelcoat	Resin infusion	Dismantled blade part
D	20 kg	Aeronautical	CF/ epoxy, protective films	Uncured prepreg	Cutting operations scraps
E	150 kg	Aeronautical	CF/epoxy, protective film	Uncured prepreg	Expired prepreg
F1 F2	8 kg 150 kg	Aeronautical	CF/epoxy	Autoclave	Defective part

atmosphere. Depending on the nature of these gases (irritability, toxicity, etc.), the optimum opening temperature is determined.

2 FiberEUse Pyrolysis Processes

2.1 Materials

In the FiberEUse project, composite material waste from different sectors were considered as study materials. The objective was to recover clean fibers after removing the polymer matrix. Table 1 shows a summary of the waste materials studied together with their main characteristics.

The above classification helps to estimate the amount of fiber that is expected to be obtained after the thermal recycling process. It was also possible to predict the final format of the fiber that would be obtained or even the degree of cleaning. In this way, from wastes that contain a high amount of inorganic fillers, it is expected to obtain dirty fibers or from wastes with fabrics in their original textile format is expected to be recovered.

2.2 Pyrolysis Equipment and Media

Different types of ovens were used for the heat treatments of the different described materials. First, small laboratory equipment was used and later bigger ovens, all

Fig. 2 a HC500 Oven for thermal treatment. **b** Semi-industrial oven Solvo Line 166

of them in TECNALIA. Finally, pilot plant was installed in RIVIERASCA. The GF materials were treated initially in ovens with the presence of air, while the CF composite materials were pyrolyzed in ovens with inert atmosphere.

Ovens with air atmosphere (Fig. 2):

- A laboratory oven HC500 with unforced air circulation and with dimension of 600*600*600 mm and maximum working temperature of 500 °C was used for small GF composite samples.
- Semi-industrial oven consisting of a VOCs (Volatile Organic Compounds) treatment system and SOLVO Line 166-Type 120 oven 600*600*1000 mm, loading capacity of 500 kg/m and equipped with an automatic unit control, able to work to 500 °C and 0.65 bar was used for large GF composite samples.

Ovens with inert atmosphere (Fig. 3):

- Small laboratory oven for the pyrolysis of small CF composite coupons. It is composed of a glass tube resistant to high temperature in which a controlled flow of argon circulates. It is jacketed with an resistors oven capable of reaching 500 °C.
- Nabetherm lab oven with a continuous nitrogen flow and a heating rate of 11.5 °C/min. The internal dimensions are 200*200*200 mm.
- Nabertherm semi-industrial oven. Argon Oven with gas extraction and filters to concentrate liquid/solid sub-products. The oven internal chamber has a dimension of 1000*500*250 mm.
- Industrial plant: oven with inert atmosphere and a capacity of 500 kg
- Pilot plant KOREC: thermochemical depolymerization in the presence of a controlled carbon dioxide environment that allows to recover both the inorganic part (carbon fibers or glass fibers) and the organic part (resin) in the form of an organic liquid fraction. This organic liquid contains C=C unsaturations and can be co-formulated with virgin polyester resins and take part in the following polymerization reactions that produce new thermosetting composites. This process has been developed and patented by Korec (EP 3114191, 2018). The innovation in

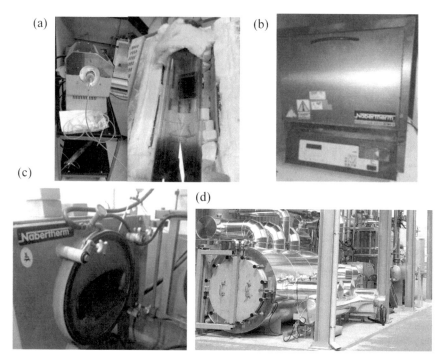

Fig. 3 **a** Pyrolysis lab oven with argon circulation. **b** Nabetherm nitrogen oven. **c** Nabertherm Argon Oven. **d** Korec industrial pilot plant

the Korec process lies in the recovery of resin fraction that, as a reactive blending component, can be reintroduced into the production chain of composites and this is what makes the process more economically sustainable. CO_2-containing environment with a predetermined CO_2 volume concentration. The capacity of the pilot plant is 950*950*2500 mm.

2.3 Methodology and Process Parameters

In order to obtain different fiber formats as a product of the thermal process or pyrolysis, the samples were introduced in the ovens with different sizes. The target format is related to the future use of the fiber. Thus, for laminating, resin infusion or RTM (Resin Transfer Moulding) processes, the preferred formats are mat or fabric, while for compounding processes (extrusion-injection processes) the preferred format is shredded fiber between 6 and 15 mm. The samples from construction (A), Automotive (B), Wind (C1) and Aeronautic (D and E) have been pyrolyzed in their original format, that is, as the waste was received. However, the samples from Wind (C2 and C3) and Aeronautic (F1 and F2) have undergone a preliminary shredding process

Table 2 Processes parameter

Equipment		HC50 oven	Solvo Line pilot plant	Glass tube lab oven	Nabertherm pilot plant	Nabertherm oven	Ind. plant	Korec pilot plant
Parameters↓								
Ramp 1	Heating rate (°C/min)	2	30	3.5	11.5	11.5	11.5	10–12
	Temperature (°C)	600	320	450	450	450	450	350–450
	Time (h)	4	0.83	2	0.62	0.62	0.62	0.55–0.60
Ramp 2	Heating rate (°C/min)	–	5.3	–	–	11.5	–	–
	Temperature (°C)	–	400	–	–	550	–	–
	Time (h)	–	5	–	–	0.15	–	–
Ramp 3	Heating rate (°C/min)	–	20	–	–	–	–	–
	Temperature (°C)	–	500	–	–	–	–	–
	Time (h)		0.1	–	–	–	–	–
Stabilisation Time (h)		2	6.5	1	2.5	30 (Ramp1) +15 (Ramp2)	6	0.2
Atmosphere		Air circulation	Vacuum + air circulation	Argon	Argon	Ramp1: Nitrogen Ramp 2: Air	Inert	CO_2
Cooling		Yes	Yes	Yes	Yes	Yes	Yes	No

(see in this chapter) to obtain shredded composite material with the defined size (6–15 mm).

The cycles used in each of the oven have been adjusted to the characteristics of the oven/technology itself, such as the rate of heating, the capacity or volume or the type of atmosphere. The parameters used are summarized in Table 2:

3 Description of the Process for Each Material

3.1 Material from Construction Waste (Ref. A1)

This waste was introduced into the oven without any previous process. An oven with air atmosphere (HC500) was used for the heat treatment. After the defined cycle, GF were obtained in mat format with no apparent residues (Fig. 4).

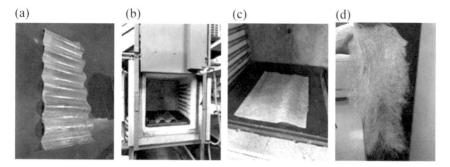

Fig. 4 **a** Waste Ref. A1, **b** Ref. A1 in HC500 oven, **c** Obtained product, **d** Mat of rGF

Fig. 5 **a** Waste Ref. A2 in Solvo Line oven tray, **b** Obtained product, **c** Clean rGF with mat format

3.2 Material from Construction Waste (Ref. A2)

In the same way, this waste was introduced into the oven without any previous process. An oven with air atmosphere (Solvo Line) was used for the heat treatment. After the defined cycle, the results were the same in both ovens, although the waste amounts treated were much higher (Fig. 5).

3.3 Material from Automotive Waste (Ref. B)

This waste was also introduced into the oven without any previous process. Again, an oven with air atmosphere (Solvo Line) was used for the heat treatment. After the defined cycle, GF were obtained in their original format, this long cut fibers. In this case, the high amount of inorganic fillers contained in the resin did the recovered fiber especially dirty and contaminated (Fig. 6).

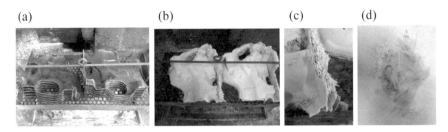

Fig. 6 **a** Waste Ref. B in Solvo Line oven tray, **b** Obtained product, **c** and **d** Long rGF with high filler content

3.4 Material from Wind Sector Waste (Ref. C1)

This material from large wind blades (length between 10 and 40 m) was previously cut by the owner company to optimal sizes for this study (maximum 0.5 m), but it was not treated with a shredding process. Again, due to the material contained GF, an oven with air atmosphere (Solvo Line) was used for the heat treatment. After the defined cycle, GF were obtained as a mix of different formats. This could be explained because the production process uses fibers in very different formats as roving, mat or fabrics. Furthermore, the presence of other materials such as foam cores and especially, gelcoats make the fiber obtained with a variable percentage of inorganic fillers. In this case, because the content of fillers is not very high, it can be removed by a mechanical process (Fig. 7).

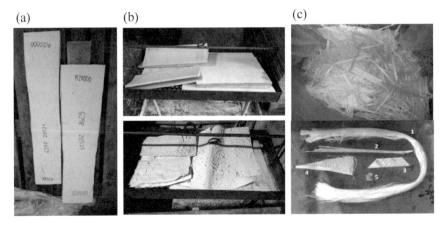

Fig. 7 **a** Waste Ref. C1, **b** Ref. C1 after and before the cycle, **c** Obtained product (mix formats)

(a) (b) (c)

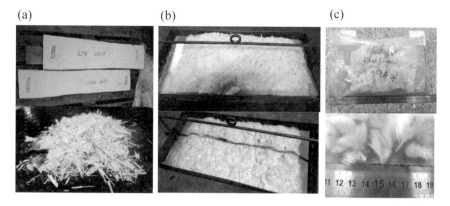

Fig. 8 **a** Waste Ref. C2, **b** Ref. C2 after and before the cycle, **c** Obtained short rGF

3.5 Material from Wind Sector Waste (Ref. C2)

This material was the same that the previous one, but in this case, it was treated with a shredding process. The same oven (Solvo Line) with air atmosphere was used for the heat treatment. In this case a tray with small holes for aeration was needed. After the defined cycle, short GF were obtained with length between 6 and 15 mm. The fiber was obtained quite clean, although some agglomerates of inorganic fillers were observed (Fig. 8).

3.6 Material from Aeronautic Waste (Ref. D)

This material was introduced into the oven without any shredding process. First, some small coupons of this material were used in the glass tube laboratory oven with inert atmosphere. The results obtained after the defined cycle were small pieces of clean CF fabric. The rest of material was introduced in Nabertherm argon oven. In this case the obtained fibers were covered of char. This effect was reduced when the waste was pyrolyzed after a curing process (Fig. 9).

3.7 Material from Aeronautic Waste (Ref. E)

This material was sent to an industrial recycling plant in its original roll format. It was unrolled, pyrolyzed and once a clean recycled CF fabric was obtained, it was rerolled. The fabric obtained was clean with any residue of char (Fig. 10).

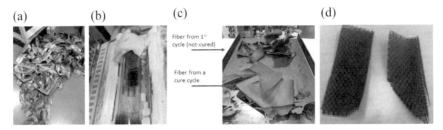

Fig. 9 **a** Waste Ref. D, **b** Ref. D in Glass tube oven, **c** Ref. D after the pyrolysis in Nabertherm oven, **d** Obtained product

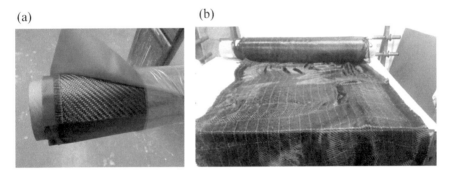

Fig. 10 **a** Waste Ref. E, **b** Obtained product

3.8 Material from Aeronautic Waste (Ref. F1)

This material was introduced into the oven after a shredding process. An oven with nitrogen atmosphere (Nabertherm) was used for the pyrolysis treatment. The results obtained after the defined cycle were short CF with length from 1 to 15 mm. The recovered CF was char covered. To recover the rCF with clean surface, the residual char was removed by mechanical processes (decompressing and sieving). The final products were those obtained in the different fractions of the sieves (Fig. 11) (Table 3).

Finally, the up scaling was carried out in the KOREC pilot plant. The C3 and F2 wastes were used for the final project demonstrators manufacturing, so large amounts of waste had to be treated. In this case, the composite waste was first shredded since the recovered fiber will be used in extrusion-compounding processes and in lamination processes from mat that will also be manufactured with short fiber.

(a) (b) (c) (d)

Fig. 11 **a** Waste Ref. F1, **b** Ref. F1 after shredding, **c** Obtained product, **d** rCF after sieving

Table 3 Mass of each fraction obtained

Fraction	Mass (kg)
L > 1 mm	3.048
0.5 < L < 1 mm	1.005
D < 0.5 mm	0.669
Total	4.053

3.9 Material from Wind Sector Waste (Ref. C3)

This material was similar to Ref. C2. The treatment involved the following steps: (i) feeding the composite materials waste into the reaction chamber; (ii) removing oxygen from the chamber until air is substantially eliminated from the reactor; (iii) creating a CO_2-containing environment in the chamber with a predetermined CO_2 volume concentration; (iv) heating the waste in the chamber and reaching a temperature set between 350 and 550 °C; (v) maintaining the CO_2-containing environment and controlling the temperature of said reactor so as to maintain said reaction temperature in the chamber, for a predetermined residence time, obtaining a gas mixture containing products of depolymerization of the resin, and a solid residue comprising the glass fibers or carbon fibers; (vi) extracting the gas mixture from the reaction chamber and cooling it down to a predetermined temperature so as to condensate the products of reaction from the gas mixture, in order to obtain a condensate liquid phase comprising a main amount of the product of depolymerization that it is separated from the uncondensed gas; (vii) extracting the solid residue from the chamber (Fig. 12).

The rGF from C3 was treated to be matrix compatible, this is, re-sized (see Chap. 7) and after this, it was transformed in plastic pellet for injection process and in mat for laminating process. Finally, the rGF was the reinforcement of two demonstrators (Chap. 16): Cowl top support and Light transmitting single skin profiled GFRP sheet.

(a)   (b)   (c)

Fig. 12 a Waste Ref. C3, **b** Organic liquid recovered from the process, **c** Obtained short rGF

(a)   (b)   (c)

Fig. 13 a Waste Ref. F2, **b** Grinded waste, **c** Obtained short rCF

3.10 Material from Aeronautic Waste (Ref. F2)

This material was similar to F1 and the process was also similar to the carried out for C3 (Fig. 13).

The rGF from F2 was also re-sized (Chap. 7) and after this, it was transformed in plastic pellet for injection process for manufacturing the Pedal bracket and the Front-end carrier demonstrators (Chap. 16).

4 FiberEUse Pyrolysis Products Characterization

The tensile strength of single fiber filaments was defined with the Fibrobotics device [9]. At least 50 fibers were tested with a gauge-length of 23.5 mm and crosshead velocity of 0.008 mm/s.

Table 4 Tensile test results

Fiber	Tensile strength (MPa)	Modulus (GPa)
Virgin CF (aeronautic sector)	1328 ± 494*	102 ± 16
rCF (Ref. F1)	534 ± 287	105 ± 25
rCF (Ref. F2)	1869 ± 747	137 ± 36
Virgin GF (construction sector)	1111 ± 485	52 ± 6
rGF (Ref. C2)	437 ± 166	51 ± 16
rGF (Ref. C3)	422 ± 204	53 ± 12

*The reference and quality of virgin fiber could be not the same of the recycled, this data is un-know for the fiber from waste

Tensile tests were carried out to characterize the different materials obtained in the pyrolysis processes. Tests were also carried out on GF and virgin CF to carry out the comparison. The results obtained are presented in Table 4.

Afterwards, the recycled fibers obtained through thermal processes have been validated through the indirect mechanical characterization of the products in which they have been reused (Chap. 7).

5 Conclusions

As a final technical conclusion of the thermal treatment study, Table 5 summarize the results obtained for each type of composite waste evaluated in FiberEUse project. This table gives a recommendation of the processes in which the recycled fiber from the different sector or applications could be integrated. This recommendation is based on the format, cleanness or properties results obtained for each recycled fiber, combined with the typical requirements of plastics and composites processes.

The up-scaled process in FiberEUse project has some sustainability advantages. The Korec technology is cost effective thanks to the recovery of the organic part of the waste, that can be re-introduced into the composite's production chain. Thanks to this technology the environmental impact is strongly mitigated:

(a) there is an elimination of the volume of FRP waste in landfills and incinerators;
(b) no hazardous chemicals are used during the process;
(c) incondensable gases and residues are used for internal energy recovery;
(d) atmospheric emissions are easily reduced by EU Best Available Techniques (the available techniques which are the best for European Commission for preventing or minimising emissions and impacts on the environment, see reference documents BREFs) and fully respect the limits imposed by European law;

Table 5 Resume of the thermal treatment, the result and de recommended re-use of the FiberEUse waste parts

Ref	Thermal treatment	Result	Recommended Re-use
A1 & A2	T^a 500 °C Time 12,5 h Air circulation	• Fiber clean • Mat format • 50% properties reduction	• Liquid molding composite processes as infusion or lamination
B	T^a 500 °C Time 12.5 h Air circulation	• Fiber + high filler content • Cut long fiber format	• Filler in a casting or liquid process
C1	T^a 500 °C Time 12.5 h Air circulation	• Fiber clean • Mixed formats • Mainly UD long fiber • 50% properties reduction	• Filler in a casting or liquid process
C2 &C3	Korec technology with CO_2 atmosp. after shredding process	• Short GF (15 mm)	• Compounding with thermoplastic
D	T^a 450 °C Time 3 h Argon	• Fiber Char covered • Small pieces of fabric	• Liquid molding composite processes as infusion or lamination
E	T^a 450 °C Time 6 h Inert gas circulation	• Fiber clean • Fabric format	• Liquid molding composite processes as RTM or T-RTM
F1	T^a 450 °C Time 2 h Nitrogen after shredding process	• Fiber clean after sieving post-treatment • Short CF (15 mm) • 50% properties reduction	• Compounding with thermoplastic • rCF mat manufacturing
F2	Korec technology with CO_2 atmosp. after shredding process	• Short CF (15 mm)	• Compounding with thermoplastic

(e) there is reduction of the process carbon footprint of the fiberglass and composites supply chain (virgin resins are products of fossil origin).

References

1. Åkesson, D. et al.: Glass fibres recovered by microwave pyrolysis as a reinforcement for polypropylene. Polymers & Polymer Composites 21 (2013)
2. Oliveux, G., Dandy, L.O., Leeke, G.A.: Current status of recycling of fibre reinforced polymers: review of technologies, reuse and resulting properties. Prog. Mater Sci. **72**, 61–99 (2015)

3. Leeke, G.A. et al.: EXHUME—Efficient X-sector use of HeterogeneoUs MatErials in Manufacturing. s.l.: UK Research and Innovation, Vol. University of Birmingham
4. Job S. et al.: Composites recycling: where are we now? (2016). https://compositesuk.co.uk/sys tem/files/documents/Recycling%20Report%202016%20-%20Light%20Background.pdf
5. Pimenta, S., Pinho, S.T.: Recycling of Carbon Fibers, Handbook of Recycling, pp. 269–283 (2014)
6. https://www.cfk-recycling.de/index.php?id=58
7. Pimienta, S., Pinho, S.T.: Recycling carbon fibre reinforced polymers for structural applications: technology review and market outlook. Waste Manage. **31**, 378–392 (2011)
8. Torres, A., et al.: Recycling by pyrolysis of thermoset composites: characteristics of the liquid and gaseous fuels obtained. Fuel **79**, 897–902 (2000)
9. Laurikainen, P. et al.: High throughput mechanical micro-scale characterization of composites and the utilization of the results in finite element analysis. In: 1st European Conference on Composite Materials, Vols. Athens, Greece (2018)

Styrene-Free Liquid Resins for Composite Reformulation

Raffaella Suriano⊙, **Andrea Mantelli**⊙, **Gianmarco Griffini**⊙, **Stefano Turri**⊙, and **Giacomo Bonaiti**⊙

Abstract Three different classes of thermosetting styrene-free resins were investigated to assess their suitability to constitute the matrix phase in the reformulation of composites reinforced with mechanically recycled glass fibers. Resin reactivity and mechanical properties after curing were compared to commercial styrene-based, unsaturated polyester resins. The polymeric resin, acting as a binder, could be properly selected depending on the desired reactivity, processability, and mechanical behavior. Some prototypal examples of reformulated composites with different types and contents of recycled glass fibers were produced and mechanically tested. The combination of the epoxy resin with up to 60 wt% of mechanically recycled glass fibers resulted in an increase of elastic modulus up to 7.5 GPa.

Keywords Polymer-matrix composite · Recycling · Glass fiber · Mechanical properties · Wind turbine blade

1 Introduction and Context of Reference

The exponential growth of the human population and the linear economy models currently adopted worldwide represent key drivers for the continuously increasing demand for new manufacturing products and for their current consumption rate, which unsustainably impact on natural resources while generating growing levels of discarded materials and components. Indeed, a great number of end-of-life (EoL) products are being disposed of, leading to several environmental problems [1]. A representative example of EoL products are wind turbine blades, whose annual production volumes are expected to increase significantly in the coming decades, reaching 200,000 tons of wind blade wastes in 2034 [2]. Wind turbine blades are

R. Suriano (✉) · A. Mantelli · G. Griffini · S. Turri
Department of Chemistry, Materials and Chemical Engineering "Giulio Natta", Politecnico Di Milano, Piazza Leonardo da Vinci 32, 20133 Milan, Italy
e-mail: raffaella.suriano@polimi.it

G. Bonaiti
Rivierasca S.P.A, Via Strasburgo 7, 24040 Bottanuco, BG, Italy

© The Author(s) 2022   99
M. Colledani and S. Turri (eds.), *Systemic Circular Economy Solutions for Fiber Reinforced Composites*, Digital Innovations in Architecture, Engineering and Construction, https://doi.org/10.1007/978-3-031-22352-5_6

mainly composed of glass fiber reinforced polymers (GFRPs) and the application of a circular economy approach to EoL GFRPs management is one of the challenges of the modern wind turbine manufacturing industry [3, 4]. Responsible treatments of EoL composite products that may include reusing, recycling, or remanufacturing are desirable and beneficial both environmentally and economically because waste is minimized, while valuable components and materials are recovered [5, 6].

Several recycling technologies have been proposed and developed concerning fiber reinforced composite materials [7]. Among them, the mechanical treatment still represents the most suitable process to reduce the size of GFRP composite parts, since glass fibers (GFs) tend to be damaged during thermo-chemical processes [6, 7].

The combination of mechanically recycled GFRPs with a polymer matrix will give shape to a composite material whose reinforcement is a waste. The matrix selection in composite materials is also quite important because it has a large influence on the final properties of the composite being constructed. Particularly for recycled GFRPs, the presence of contamination in wastes as well as some degradation components may affect the behavior of composites reinforced with recycled GFRPs. The main roles of a matrix in both the pristine and recycled composites are:

- to hold and protect reinforcing materials from environmental and physical damage;
- to distribute and transfer the load to the reinforcements;
- to define external characteristics, such as the final shape of composites;
- to prevent the buckling of the fibers;
- to ensure the chemical and thermal compatibility with the reinforcement;
- to avoid rapid propagation of composite cracks [8, 9].

Depending on their final applications, polymeric composites can employ matrices that are thermoplastic or thermosetting. Both of them present advantages and disadvantages. Polymer matrices used for structural applications are almost exclusively based on thermosetting systems, thanks to the superior thermal stability, chemical resistance, low creep and relaxation properties and better ability to impregnate the fiber reinforcements of thermosetting in comparison with thermoplastic matrices [10].

2 State of the Art for Composite Reformulation from Mechanical Recycling

2.1 Brief Description of the Scientific and Industrial State-Of-The-Art

Unsaturated Polyesters. One of the most popular class of resins, as well as one of the most versatile group of synthetic polymers in composites manufacturing, is represented by unsaturated polyester (UP) resins. They can be found in a very wide range of engineering applications. Moreover, they are characterized by a highly favorable performance/cost ratio. Important product areas for UPs are marine, automotive, electric and electronic, building, constructions, sport and leisure, domestic and sanitary appliances, furniture as well as military applications [9]. The processing of a UP resin into a composite product can be done using several technologies: hand lay-up and spray lay-up, lamination, casting, compression molding, pultrusion, resin transfer molding (RTM), vacuum infusion and filament winding.

The advantages of UPs are their very good structural properties together with affordable costs. Other advantages include easy handling, processing, fabrication and a good balance of mechanical, electrical and chemical properties. Some special formulations offer high corrosion resistance and fire retardancy.

Vinyl Esters Resins. Vinyl esters are ester resins with unsaturated C=C terminal groups. The terminal unsaturation makes them very reactive. Moreover, the low number of esters group in the main chain of the polymer gives good chemical resistance to this class of resins. In terms of properties and cost, they present characteristics in between those of UP resins and epoxy resins, maintaining the processability and versatility of polyesters, while having mechanical performances closer to the epoxies. This range of properties fits the requirements of corrosion resistance products and for this reason, they are largely employed in this field. Other main markets include pultruded construction and electrical components, automotive structural applications, polymer concrete vessels for mining and chemical operations, high-performance marine applications and sporting goods.

Epoxy Resins. An epoxy resin is a formulation comprising an epoxy-functional oligomer combined with other molecules like polyamines and anhydrides able to cure into a rigid network. As for polyesters, epoxy resins exhibit tremendous versatility because they can be designed to meet a wide range of specific performance requirements. In addition, all epoxy systems involved in composite manufacturing are characterized by the presence of aromatic rings giving mechanical rigidity and thermal stability to the final solidified crosslinked polymer [11].

The main advantages of epoxy resins are:

- the relatively high polarity that confers excellent adhesion to a wide variety of fibers;

- dimensional accuracy of fabricated structures due to low cure shrinkage;
- absence of bubbles and voids due to the addition curing reaction without volatile byproducts;
- very stable and resistant crosslinked structure.

The epoxy composite market was valued at USD 21.6 billion in 2015 and is projected to reach USD 33.1 billion by 2021 driven by the increasing wind energy installations and the increasing use of epoxy composites in structural applications. Moreover, in the aerospace and defense industry, the need for lightweight materials to produce fuel-efficient vehicles due to the increasing environmental concerns, especially in Europe, is constantly increasing together with the growing demand from the pipe and tank and oil and gas industries. In addition, the use of epoxy composite is also increasing in the sporting goods industry for the manufacturing of products, such as tennis rackets, bicycles, fishing rods, and golf shafts [12].

Polyurethane Resins. Polyurethanes (PUs) are generally obtained by the reaction between polyisocyanates and polyols. Tuning the molecular weight, composition and functionality of the two components, a wide range of products can be obtained, like adhesives, coatings, elastomers and foams. Polymeric 4,4′-methylene diphenyl diisocyanate (MDI) is the polyisocyanate of choice for composite manufacturing due to its low viscosity and thereby good processability even at high filler loading, while polyethers and polyols are typically used as hardeners. As cast materials, PUs have gained importance as solvent-free systems, in combination with other components including desiccants, fillers, plasticizers, pigments, flame retardant additives, and catalysts.

The hardened PU materials range from hard and tough to soft and elastic. A general advantage when compared to other matrices is represented by the large toughness and elasticity. Making a comparison with other thermosetting resins, it is possible to state that PUs can be crosslinked as fast as polyesters but yield mechanical and thermal performances comparable to epoxies [13], but with a lower thermal stability.

For decades, crosslinked PUs have occupied a strong position as electrical insulation materials providing notable protection of electrical equipment from aggressive media such as water, chemicals, and dust [14]. Nowadays, the excellent mechanical properties of PU composites open the way to redesign existing profiles, in some cases allowing a redesign of the reinforcement package.

2.2 Limitations, Open Issues and Needs

In the field of UP resins, the incorporation of low molecular weight monomers as reactive diluents, most commonly styrene, enables to tune the final properties of the cured material, especially reducing the viscosity and improving processability. The dilution of UPs and vinyl ester resins in reactive monomers like styrene, results in the formation of a thermosetting resin with good heat resistance, excellent mechanical properties (particularly high tensile elongation) and chemical resistance [4].

However, styrene has several health and safety issues, because of its flammability, volatility and hazardous character. Its replacement is therefore needed for a better environmental sustainability of fiber reinforced polymer composites.

3 Development of Sustainable Styrene-Free Resins for Composite Reformulation

To select an adequate matrix material for remanufacturing, in terms of processability and mechanical properties, three styrene-free, thermosetting binders were investigated and characterized: an acrylic system; an epoxy resin; and a polyurethane resin. Subsequently, the epoxy resin was selected for further formulation studies and filled with different concentrations of mechanically recycled GFRPs. The corresponding composites were tested through thermogravimetric analysis (TGA) and tensile tests to measure the real content of the reinforcement phase contained in each sample and their mechanical properties. Some limitations regarding the ultimate tensile properties and the adhesion between the matrix and the recycled fibers as evidenced by scanning electron microscopy (SEM) images were found, and still need to be fully optimized. Nevertheless, the possibility of using recycled GFRPs as reinforcement phase to obtain new polymer-based composite materials with high values of elastic modulus was demonstrated.

4 Description of Materials and Methods

4.1 Liquid Resin Preparation

Acrylic Resin. The base acrylic resin was mainly composed of a bifunctional acrylate oligomer, an ethoxylated bisphenol A diacrylate (Fig. 1), named SR349 (Sartomer, Arkema).

As the viscosity of SR349 is too high, a methacrylic reactive solvent was added for dilution. The selection of the reactive diluent for SR349 was based on literature data [15]. Considering products, which could be at least partially obtained from renewable sources, four potential candidates, were identified and their chemical structures are shown in Fig. 2:

- isobornyl methacrylate (IBOMA);
- methyl methacrylate (MMA);
- lauryl methacrylate (LMA);
- 1,4-butanediol dimethacrylate (BDDMA).

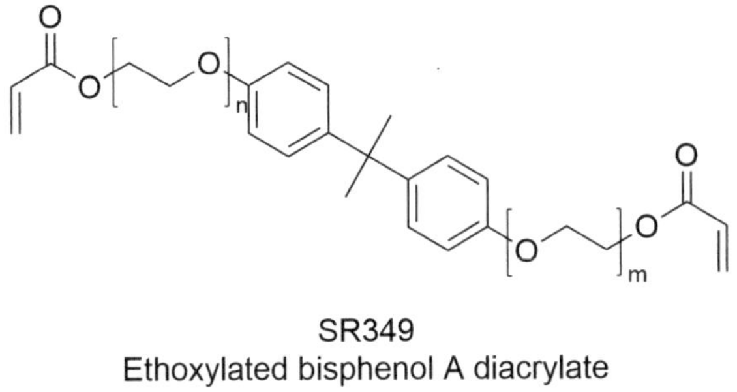

SR349
Ethoxylated bisphenol A diacrylate

Fig. 1 Chemical structure of the bifunctional acrylate oligomer, named SR349

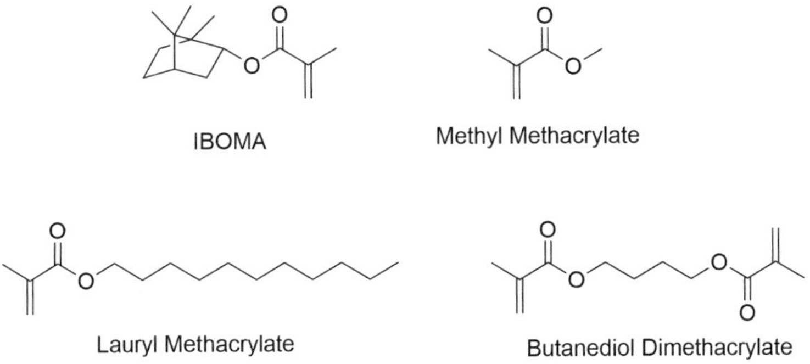

IBOMA   Methyl Methacrylate

Lauryl Methacrylate   Butanediol Dimethacrylate

Fig. 2 Chemical structures of some bio-based reactive diluents

After the rheological characterization shown in Sect. 5.2, BDDMA was selected as target reactive diluent and the acrylic resin formulation was developed according to the weight percentages reported in Table 1.

For the preparation of the acrylic resin, the following procedure was followed. SR349 and the reactive diluent were mixed in a beaker. An initiator/catalyst system was then added, according to the following order: (1) dicumyl peroxide; (2) cobalt(II) naphthenate; (3) 2-butanone peroxide. Once the initiator/catalyst system was mixed, the mixture turned from an uncolored state to a purple color and then to a dark green color. The mixing process generated gasses, resulting in the formation of a layer of

Table 1 Acrylic resin formulation

Component	Weight fraction (%)
SR349	78.44
BDDMA	19.61
Dicumyl peroxide	0.29
Cobalt(II) naphthenate	0.19
2-Butanone peroxide	1.47

bubbles on top of the liquid resin. For this reason, a degassing process was performed, putting the resin under vacuum at 50 °C. Afterward, the resin was poured into a mold for the crosslinking. The thermal curing treatment, required for the crosslinking of the acrylic resin, consisted of:

- 2 h at room temperature;
- 3 h at 80 °C;
- 2 h at 140 °C.

Epoxy Resin. The epoxy resin is a bisphenol-A (epichlorohydrin) resin, containing an epoxy-terminated oligomer based on bisphenol-A (Fig. 3) and 1,4-bis(2,3-epoxypropoxy)butane (Fig. 4, Araldite®BY158, Huntsman Corporation). To tune the reactivity of the system, Araldite®BY158 was combined with either Aradur®2992 or Aradur®21 or both hardener mixtures, supplied by Huntsman. Aradur®2992 is a mixture of trimethylhexane-1,6-diamine (Fig. 5) and toluene-4-sulphonic acid, while Aradur®21 is a pure trimethylhexane-1,6-diamine curing agent.

Three different formulations (Table 2) were supplied by Huntsman and were used for the epoxy resin, depending on the desired gel time of the system (Table 3).

Depending on which formulation was chosen, the selected components were properly mixed in a beaker with a spatula. During mixing, a transition from clear to opaque

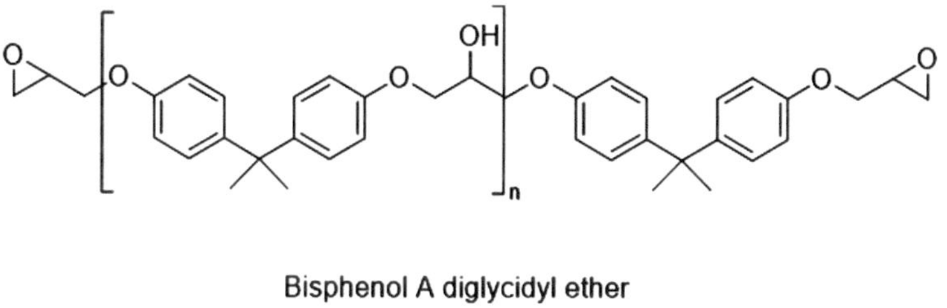

Bisphenol A diglycidyl ether

Fig. 3 Chemical structure of the epoxy-terminated oligomer-based bisphenol-A

Fig. 4 Chemical structure of 1,4-butanediol diglycidyl ether, which is a reactive diluent

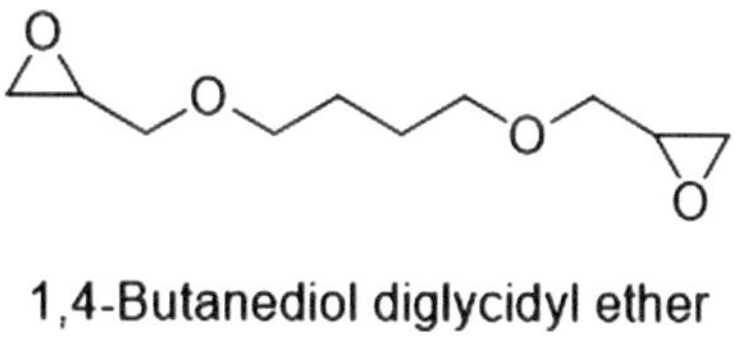

1,4-Butanediol diglycidyl ether

Fig. 5 Chemical structure of trimethylhexane-1,6-diamine curing agent

Trimethylhexane-1,6-diamine

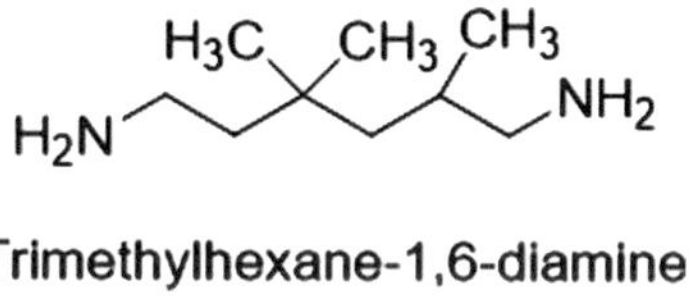

Table 2 Weight fractions and name for three different epoxy resin formulations

Component	Weight fraction (%)		
Araldite BY158	74.63	77.52	78.12
Aradur 2992	25.37	5.43	/
Aradur 21	/	17.05	21.88
Formulation name	EPO10	EPO	EPO80

Table 3 Epoxy resin formulations with their specific viscosity and gel time

Formulation name	EPO10	EPO	EPO80
Viscosity at 25 °C (mPa s)	150	100	90
Gel time (min)	10	30	80

of the system was observed. When the resin became clear again, the mixing was interrupted and the resin was employed for further steps. The thermal curing treatment, required by the epoxy resin to cure in a solid material, was conducted as follows:

- 24 h at room temperature;
- 1 h at 100 °C.

As shown in Tables 2 and 3, three different formulations of the epoxy resin were considered. For this reason, the most catalyzed one was characterized by 10 min of gel time, the intermediate one by 30 min of gel time and the one without catalysts by 80 min of gel time. They were respectively named 'EPO10', 'EPO' and 'EPO80'.

Polyurethane Resin. The PU formulation consisted of an isocyanate and a polyol. Different polyols were tested to determine the optimal formulation, in terms of reactants miscibility, matrix rigidity, and the presence of voids in the cured material. For all the formulations, a molar ratio between hydroxyl groups and isocyanates of 1.05 was selected. The tested reactants were:

- Poly[(phenyl isocyanate)-co-formaldehyde] (PMDI) with average Mn ~ 400 (Fig. 6, Sigma Aldrich);
- Tripropylene glycol (purity ≥97%, Fig. 7, Sigma Aldrich);

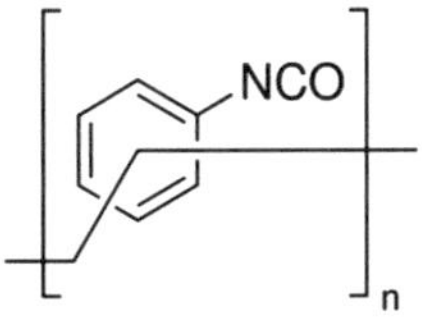

Fig. 6 Chemical structure of poly[(phenyl isocyanate)-co-formaldehyde] monomer

Fig. 7 Chemical structure of tripropylene glycol

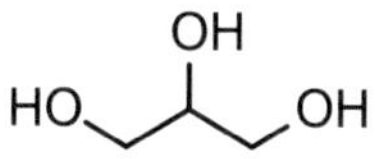

Fig. 8 Chemical structure of glycerol

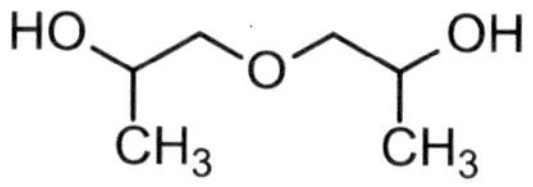

Fig. 9 Chemical structure of dipropylene glycol

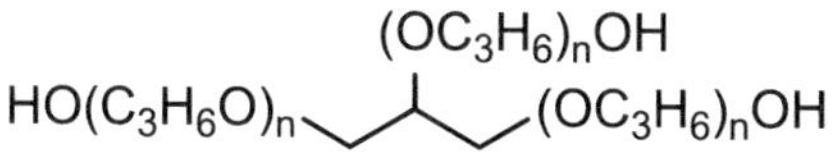

Fig. 10 Chemical structure of glycerol propoxylate

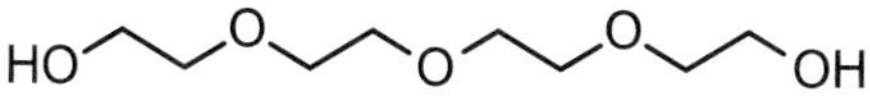

Fig. 11 Chemical structure of tetraethylene glycol

- Glycerol (purity ≥99.5%, Fig. 8, Sigma Aldrich);
- Dipropylene glycol (purity ≥99%, Fig. 9, Sigma Aldrich);
- Glycerol propoxylate with average Mn ~ 266, Fig. 10, Sigma Aldrich);
- Tetraethylene glycol (purity ≥99%, Fig. 11, Sigma Aldrich);
- UOP-L powder, 3A zeolite type molecular sieves for high-quality PU system, Obermeier GmbH & Co. KG Company.

After the measurement of glass transition temperature Tg shown in Sect. 5.2, the final formulation for the PU resin was selected and prepared according to the weight percentages reported in Table 4.

Processing of polyisocyanates, such as PMDI, requires the total absence of water, because the isocyanate groups can react with water in competition with the polyol, resulting in an incomplete and inefficient curing of the resin. To avoid this, tripropylene glycol was dehydrated under vacuum at 90 °C for 16 h under magnetic stirring in a flask. To deactivate the remaining water in the polyol, a zeolite-type molecular sieve (UOP-L powder, ten parts by weight with respect to the polyol) was added to the stirring polyol. Before the addition of tripropylene glycol, the zeolites needed to

Table 4 PU resin formulation

Component	Weight fraction (%)
PMDI	52.89
Tripropylene Glycol	42.83
Zeolite powder	4.28

Table 5 Styrene-based conventional UP resin formulation

Component	Weight fraction (%)
SIRESTER FS 0910/M	98.33
Cobalt(II) naphthenate	1.47
2-butanone peroxide	0.20

be activated with a heat treatment at 350 °C for 4 h under vacuum. The mixture of tripropylene glycol and zeolites was stirred for 1 h under vacuum at 90 °C. Afterward, the proper quantity of PMDI was mixed with the glycol and the mixture was stirred for 15 min under vacuum at room temperature. The thermal curing treatment required by the PU resin to cure into a solid material was performed according to the following cycle:

- 1 h at 50 °C (during molding);
- 5 h at 80 °C.

Commercial styrene-based orthophthalic polyester resin. The UP resin, commercially named SIRESTER FS 0910/M (Sir Industriale, S.p.A.) was composed of a polyester binder, styrene (around $40 \div 50$ wt%), and octabenzone ($\leq 0.25\%$). The conventional styrene-based UP resin was provided in a ready-to-be-used liquid phase. However, the crosslinking required the addition of a catalytic system, composed of cobalt(II) naphthenate and 2-butanone peroxide. The formulation used for the conventional styrene-based UP resin is shown in Table 5.

4.2 Casting and Molding

Acrylic and epoxy liquid resins were cast into a rectangular aluminum mold. The resins undergoing this process were fluid enough to wet the whole mold surface and to generate a smooth surface. PU resins required a faster molding and a curing method to avoid contact between the mixture and the humidity of the air. For this reason, a hydraulic press was chosen to perform the compression molding of PU samples for 1 h at 50 °C. To assure that the cured material could be pulled-off after crosslinking without being damaged, a PTFE treated mold was designed and used.

4.3 Curing Cycle

Thermosetting polymers can achieve gelation with room temperature curing. However, a post-curing process is normally required to complete crosslinking. Different post-curing treatments were designed for the different resins and they are shown in Table 6.

Table 6 Curing cycles designed for the developed resins

Resin	Curing process
Acrylic	(i) 2 h at room temperature (ii) 3 h at 80 °C (iii) 2 h at 140 °C
Epoxy	(i) 24 h at room temperature (ii) 1 h at 100 °C
Polyurethane	(i) 1 h at 50 °C (during molding) (ii) 5 h at 80 °C
Styrene–based UP	(i) 24 h at room temperature (ii) 3 h at 80 °C

4.4 Composite Reformulation

The reformulation process of mechanically recycled composites with a fresh polymer matrix was mainly studied using the epoxy resin, due to the better mechanical performances showed by this type of resin (as described in Sect. 5.5).

Before the formulation with the resin, the shredded recyclate was dried at 50 °C in an oven for 2 h to remove entrapped humidity and improve the composite behavior [16]. Afterward, the liquid resin was poured into the beaker containing the recyclate and manually mixed with a spatula. Compression molding technology was selected to cure the very viscous mixture. Compression molding also enabled the realization of composite samples with a smooth surface and controlled thickness. A mold composed of stainless steel and poly(methyl methacrylate) plate was chosen to allow an easy detachment of the epoxy-based composites from the mold. Compression molding was carried out at room temperature for 24 h, to make the sample cure with the desired shape. Then, an off-mold post-curing treatment at 100 °C for 1 h was performed.

4.5 Characterization Methods

TGA was performed for mechanically recycled shredded composites based on glass fiber and UP, to measure the exact content of GFs in the recyclate flakes. All the tests were performed with a Q 500 instrument (TA Instruments®), setting a heating ramp from 25 to 800 °C at 20 °C/min in air atmosphere, to degrade the resin present in the recycled material and measure the glassy inorganic residue in the sample chamber.

Differential scanning calorimetry (DSC) analyses were performed using a DSC 823e (Mettler-Toledo®) instrument. To identify the glass transition temperature (Tg) of the crosslinked material, the tests consisted of the following three runs with a rate of 20 °C/min:

- first heating run from 25 to 250 °C to check the absence of any unreacted groups;
- cooling run from 250 to 0 °C; and
- second heating run from 0 to 250 °C to identify the Tg.

To determine the total crosslinking enthalpy of the resin, starting from a liquid sample of resin, a single heating DSC ramp from 0 °C to 250 °C was employed with a rate of 10 °C/min.

Rheological properties of liquid resins were measured by means of a rotational rheometer (Melvern instruments Ltd.) with a parallel-plate geometry, with a diameter equal to 20 mm and a gap of 0.5 mm between the plates.

The gel content method was used to evaluate the extent of crosslinking by measuring the effective gel percentage present in the crosslinked resin. The solid crosslinked sample was immersed into an adequate solvent, to allow the dissolution of any residual unreacted components. Acetone or tetrahydrofuran (THF) were selected as solvent, according to the solubility of the liquid resins. 70 mL of solvents were used for each gram of sample. The immersion lasted 24 h when using acetone and 7 h with THF.

At the end of the immersion, the residual, not solubilized, sample was introduced into a beaker. The solution was carefully filtered to collect every residual insolubilized particle in the filter itself, while the solution was poured into the flask. The filter was then introduced into the beaker, while the whole amount of solvent in the flask was evaporated. Afterward, both the flask and the beaker (with the filter inside) were inserted into a vacuum oven at 50 °C and left there for approximately 48 h, until their weights were constant. The final weights were used to calculate the final mass of the solubilized residue (m_{SOL}). Gel percentage (Gel%) was then evaluated with the Eq. (1), which defines the percentage of insolubilized material:

$$Gel\% = 100 \times \left(m_{sample} - m_{SOL}\right)/m_{sample} \tag{1}$$

where m_{sample} is the initial mass of the sample.

The tack-free time determination is a test that allows the estimation of the time required by a liquid resin to reach a correspondingly robust state for safe handling. This means that, from that moment, the partially-cured resin may be removed from the mold. The test involved the casting of a thin layer of the liquid polymeric resin in a round mold of PP, featuring a diameter of 1 cm. The layer, then, was periodically subjected to a scratch and the time required by the resin to recover was measured. The test was repeated until it was not possible to scratch the surface of the layer anymore.

Mechanically recycled fibers from composite materials are not completely liberated but are partially embedded in the cured matrix that was binding them in the 1st life application product. Therefore, it is important to measure the adhesion between the polymer resin constituting the former matrix and the "new" resins, employed in this work as matrices for remanufactured products. For this purpose, a PosiTest Pull-Off Adhesion Tester (DeFelsko) was used to measure the perpendicular force required to pull a material layer of specified area away from its substrate using hydraulic pressure connected to a metallic dolly. The highest value of tensile pull-off force that the layer can bear before detaching was measured and it represented the strength of adhesion to the substrate. In this specific case, a substrate made of the styrene-based polyester commercial resin was prepared. Then, a thin layer of the chosen "styrene-free" reformulation resin was applied on the dolly surface. After that, the dolly was attached to the substrate and the pull-off test was performed. The test was performed for each of the three resins developed in this work.

For the tensile tests, a Zwick/Roell GmbH & Co. dynamometer (10 kN Allround Tisch), equipped with a 10 kN load cell was employed. The dimensions and the operative parameters were selected from the international standard related to testing conditions for isotropic and orthotropic fiber-reinforced plastic composites (ISO 527-4). In this case, rectangular shape specimens without end tabs were used. The dimensions indicated in the international standard ISO 527-4 were mostly scaled with a factor of 0.4.

5 Results and Validation

5.1 Characterization of the Recycled Reinforcement Materials

Two types of shredded flakes, obtained from EoL GF reinforced composites, were used. The first type consisted of coarse white flakes with a maximum dimension of around 4 mm, deriving from the EoL GF reinforced UP, produced for construction applications, shredded and supplied by Rivierasca S.p.A. The second type of recyclates was obtained by grinding EoL wind turbine blades, provided by Siemens Gamesa Renewable Energy S.A and shredded by STIIMA—CNR (Istituto di Sistemi e Tecnologie Industriali per il Manifatturiero Avanzato—Consiglio Nazionale delle Ricerche). EoL wind blades were made of an epoxy resin reinforced with a GF woven cloth.

Mechanically recycled glass fibers from composites for construction applications. Mechanically recyclates consisting of EoL GF reinforced polyester composites showed an average content of GFs of 30.5 ± 2.3 wt%, determined by TGA. These composites were originally composed of styrene-based orthophthalic polyester resin and 5 cm long fibers with a diameter of 13 μm. After the mechanical treatment, recycled composites consisted of matrix particles and GFs still partially embedded in the original matrix, as shown by representative SEM images of the recyclates (Fig. 12a). The average length of GFs turned out to be around 500 μm, after the elaboration of SEM images by ImageJ software.

Mechanically recycled glass fibers from composites for wind turbine blades. TGA tests were performed on different batches of the same recyclate and an average GF content of $71.5 \pm 3.0\%$ was measured. In order to estimate the average diameter of the fibers, SEM images were used and elaborated on through ImageJ software. The average diameter turned to be 20.5 ± 4.5 μm (Fig. 12b). The average length of fibers was determined by the elaboration of optical microscope images and photos taken by a common camera. Considering the median value of the distribution as an

Fig. 12 Bundle of GFs present in the mechanically composite recyclates obtained by grinding **a** GF reinforced polyesters for building applications and **b** GF reinforced epoxy resins for wind turbine blades

average value, the fibers showed an average length of 157.7 μm. As an average, the degree of liberation of glass fibers from the cured matrix seemed much better than in the previous case.

5.2 Formulation and Crosslinking of Styrene-Free Liquid Resins

Acrylic Resin. To select the reactive diluent suitable for the acrylic resin, four candidates were identified (MMA, IBOMA, LMA, and BDDMA) and characterized considering the following properties: volatility; miscibility; viscosity cut; handling safety.

As shown in the literature, MMA showed a level of volatility comparable to styrene and higher than the that exhibited by the other systems [15]. IBOMA and LMA showed poor miscibility with SR349 and low viscosity cut. These considerations led to the selection of BDDMA as target reactive diluent. To determine the suitable amount of reactive diluent, different weight fractions of BDDMA and SR349 were analyzed through rheological analyses to assess the viscosity of the system. The rheological analysis shows a very weak shear-thinning behavior for the resin diluted with BDDMA (Fig. 13). Considering the commercial UP resin viscosity as a reference value (0.39 ± 0.04 Pas), an optimal percentage of BDDMA was found to be around 20 wt% (viscosity0.30 ± 0.01 Pa·s). A further increase of BDDMA percentage was considered unnecessary.

The Tg of the crosslinked acrylic resin obtained with 80 wt% of SR349 and 20 wt% of BDDMA was evaluated via DSC. The Tg value is shown in Table 7, together with the gel content percentage for the solid resin, evaluated gravimetrically after solvent immersion and extraction.

Fig. 13 Viscosity curve of SR349 + BDDMA resin with different weight concentrations of reactive diluent (BDDMA, also called with the commercial name SR214)

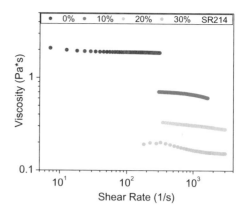

Table 7 Glass transition temperature (Tg) and gel content of the acrylic solid resin, obtained with 80 wt% of SR349 and 20 wt% of BDDMA

Properties	
Tg	$123 \pm 1 \, °C$
Gel content	$98 \pm 1\%$

Epoxy Resin. DSC analyses were performed for all three epoxy systems and the results were found to be comparable. The Tg obtained by DSC and gel content evaluation tests for the solid epoxy resin (EPO) are shown in Table 8.

Polyurethane Resin. DSC analyses were performed to evaluate the Tg obtained by combining the polymer bearing isocyanate side groups with different diols. The results are listed in Table 9.

The formulation including the glycerol resulted too brittle, showing a Tg of 153 °C. Those involving dipropylene glycol and tetraethylene glycol were characterized by limited miscibility that produced biphasic materials (two Tg values were observed), not suitable for binders. On the other hand, formulation #3, with glycerol propoxylate, resulted too soft. Finally, formulation #5, obtained with MDI and tripropylene glycol, led to a cured resin with a Tg of 92 °C. Furthermore, this resin showed good miscibility and properly low viscosity, during processing. In light of the above, the formulation involving MDI and tripropylene glycol was selected as the PU system in further steps. The properties for the solid PU resin obtained with tripropylene glycol are provided in Table 10.

Table 8 Glass transition temperature (Tg) and gel content of the epoxy solid resin named EPO

Properties	
Tg	$72 \pm 1 \, °C$
Gel content	$98 \pm 1\%$

Table 9 Glass transition temperature (Tg) of the polyurethane solid resin crosslinked with different glycols or polyols

Number	Isocyanate (Part A)	Polyol (Part B)	Molecular sieve (Part C)	Parts by Weight (A:B:C)	Tg (°C)
1	MDI	Glycerol	/	100:26:00	153
2	MDI	Dipropylene glycol	/	100.56:00	63.5; 143
3	MDI	Glycerol propoxylate	Zeolite A type	100:74:7.4	<80
4	MDI	Tetraethylene glycol	Zeolite A type	100:81.6:8.2	49; 104
5	MDI	Tripropylene glycol	Zeolite A type	100:81:8.1	92.5 ± 2.5

Table 10 Glass transition temperature (Tg) and gel content of the polyurethane solid resin obtained with tripropylene glycol

Properties	
Tg	$93 \pm 3 °C$
Gel content	$96 \pm 1\%$

Table 11 Glass transition temperature (Tg) and gel content of solid commercial styrene-based unsaturated polyester resin

Properties	
Tg	$121 \pm 2 °C$
Gel content	$99 \pm 1\%$

Commercial styrene-based orthophthalic polyester resin. The viscosity of the liquid commercial styrene-based UP resin was found to be 0.39 ± 0.04 Pa s. This value was used as a reference value for the other resins. The results obtained with DSC tests and gel content evaluation tests for the solid resin are provided in Table 11.

5.3 Reactivity of the Resins

A key property is the reactivity because it is crucial for the understanding of the time available for the molding process. For this reason, an evaluation of the de-molding time and an analysis with DSC measurements were conducted. The de-molding time determination method is a test that allows the estimation of the time required for a liquid resin to reach a correspondingly robust state for safe handling. This means

Table 12 De-molding time
results of the different resins

Resin	Phase transition (min)*	De-molding time (min)
Styrene-based polyester	10	15
EPO10	15	20
EPO	90	240
EPO80	270	600
Acrylic	180	360
Polyurethane	30	180

*Liquid to a viscous phase transition

that the semi-cured resin may be removed from the mold for times longer than the de-molding time.

The test involved the casting of a thin layer of the liquid polymeric resin in a round mold of PP, featuring a diameter of 1 cm. The layer, then, was periodically subjected to a scratch and the time required by the resin to recover the scratch was measured. The test was repeated until it was not possible to scratch the surface of the layer anymore. The results of the de-molding time tests of the resins are shown in Table 12.

DSC analysis was instead used to determine the total crosslinking enthalpy of the resin. To this end, starting from a liquid sample, a single heating ramp was performed from 0 to 250 °C with a rate of 10 °C/min. The styrene-based polyester resin was considered as a benchmark, because it is currently used in a large number of applications, especially for composite manufacturing. The reactivity of resins for composite reformulation should be moderately high to guarantee good processability together with fast curing, to achieve larger productivity. The results of the reactivity of the resins obtained by DSC analysis are in good agreement with the gel time indications (Tables 11 and 12). According to the DSC curve, the styrene-based resin started to react at room temperature and a de-molding time of 15 min was measured in this case. A total exchanged enthalpy of 225.3 J/g was also measured for the styrene-based resin (Table 13). DSC curves for the epoxy systems are shown in Fig. 14. EPO10 showed a shorter curing time with a de-molding time of 20 min and values of onset and end temperatures comparable to those of reference resin. On the other hand, EPO and EPO80 respectively exhibited a de-molding time of 4 and 10 h and higher values of onset and end temperatures. All the epoxy binders started to react above room temperature and were characterized by almost the same large heat flux. Table 13 also shows that the onset temperature is a good indicator of the reactivity of the resins. Indeed, the resins with faster curing (lower gel and de-molding time) were characterized by lower onset temperature. By increasing the onset temperature, the reaction was slower (lower gel and de-molding time). Thus, the epoxy formulation could be tuned to achieve fast curing or, otherwise, a slow curing system. Moreover, EPO10 can be considered the system more similar to the benchmark in terms of reactivity.

Table 13 Peak temperature (T_{peak}), onset temperature (T_{onset}), end temperature (T_{end}), and peak area underneath the peaks measured by DSC for the liquid resins under investigation

Resin	T_{Peak} (°C)	T_{onset} (°C)	T_{end} (°C)	Peak area (J/g)
Styrene-based polyester	84	40	132	225
EPO10	91	55	146	424
EPO	101	66	178	458
EPO80	112	79	178	458
Acrylic	76	68	100	226
Polyurethane	108	58	192	153

Fig. 14 DSC curves of all three liquid epoxy formulations

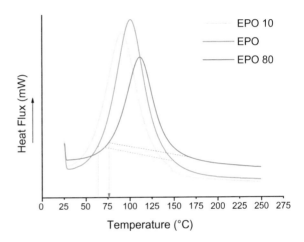

Regarding the acrylic resin, the measured amount of heat flux was comparable to the one of the commercial styrene-based polyester resin, but the onset temperature was found to be 68.4 °C. A much larger de-molding time of 6 h was also detected and this suggests that this resin has a slow crosslinking kinetics.

Characteristics similar to the previous one were observed for the PU system. A de-molding time of 3 h was observed and the onset temperature was higher than room temperature (T_{onset} = 57.6 °C). A heat flux of 153.39 J/g, which is lower than that observed on all three resins, was measured.

5.4 Adhesion Tests

Mechanically recycled GFs usually consist of GFs still partially embedded in the original matrix material (see Fig. 12). This could be a problem for the adhesion strength between the fresh matrix and the recycled reinforcement. A poor adhesion strength could result in unsatisfying mechanical performances of the final re-manufactured

Table 14 Pull-off tests results

Resin	Adhesion strength (MPa)	Failure type
Acrylic	5.6 ± 0.3	Cohesive
Epoxy	2.0 ± 0.6	Adhesive
Polyurethane	4.2 ± 0.4	Cohesive

product since the physical and mechanical properties of GFRP depend essentially on how tightly the resin adheres to GFs. This is related to how efficiently the mechanical stresses can be transferred from the matrix to the reinforcing fiber. It was important, therefore, to reliably estimate the adhesion of the different resins to the recycled material through the pull-off test. A styrene-based polyester substrate was prepared and the candidate resins were the adhesives between the dolly and the substrate. The pull-off test results are shown in Table 14.

The best adhesion was achieved by the acrylic resin with a value of adhesion strength of 5.6 ± 0.3 MPa, while the epoxy and the PU showed lower results, respectively characterized by an adhesion strength of 2.0 ± 0.6 and 4.2 ± 0.4 MPa. If the failure occurred along with the interface between the adhesive and the substrate, the specific value of load required to peel off the dolly from the substrate was set as the level of the adhesion strength. On the contrary, if the failure occurred in the substrate bulk, the level of cohesive strength in the substrate was measured. When the substrate failed and thus the cohesive strength was established, the adhesive strength could not be determined, but it was considered to be greater than the cohesive strength value. This was the case with acrylic and PU resins.

5.5 Tensile Tests

Mechanical properties of the crosslinked resins under investigation were determined by tensile tests. As a reference system for commercial applications, tensile tests were also performed on the commercial styrene-based polyester resin. Young's modulus (E), tensile strength (σ_b), and deformation at break (ε_b) are reported in Fig. 15.

Considering the high value of Tg (123 °C) and the gel content (98%), the acrylic resin was expected to be a very rigid material with a high degree of crosslinking. Tensile tests confirmed these expectations (Fig. 15). When compared to the reference styrene-based resin, the acrylic one showed a lower Young's modulus (about 1 GPa lower) together with a slight increase of elongation at break. On the contrary, concerning the tensile strength (including the standard deviation), the average value of the acrylic proved to be almost the same as the reference one. This means that the acrylic resin was a less stiff material, capable of larger deformation at break when compared to the styrene-based resin.

The epoxy resin under investigation was characterized by a Tg of 73 °C and a gel content of 98%. Due to the lower Tg and the presence of a few residual reactive

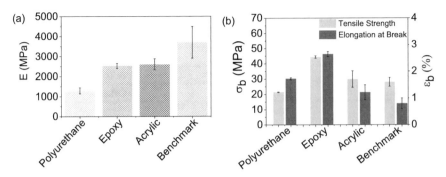

Fig. 15 Young's modulus (E), tensile strength (σ_b), and deformation at break (ε_b) for the three styrene-free resins under investigation, compared with the styrene-based reference resin

groups, a rigid material with good elongation capacity was expected. Indeed, the resin showed almost the same average Young's modulus as the acrylic resin, and an average value of elongation at break doubled (Fig. 15). Furthermore, a very high tensile strength was measured (Fig. 15b). Also, the epoxy resin showed a lower elastic modulus and a higher average tensile strength and elongation at break than the reference.

A PU resin was selected to find an alternative type of binder with slightly different characteristics. More specifically, it should provide a more ductile and tougher matrix, able to bear higher deformations. According to the data obtained, the desired objectives were only partially achieved. Indeed, the PU resin showed a low Young's modulus and tensile strength (respectively 1.7 GPa and 21.5 MPa), in agreement with the desired mechanical behavior (Fig. 15). However, the determined average elongation at break was too low. This could be due to the presence of some defects in the tested specimens. Those defects likely originated from humidity trapped in the tripropylene glycol reagent, during the manufacturing process. Defects gave rise to stress intensification and led to the premature fracture of the sample.

A comparison between the mechanical properties of the crosslinked resins developed was performed and the one showing the best performances was selected for the following tests. The average Young's modulus comparison is provided in Fig. 15a, whereas tensile strength and deformation at break of the different systems are compared in Fig. 15b. The PU resin was characterized by lower mechanical properties. The epoxy and acrylic resins also exhibited an elastic modulus lower than the styrene-based polyester resin, but ultimate properties at break were found comparable for the acrylic resin and even higher for the epoxy resin, in comparison with to the standard styrene-based system. The acrylic and epoxy systems were analogous in terms of stiffness, considering the very similar average Young's modulus. However, the epoxy material showed a larger capability to sustain more intensive loads and deformations. Thus, the epoxy resin seemed the most appropriate candidate for the combination with the recycled fibers. Furthermore, most of the GFRP products from the wind energy sector employ an epoxy matrix. Hence, a large stream of mechanically recycled fibers, still partially embedded in an epoxy resin, is expected to affect the composite market. For this reason, the composite remanufacturing, involving a new epoxy binder, should result in an improved adhesion between matrix and fibers, ultimately yielding better mechanical properties.

6 Feasibility Evaluation of Reprocessed Composites Using Mechanical Recyclates

6.1 Polymer Composites with Mechanically Recycled Glass Fibers from the Construction Sector

The reformulation of mechanically recycled composites with a polymer matrix was mainly performed with the epoxy resin, due to the higher mechanical performances showed in the previous section. Three formulations, which were different in terms of weight concentration of recycled coarse flakes added to the liquid resin, were mechanically tested. More specifically, they were characterized by 30, 35, and 40 wt% of reinforcement material and named 'EPO30', 'EPO35' and 'EPO40', respectively. The results of tensile tests are reported in Table 15.

It is clear that the remanufacturing process resulted in an improvement of modulus and therefore of stiffness (Table 15). However, no significant changes were found by increasing the percentage of recyclate from 30% to 40 wt%. This can be due to the small variation in GF content, going from 9 wt% to 12 wt% in the case of fibers weight fraction of 30% and 40%, respectively.

On the other hand, the remanufacturing process led to a large reduction of tensile strength and elongation at break, which may be due to the poor interfacial adhesion and to the presence of defects, originated during the manufacturing operations.

Table 15 Young's modulus (E), tensile strength (σ_b), and deformation at break (ε_b) for the three differently epoxy systems reinforced with different amounts of mechanically recycled composites from the construction sector

	E (GPa)	σ_b (MPa)	ε_b (%)
EPO	2.53 ± 0.13	44.70 ± 0.74	2.65 ± 0.09
EPO30	3.16 ± 0.38	26.02 ± 4.18	0.91 ± 0.13
EPO35	3.41 ± 0.40	28.29 ± 5.07	0.92 ± 0.16
EPO40	3.45 ± 0.31	30.82 ± 5.29	0.96 ± 0.13

6.2 Polymer Composites with Mechanically Recycled Glass Fibers from the Wind Energy Sector

The remanufactured composites were obtained firstly by mixing the epoxy resin with the recyclates, then by applying vacuum to enhance the wettability of the new resin with the recyclates. Finally, compression molding was used to obtain the final object. In this case, the use of the epoxy resin as a matrix should improve the adhesion between the new resin and the GFs, because the original resin used for the wind blade was also an epoxy system.

Different percentages of recycled materials were mixed with the epoxy resin and good results for recyclate feed concentration up to 60% were achieved, as the final object did not show any evident defect. As shown in Fig. 16, the samples loaded with the 60% of recyclates showed a high elastic modulus of 7.4 GPa, comparable to that measured for standard GF composites for building applications. For percentages higher than 60 wt%, the final specimens showed voids and defects and no increase in Young's modulus, compared to samples reinforced with 60 wt% of recyclates. The presence of defects in the specimens also explained the values of tensile strength and elongation at break for samples reinforced with 65% and 70 wt% of recyclates that are lower than those loaded with 60 wt%

The decrease of values measured for ultimate mechanical properties by increasing the percentage of recyclates added in the epoxy resin can also be due to small contaminations present on the surface of GFs and acting as defect, as shown in Fig. 17d. SEM images of a single mechanical recycled GF were acquired to understand if these pieces are residues of the old matrix or parts of the new one attached to the fiber surface. These small pieces were assigned to residues of the old matrix because they were found only in the recycled materials. In Fig. 17b and c, examples of frictional sliding and fiber pull-out are represented. Frictional sliding constitutes one of the most significant sources of fracture for most composites with fibers. Indeed, this

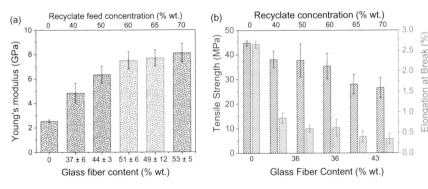

Fig. 16 a Young's modulus (E), **b** tensile strength (σ_b), and deformation at break (ε_b) for the epoxy systems reinforced with different amounts of mechanically recycled composites from the wind energy sector

Fig. 17 SEM images with a magnification of 1000× of epoxy systems reinforced with **a** 40%, **b** 50%, **c** 60%, and **d** 70 wt% of mechanically recycled composites from the wind energy sector

process can absorb large quantities of energy depending on the superficial rough-ness, contact pressure, and sliding distance, which result in the pull-out of fibers from their socket in the resin. The voids, observable in many regions and indicated by the arrows, represented signs of the previously described phenomena. Moreover, also the voids present in the matrix surrounded by red circles represent regions subjected to the fiber pull-out [17].

7 Conclusions and Future Research Perspectives

Three different thermosetting resins were developed and tested to assess their suit-ability to constitute the matrix phase in a composite material. Their reactivity and mechanical properties were compared to those of resins currently used in commercial applications. Moreover, the polymeric resin, acting as reinforcement binder, could be properly selected depending on the desired reactivity, processability, and mechanical behavior. The epoxy resin seemed the most appropriate candidate for incorporation of mechanically recycled fibers. On the other hand, a poor level of adhesion was found between two of the resins under investigation and the unsaturated polyester resin, which was used as a matrix in composites for building applications and probably will present as a residue in the corresponding mechanically recycled GFs. This could represent a limitation because it could lead to a poor level of adhesion and defects in final remanufactured composites.

The feasibility of the composite remanufacturing process has been evaluated so far combining the epoxy resin with mechanically recycled GFs obtained from EoL GFRP for construction applications. Promising results, showing an improvement of elastic modulus and therefore of stiffness in comparison with neat resin, were obtained. However, no significant changes were observed by increasing the recyclate percentage from 30 to 40%, due to the low corresponding variation (only 3 wt%) of the GF content in the tested range of recyclate percentage. Moreover, an increase of 25–35% in the elastic modulus of remanufactured composites was only observed, because the maximum recyclate percentage used so far corresponds to 12% of GF weight fraction in the composites. A reduction of mechanical properties at break was observed for remanufactured composites when compared to the neat resin, probably due to some defects and flaws present in the remanufactured samples.

A viable opportunity can be provided by the use of a different type of recyclate with a much larger content of GFs. This opportunity was given by the recyclate coming from the wind energy sector. Recyclate materials obtained from the mechanical recycling of wind turbine blades showed a GF weight fraction of nearly 70 wt% This type of recycled reinforcement phase allows the addition of GF contents comparable or even larger than the ones measured in commercial products for construction applications. A tensile modulus of 7.5 GPa was measured for samples reinforced with the 60 wt% of mechanically recycled GFs. However, during mechanical tests, a decrease of the tensile strength and elongation at break was registered by increasing the amount of recycled GFs. SEM analysis of the fractured specimens of the recycled GF epoxy composites was exploited to gain some insight into some phenomena related to the matrix-fiber adhesion, such as frictional sliding and fiber pull-out.

This study suggests that composites recovered from the wind energy sector can be recycled for the remanufacturing of products for application fields requiring lower mechanical performances (i.e., building and sanitary sectors) in a cross-sectorial approach. Therefore, the obtained results provide a clear evidence of the possibility to reuse mechanically recycled GFs as reinforcement in styrene-free composites materials.

References

1. A European Strategy for Plastics in a Circular Economy. https://ec.europa.eu/environment/cir cular-economy/pdf/plastics-strategy-brochure.pdf. Last accessed 20 June 2020
2. Liu, P., Barlow, C.Y.: Wind turbine blade waste in 2050. Waste Manage **62**, 229–240 (2017)
3. Ayre, D.: Technology advancing polymers and polymer composites towards sustainability: a review. Current Opinion in Green and Sustainable Chemistry **13**, 108–112 (2018)

4. Hao, S., Kuah, A.T.H., Rudd, C.D., Wong, K.H., Lai, N.Y.G., Mao, J., Liu, X.: A circular economy approach to green energy: wind turbine, waste, and material recovery. Sci. Total Environ. **702**, 135054 (2020)
5. Oliveux, G., Dandy, L.O., Leeke, G.A.: Current status of recycling of fibre reinforced polymers: review of technologies, reuse and resulting properties. Prog. Mater Sci. **72**, 61–99 (2015)
6. Rybicka, J., Tiwari, A., Leeke, G.A.: Technology readiness level assessment of composites recycling technologies. J. Cleaner Prod. **112**, 1001–1012 (2016)
7. Naqvi, S.R., Prabhakara, H.M., Bramer, E.A., Dierkes, W., Akkerman, R., Brem, G.: A critical review on recycling of end-of-life carbon fibre/glass fibre reinforced composites waste using pyrolysis towards a circular economy. Resour. Conserv. Recycl. **136**, 118–129 (2018)
8. Callister, W.D., Rethwisch, D.G.: Fundamentals of Materials Science and Engineering: An Integrated Approach (2012)
9. Bagherpour, S.: Fibre reinforced polyester composites. In: Saleh H.E.-D.M. (ed.) Polyester (2012)
10. Benzarti K., Colin X.: Understanding the durability of advanced fibre-reinforced polymer (FRP) composites for structural applications. In: Bai, J. (ed.) Advanced Fibre-Reinforced Polymer (FRP) Composites for Structural Applications. Woodhead Publishing, pp. 361–439 (2013)
11. Penn, L.S., Wang, H.: Epoxy resins. In: Peters, S.T. (ed.) Handbook of Composites. Springer, Boston (1998)
12. MARKETS and MARKETS: Epoxy composite market. https://www.marketsandmarkets.com/Market-Reports/epoxy-composite-market-90247237.html. Last accessed 07 Dec 2020
13. Bramante, G., Bertucelli, L., Benvenuti, A., Meyer, K.: Polyurethane composites: a versatile thermo-set polymer matrix for a broad range of applications—mechanical analysis on pultruded laminates. In: Proceedings of Polyurethanes 2014 Technical Conference. Dallas, TX, US (2014)
14. Oertel, G. (ed.): Polyurethane Handbook. Carl Hanser Verlag, Leverkusen (1985)
15. Cousinet, S., Ghadban, A., Fleury, E., Lortie, F., Pascault, J.-P., Portinha, D.: Toward replacement of styrene by bio-based methacrylates in unsaturated polyester resins. Eur. Polym. J. **67**, 539–550 (2015)
16. Mamanpush, S.H., Li, H., Englund, K., Tabatabaei, A.T.: Recycled wind turbine blades as a feedstock for second generation composites. Waste Manage **76**, 708–714 (2018)
17. Beauson, J., Madsen, B., Toncelli, C., Brøndsted, P., Ilsted Bech, J.: Recycling of shredded composites from wind turbine blades in new thermoset polymer composites. Compos. A Appl. Sci. Manuf. **90**, 390–399 (2016)

Fiber Resizing, Compounding and Validation

Pekka Laurikainen⬡, Sarianna Palola⬡, Amaia De La Calle⬡,
Cristina Elizetxea⬡, Sonia García-Arrieta⬡, and Essi Sarlin⬡

Abstract The mechanical performance of a composite is greatly related to the load transfer capability of the interface between the matrix and the reinforcing fibers, i.e. the fiber/matrix adhesion, which is enhanced by a surface treatment called sizing. The original sizing of reinforcing fibers is removed during recycling process, which is recognized to contribute in typical issues of recycled fibers, namely uneven fiber properties and poor fiber/matrix adhesion. Applying a new sizing, a process denoted here as resizing, can help mitigate the issues. Furthermore, the sizing has a major role in improving the processability of the fibers as it contributes to the distribution of the fibers in the matrix. Proper distribution, along with the fiber fraction, are highly important for the composite performance. These properties are ensured by proper compounding. Here we demonstrate and validate the process steps to resize and compound recycled glass and carbon fibers with thermoplastic matrices. We found that at a relatively high sizing concentration, the compounding of all tested material combinations was possible. The resizing of the recycled fibers improved the compatibility at the fiber/matrix interface. It was concluded that recycled fibers can be used to replace virgin fibers in automotive industry to allow weight reductions and to promote circularity.

Keywords Recycled fibers · Sizing · Compounding · Composite performance

1 Introduction

The mechanical performance of a composite material is defined by the properties of its constituents: the fiber, the matrix and the adhesion between them. While the matrix

P. Laurikainen · S. Palola · E. Sarlin (✉)
Materials Science and Environmental Engineering, Tampere University, Korkeakoulunkatu 6, 33270 Tampere, Finland
e-mail: essi.sarlin@tuni.fi

A. D. L. Calle · C. Elizetxea · S. García-Arrieta
TECNALIA, Basque Research and Technology Alliance (BRTA), Mikeletegi Pasealekua, 2, 20009 Donostia-San Sebastián, Spain

© The Author(s) 2022
M. Colledani and S. Turri (eds.), *Systemic Circular Economy Solutions for Fiber Reinforced Composites*, Digital Innovations in Architecture, Engineering and Construction, https://doi.org/10.1007/978-3-031-22352-5_7

125

provides the shape and the integrity for the composite, the actual load bearing capacity of the structure is provided by the fibers. Good load-transfer between the components and, therefore, the full potential of the mechanical performance of the composite structure, can be achieved only if adequate fiber/matrix adhesion is ensured.

Virgin fibers are coated with a multicomponent thin coating called *sizing*, *size* or *finish*, both to ensure wetting and adhesion at the fiber/matrix interface, and to enable good handling and processing properties for the fiber tows or fabrics. Therefore, the sizing is the key to achieve the desired interfacial properties. The sizing formulation can be designed for a specific fiber and matrix combination or it can be a general one. The main components of these polymer-based aqueous solutions or emulsions are the film former and the coupling agent, the latter of which is not typically used for carbon fibers [1]. Other possible components are lubricants, antistatic agents, wetting agents and cross-linking agents [1]. The film former provides good processing characteristics, protects against fiber damage, bonds the fibers into a bundle and affects the interfacial properties. The coupling agent improves the fiber/matrix adhesion by being able to bond chemically to the fiber surface [2], which is why the coupling agent might also bridge minor defects on the glass fiber surface. As polymeric materials, the sizing components can generate an interpenetrating network at the fiber/matrix interphase through interdiffusion. Simultaneously, the interfacial properties generate a gradient in the radial direction and become approaching the bulk matrix properties when moving away from the fiber [3].

During thermal and chemical recycling processes, the original sizing is lost from the fiber surface. Furthermore, the fiber surface characteristics cannot be fully controlled in recycling due to the uncertainties related to the origin and state of the fibers. The recycling process itself also affects these characteristics. Ultimately, this leads to non-uniform and degraded fiber properties, which hinders the potential application possibilities of the recovered fibers [4]. To mitigate the issue of poor performance of recycled glass and carbon fibers, the sizing should be renewed, a fact that has also been acknowledged in the literature [5]. However, due to the nature of recycled fibers, a similar process as used for virgin fibers, is not applicable. The sizing can be applied by spraying or by various electrodeposition, electropolymerization or plasma polymerization methods [2]. The process is applied to continuous virgin fibers in a very fast roll-to-roll process. In turn, recycled fibers are typically chopped and vary in their length, which makes it impossible to follow similar sizing process, requiring instead a batch-type process. Naturally, the application of sizing on recycled fibers, or *resizing* as we call the process here, depends also on the use of the recycled fibers. For example, in the spray-up composite fabrication process, the resizing could be applied together with the binder or as the binder. However, if the composite structure fabrication is to be done with injection molding, the resizing should be done prior to compounding to ensure good processability and wetting of fibers. In that case, a dipping process is an intuitive choice.

Compounding with a thermoplastic matrix and injection molding are considered as very suitable processing methods for recycled fibers, which are usually obtained in chopped form. The optimization of the resizing chemistry and the process parameters must be done by also considering the requirements of the compounding process.

For best mechanical performance, a very thin sizing layer and lack of fiber clusters are preferable, whereas a higher specific density and fiber agglomeration are advantageous for the compounding process, as agglomerated fiber bundles are easier to feed into the compounder. In addition, special design might be required for the compounding device geometry.

Here the process steps to resize and compound chopped recycled fibers are demonstrated and the effects of the process parameters on the achievable composite performance are discussed. Recycled glass and carbon fibers were used together with thermoplastic polypropylene and polyamide matrices. The properties of injection molded composites were characterized by mechanical and thermal methods. The main aim of the work was to validate the applicability and the technical viability of recycled fibers in automotive mass production components. The work was divided into optimization of the resizing process and the compounding process.

2 Resizing Recycled Fibers

The aim of the resizing process optimization was to understand the effects of resizing process parameters on the composite properties, as resizing differs fundamentally from the application of sizing on virgin fiber. The following variables were studied: resizing solution concentration, immersion time and rinsing to remove excess sizing. These variables mainly affect the thickness of the sizing layer.

2.1 Materials and Processing

Thermally recycled glass (rGF) and carbon (rCF) fibers were resized in batch-type dipping process. The recycling was done as a part of the FiberEUse project (described in Chap. 5). Prior to resizing, the fibers were analyzed with Fourier Transformation Infrared Spectroscopy (FTIR, Bruker Tensor 27 with PIKE Technologies GladiATR attenuated total reflectance unit), Thermogravimetric analysis (TGA, Netzsch TG 209 F3 Tarsus) and Scanning Electron Microscopy (SEM, Zeiss ULTRAplus) to verify that the fibers are clean without matrix residues.

A polypropylene copolymer (copoPP) and a polyamide 6 (PA) were used when optimizing the resizing process and the chemistry of each resizing formulation was selected according to the corresponding matrix. Due to the FiberEUse project objectives, only copoPP matrix was used with rGF whereas rCF compounds were prepared from both matrix types. For the PP matrix, a nonionic maleated PP dispersion (Hydrosize® PP2-01, Michelman Inc.) was used as film former. For the PA matrix, a nonionic polyurethane dispersion (Hydrosize® U2022 Michelman Inc.) was used. According to the supplier, both matrices are suitable for chopped strands and provide better adhesion with the matrix. In the case of rGF, (3-aminopropyl)triethoxysilane (Sigma Aldrich) was used as coupling agent and it was expected to generate covalent bonds with the glass fiber surface [6]. The resizing solution was prepared by mixing the chemicals with deionized water to generate solutions with specific total solid content. The pH of the rGF solution was adjusted with acetic acid to 7 to ensure proper silane reactions. Due to the scientific nature of the study, the simplified formulation for the resizing formulation was accepted. Rinsing the fibers in deionized water after dipping in the sizing solution was used as one of the processing parameters. The resized fibers were dried at 80 °C for at least 12 h.

The resized fibers were compounded, i.e. melt blended, with the copoPP and PA matrices in a laboratory-scale twin-screw microcompounder (DSM Xplore). The material was then injection molded with the microinjection unit of the compounder into dog-bone shaped tensile specimens and tested with a universal testing device [ISO 1172 (method A) standard, Instron 5067 device]. Prior to the compounding, the fibers were chopped to approx. 5 mm length, which were aligned relatively well in the injection molding process. The fiber fraction of the compounds was 20 wt% and verified by the burning-off method according to the ISO 1172 (method A) standard. The compounds were denoted as copoPP/rGF20, copoPP/rCF20 and PA/rCF20. Reference compounds were prepared with recycled fibers without resizing.

2.2 Characterization of the Resized Fibers and Compounds

FTIR, TGA and SEM characterization was used to study the resized fibers. The FTIR analysis verified that the resized rGF surface contained the film former and coupling agent (data not shown). Increasing the relative concentration of coupling agent induced a slight increase in the corresponding absorption bands at a specific total solid content. However, at highest total solid content this was not observed, and it was assumed that the high concentration of silane led to self-polymerization reaction instead of silane reaction with the reactive sites at the fiber surface followed by removal of coupling agent during rinsing. In general, higher total solid content of the resizing solution strengthened the corresponding absorption bands. Similar behavior was seen in the total mass loss of the thermogravimetric analysis. In the micrographs of the resized rGF surfaces, a slightly uneven sizing layer was observed (Fig. 1). However, the appearance of the resized rGF was not affected by the relative concentration of the coupling agent and the film former, the solid content of the resizing solution, the immersion time or the rinsing.

In the case of rCF, the sizing layer was more homogeneous with low resizing solution solid contents and both the PP and PU based sizings smoothed out some of the surface roughness (Fig. 2). Thermal analysis revealed that the mass loss, i.e. the average sizing layer thickness, followed the resizing solution concentration if the fibers were rinsed after the dipping, whereas without rinsing the amount of sizing in the fibers was higher and more random. More detailed characterization of the resized fibers can be found from [7].

According to the tensile test results of the copoPP/rGF20 compounds (Fig. 3), the tensile strength seems to reach a maximum value around 2 wt% sizing concentration,

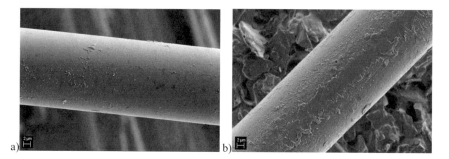

Fig. 1 Examples of the recycled and resized glass fibers: **a** rGF prior resizing, **b** resized rGF

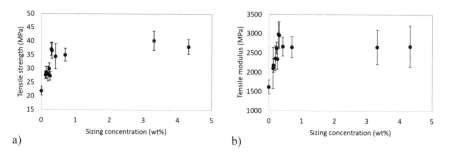

Fig. 2 Examples of the recycled and resized carbon fibers: **a** rCF prior resizing and **b** resized rCF

whereas for the modulus the maximum is reached already earlier, around 0.5 wt%. A further increase in the concentration did not improve the tensile properties. The SEM micrographs from the fracture surfaces (Fig. 4) supported the tensile test results showing a better fiber/matrix contact for the resized rGF when compared to the fibers prior resizing.

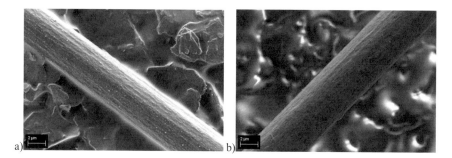

Fig. 3 Tensile test results of the copoPP/rGF20 compound: **a** strength and **b** modulus. Here, the tensile properties are expressed as a function of sizing concentration of the fibers (total mass loss in thermal analysis)

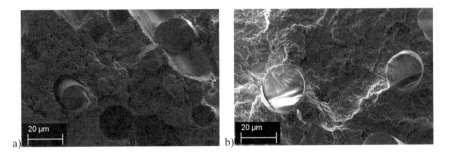

Fig. 4 Fracture surface of the copoPP/rGF20 compounds: **a** without sizing and **b** with resized fibers

For the rCF compounds, no significant difference was observed in the tensile properties as a function of sizing concentration (Fig. 5). This may indicate that the surface of recycled carbon fibers is more compatible than glass with thermoplastic matrices and do not need any pre-treatments from the adhesion point of view. This was also visible in the fracture surface micrographs (Fig. 6). However, the fractography revealed that the distribution of the fibers was improved when resizing was used. Further, the higher specific density introduced by the resizing induced a major improvement in the processability of the fibers.

To define the resizing process parameters to be used later in the FiberEUse project, the tensile properties of the thermoplastic rGF and rCF compounds were evaluated together with the processability experience gained during the study. To enable an industrially feasible processing of the recycled fibers, a high enough specific density must be obtained to ease the handling of the fibers. If the specific density is too low, the feeding the fibers into the mixer becomes difficult or even impossible. For a larger scale process, this aspect proved more critical than the exact mechanical properties, and therefore a relatively high sizing concentration was chosen for the next steps.

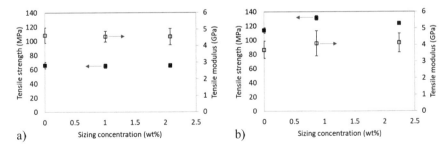

Fig. 5 Tensile test results of **a** the copoPP/rCF20 compound and **b** the PA/rCF20 compound. Here, the tensile properties are expressed as a function of sizing concentration (total mass loss in thermal analysis)

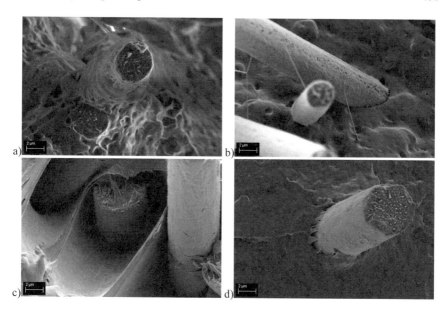

Fig. 6 Fracture surface of **a** the copoPP/rCF20 compound without sizing, **b** the copoPP/rCF20 compound with resized fibers, **c** the PA/rCF20 compound without sizing, and **d** the PA/rCF20 compound with resized fibers

For the compounding experiments and the FiberEUse project demonstrators, larger batches of resized fibers were prepared. While the resizing solution preparation was kept similar, inevitable differences—between the first small-scale resizing experiments described here and the larger scale batches—were introduced in terms of the origin of the fibers (different batch of original waste) and the recycling process of fibers (more advanced recycling process). This also affected the average length of the recycled fibers (reduction from >10 cm to >5 cm). The fiber fraction of the compounds was also different from the resizing process optimization step. The changes were assumed to influence the final properties of the recycled fibers and therefore the results of the resizing process optimization study and the compounding study are not directly comparable.

3 Compounding Recycled Fibers

Industrial scale compounding starts with a base resin or polymer, i.e. the matrix. The polymer is melt and mixed with the reinforcement and possible other additives in a twin-screw extruder at a high temperature, after which it is extruded and pelletized. The used additives can be, e.g., compatibilizers, antioxidants or pigments, which ensure that the matrix has a good compatibility with the reinforcement, does not degrade during the extrusion and injection processes and has the desired optical

Fig. 7 Specific hopper and
feeder for short fiber feeding

properties. Compounding process changes the physical, thermal, electrical and/or optical characteristics of the plastic. The final product is called a *compound* and the pellets can be the further processed, e.g., by injection molding.

The most important difference between compounding fibers and additives which are in powder or pellet format, is the importance of maintaining the maximum possible fiber length for the optimization of mechanical properties. For this, the process conditions and the geometry of the screws are important parameters. The specific characteristic of compounding recycled fibers is, that instead of a continuous rowing as a starting material, the recycled fibers are obtained in chopped form. Therefore, the specific density of the fibers is particularly important for successful feeding of the fibers. These, in turn, set new requirements for the feeder and hopper of the feeding system (Fig. 7).

In the rCF compounds of the FiberEUse project, the same PA and copoPP matrices were used as in the resizing optimization step. In addition, two other types of polypropylene grades were used: a homopolymer for the rCF compounds (denoted here as homoPP) and another polypropylene copolymer grade for the rGF compounds (denoted simply as PP). In addition to the rGF and rCF fibers recycled in the FiberEUse project, commercial recycled fibers Carbiso CT from ELG were used as s reference. However, it must be noted that the commercial fibers have not been obtained by pyrolysis, and that the product is based on production waste with general purpose epoxy sizing, which is not optimized for PP or PA. The origin of the recycled fibers of the FiberEUse project are various components from aeronautics sector, in which several carbon fiber grades are presumably used. The FiberEUse project fibers were used with and without resizing. All compound formulations are listed in Table 1.

The compounding process was carried out with a twin screw extruder (Coperion Werner & Pfleiderer ZSK 26 P 10,6) using a Brabender feeder to feed the matrix and a double screw side feeder to feed the fibers (Fig. 8). The parameters used in the compounding process are listed in Table 2. The final fiber fraction of the compounds was verified with thermogravimetric analysis (TGA, TA Instrument Universal V4) and the results are listed in Table 1. As can be seen, the final fiber fractions were within ± 2 wt% from the intended value.

Table 1 Description of the different compounds

Code	Matrix	Fiber	Fiber origin	Resizing	Fiber fraction (wt%)	
					Aim	Verified
A	PA	rCF	FiberEUse project	No	20	19.05
B	PA	rCF	FiberEUse project	Yes	20	21.68
C	PA	rCF	Commercial	Commercial	20	21.81
D	homoPP	rCF	FiberEUse project	No	20	19.58
E	homoPP	rCF	FiberEUse project	Yes	20	19.28
F	homoPP	rCF	Commercial	Commercial	20	20.02
G	copoPP	rCF	FiberEUse project	No	20	19.85
H	copoPP	rCF	FiberEUse project	Yes	20	20.65
I	copoPP	rCF	Commercial	Commercial	20	20.00
J	PP	rGF	FiberEUse project	No	30	30.72
K	PP	rGF	FiberEUse project	Yes	30	30.32
L	PP	rGF	Commercial	Commercial	30	29.54

The fiber fractions were verified with thermogravimetric analysis

Fig. 8 Recycled glass and carbon fibers and lateral feeding addition process

4 Properties of Recycled Fiber Reinforced Composites

The materials manufactured through the compounding process were characterized by tensile tests (ISO 527-2 standard, Instron 5500 R equipment) to compare the two main factors: the influence of sizing in the fiber and the influence of the origin of

Table 2 Compounding parameters for different formulations

Code	Matrix	Fiber	Main drive speed (RPM)	Pelletizer speed (RPM)	Melting temperature range (°C)	Matrix feeder (kg/h)	Fiber feeder (kg/h)
A-C	PA	rCF	50	200	245–260	10	2
D-F	homoPP	rCF	80	200	200–220	10	2
G-I	copoPP	rCF	80	200	200–220	10	2
J-L	PP	rGF	50	200	195–220	10 ·	3

the fiber (recycling process). Naturally, the specific mechanical design of the automotive component is determined by the properties fulfilled by the current material used in that design and these values were also used here as reference for the developed compounds: For the FiberEUse formulations based on PA with 20 wt% rCF (compounds A-C in Table 2), the reference value corresponded a PA compound with 40 wt% virgin glass fibers. For the FiberEUse formulations based on homoPP and copoPP compounds with 20 wt% rCF (compounds D-I in Table 2), the reference value corresponded a PP compound with 40 wt% virgin glass fibers. For the FiberEUse formulations based on PP with 30 wt% rGF (compounds J-L in Table 2), the reference value corresponded a similar compound but with virgin fibers. After tensile testing, composite fracture surfaces were studied with SEM (JEOS JSM-5910LV).

4.1 PA/rCF Compounds

As can be seen from Fig. 9a, the PA/rCF compound with resized fibers (compound B) had 100% higher modulus compared to the material without sizing (compound A). This material B was also only 3% below the automotive reference. The commercial recycled fiber compound C had 22% higher modulus than the automotive reference.

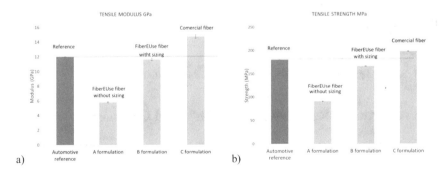

Fig. 9 **a** Tensile modulus and **b** tensile strength of the PA/rCF compounds A-C

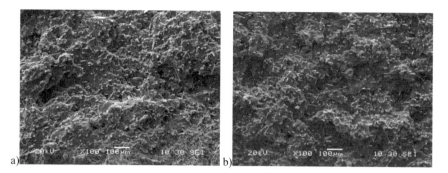

Fig. 10 Good dispersion of the fibers in the PA matrix: **a** the compound B with rCF from the FiberEUse project and **b** the compound C with commercial rCF

Regarding the tensile strength (Fig. 9b), the material without sizing (compound A) had 45% lower strength than the one with resized fibers (compound B). With respect to the automotive reference material, the strength of the compound B was 8% lower while the strength of the commercial recycled fiber compound C was 9% above the reference. The relatively minor difference between the developed PA/rCF compound B and the automotive reference material could be overcome by redesigning the automotive component so that slightly lower material properties are acceptable.

The fractography of the PA/rCF compounds (Fig. 10) exhibited a good distribution of the resized fibers, similar to the compound with commercial recycled fibers. When the compound based on commercial recycled fibers was compared with the corresponding FiberEUse material (Fig. 11), it was observed that the recycled carbon fibers from the FiberEUse project varied in diameter whereas the size of the commercial rCF was more homogenous. This was assumed to reflect the variability of original carbon fiber grades in the waste material: in the FiberEUse project the fibers originate from various components from aeronautic sector whereas the origin of the commercial recycled fiber is assumed to be homogenous.

4.2 PP/rCF Compounds

In Fig. 12 the tensile test results for the homoPP/rCF compounds D-F are shown. The material with sizing (compound E) had 37% higher modulus compared to the material without sizing (compound D). The resized compound E had also 23% higher modulus than the automotive reference. The commercial fiber material (compound F) had 70% higher modulus than the automotive reference.

Regarding tensile strength, it was observed that the material without sizing (compound D) had 40% lower strength than the compound E with sizing. With respect to the strength of the automotive reference material, compounds E and F were 15% and 13% lower, respectively. Similar to the PA/rCF compounds, the lower

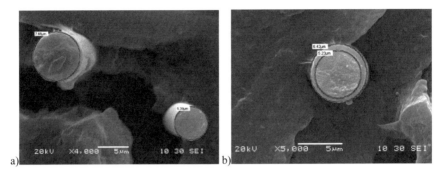

Fig. 11 Diameter of recycled carbon fibers: **a** the compound B based on recycled fibers from the FiberEUse project and **b** the compound C based on commercial recycled fibers

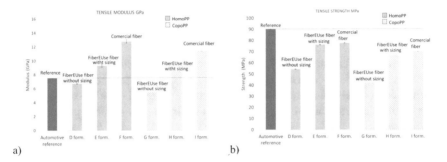

Fig. 12 **a** Tensile modulus and **b** tensile strength of the PP/rCF compounds D-I

strength level could be compensated by redesigning the components in which the recycled material is be used.

From the fractographies of the homoPP/rCF compounds (Fig. 13) it is evident that the fibers without sizing show poor compatibility with the matrix whereas the same fiber with sizing had better fiber/matrix adhesion. Also, the commercial fiber with a general sizing did not create a good fiber/matrix adhesion. The dispersion of the homoPP/rCF compound was proven to be good even without resizing (Fig. 14). When comparing the results with Fig. 5, it is evident that resizing improves especially the processability of the rCF compounds having simultaneously positive effect on the mechanical properties.

When considering the copoPP/rCF results, the material with sizing (compound H) had 54% higher modulus when compared to the material without sizing (compound G). The modulus of the resized compound H was also 13% higher than the automotive reference. The commercial recycled fiber material (compound I) had 51% higher modulus than the automotive reference. Compound G without sizing had

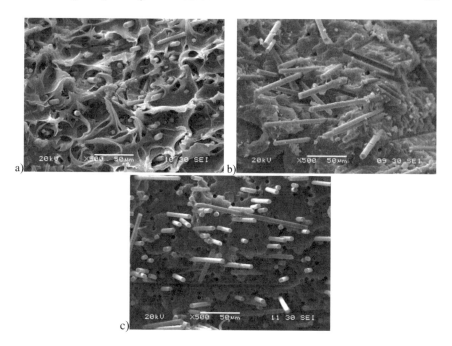

Fig. 13 Varying fiber/matrix adhesion when comparing the homoPP/rCF compounds: **a** poor adhesion in the compound D without sizing, **b** good adhesion in the compound E with resized fibers from the FiberEUse project, **c** poor adhesion in the compound F with commercial fibers and a general sizing

Fig. 14 Good dispersion of fibers in the homoPP matrix in compound D without resizing

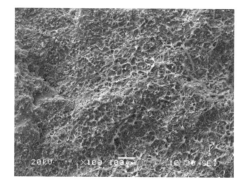

58% lower tensile strength than compound H with sizing. Compared to the automotive reference material, the strength of compounds H and I were 25% and 22% lower, respectively. The fracture surfaces were very similar to the corresponding homoPP/rCF compounds.

4.3 PP/rGF Compounds

The results of rGF compounds were not as close to the automotive reference values as the rCF compounds (Fig. 15). The PP/rGF compound K with sizing had 38% higher modulus when compared to the material J without sizing. However, the modulus of the resized material K was 45% lower than the automotive reference. Also, the modulus of the commercial recycled fiber material L was 22% lower than the automotive reference. The tensile strength of the compound J without sizing was 58% lower than the resized compound K and the strength of compounds K and L were 59% and 58% lower than the automotive reference, respectively.

Fractography of the PP/rGF compounds with and without sizing (Fig. 16, compounds J and K) revealed the positive effect of the resizing on the fiber/matrix adhesion. Considering the good adhesion between resized fibers and the PP matrix and the relatively low mechanical properties of the compound, the main issue seems to be the cohesive properties of the recycled glass fibers [8].

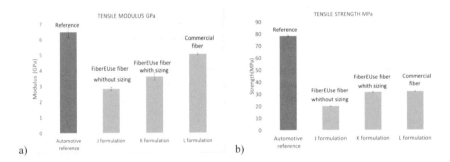

a)   b)

Fig. 15 a Tensile modulus and **b** tensile strength of the PP/rGF compounds J-L

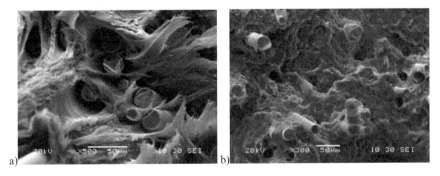

a)   b)

Fig. 16 Improved fiber/matrix adhesion when comparing the PP/rGF compounds: **a** poor adhesion in the compound J without sizing, and **b** good adhesion in the compound K with resized fibers from the FiberEUse project

5 Conclusions

The aim of this study was to demonstrate the technical feasibility of using thermoplastic compounds with recycled reinforcement in components of automotive industry. The study was divided in two steps: optimization of the resizing process of recycled glass and carbon fibers, and optimization of the compounding process of the thermoplastic composites.

The resizing of the recycled fibers should ensure good compatibility between the fibers and the matrix as well as industrially feasible processability, which was noted to correspond with high enough specific density. The latter aspect became more critical in this project and therefore a relatively high sizing concentration was selected, although slightly better mechanical properties were achieved at lower concentrations.

The compounding at industrial level was successful for all tested material combinations. Despite all compounds based on recycled carbon fibers did not satisfy the requirements for tensile strength and modulus used in designing automotive industry components, it was evident that recycled carbon fibers compounds can be used to replace virgin glass fiber compounds allowing weight reductions and promotion of circularity. The feasibility of the application of recycled glass fibers in automotive industry will depend greatly also on the cost of the materials.

References

1. Kim, J., Mai, Y.: Engineered Interfaces in Fiber Reinforced Composites, 1st edn. Elsevier, Amsterdam (1998)
2. Park, S., Hey, G.: Surface treatment and sizing of carbon fibers. In: Park, S. (ed.) Carbon Fibers, 2nd edn., pp. 101–133. Springer, Dordrecht (2015)
3. Park, S., Seo, M.: Modeling of fiber–matrix interface in composite materials. In: Park, S., Seo, M. (eds.) Interface Science and Technology, vol. 18, pp. 739–776. Elsevier Science, Amsterdam (2011)
4. Oliveux, G., Dandy, L.O., Leeke, G.A.: Current status of recycling of fiber rein-forced polymers: review of technologies, reuse and resulting properties. Prog. Mater Sci. **72**, 61–99 (2015)
5. Yang, L., Jenkins, P., Liggat, J., Thomason, J.: Strength of thermally conditioned glass fiber degradation, retention and regeneration. In: Proceedings of 20th International Conference on Composite Materials ICCM20, Copenhagen (2015)
6. Acres, R.G., Ellis, A.V., Alvino, J., Lenahan, C.E., Khodakov, D.A., Metha, G.F., Andersson, G.G.: Molecular structure of 3-aminopropyltriethoxysilane layers formed on silanol-terminated silicon surfaces. The Journal of Physical Chemistry C **116**(10), 6289–6297 (2012)
7. Matrenichev, V., Lessa Belone, M.C., Palola, S., Laurikainen, P., Sarlin, E.: Resizing approach to increase the viability of recycled fiber-reinforced composites. Materials **13**(24), 5773 (2020)
8. Ginder, R.S., Ozcan, S.: Recycling of commercial e-glass reinforced thermoset composites via two temperature step pyrolysis to improve recovered fiber tensile strength and failure strain. Recycling **4**(2), 24 (2019)

Additive Manufacturing of Recycled Composites

Andrea Mantelli⊙, **Alessia Romani**⊙, **Raffaella Suriano**⊙,
Marinella Levi⊙, **and Stefano Turri**⊙

Abstract An additive remanufacturing process for mechanically recycled glass fibers and thermally recycled carbon fibers was developed. The main purpose was to demonstrate the feasibility of an additive remanufacturing process starting from recycled glass and carbon fibers to obtain a new photo- and thermally-curable composite. 3D printable and UV-curable inks were developed and characterized for new ad-hoc UV-assisted 3D printing apparatus. Rheological behavior was investigated and optimized considering the 3D printing process, the recyclate content, and the level of dispersion in the matrix. Some requirements for the new formulations were defined. Moreover, new printing apparatuses were designed and modified to improve the remanufacturing process. Different models and geometries were defined with different printable ink formulations to test material mechanical properties and overall process quality on the final pieces. To sum up, 3D printable inks with different percentages of recycled glass fiber and carbon fiber reinforced polymers were successfully 3D printed.

Keywords 3D printing · Rheology · Glass fiber · Carbon fiber · Mechanical properties · Composites

1 Introduction to Composites Additive Re-Manufacturing: Challenges and Objectives

Over the last years, additive manufacturing (AM) has been emerging as a valid technology for the cost-effective realization of complex and customized components made of polymer composites through a layer-by-layer process [1, 2]. However, AM

A. Mantelli (✉) · A. Romani · R. Suriano · M. Levi · S. Turri
Department of Chemistry, Materials and Chemical Engineering "Giulio Natta", Politecnico Di Milano, Piazza Leonardo da Vinci 32, 20133 Milano, Italy
e-mail: andrea.mantelli@polimi.it

A. Romani
Department of Design, Politecnico Di Milano, Via Durando 38/A, 20158 Milano, Italy

© The Author(s) 2022
M. Colledani and S. Turri (eds.), *Systemic Circular Economy Solutions for Fiber Reinforced Composites*, Digital Innovations in Architecture, Engineering and Construction, https://doi.org/10.1007/978-3-031-22352-5_8

of composites shows some weaknesses such as the optimization of the processability, which changes according to the materials, and the resulting properties associated with the adhesion between layers as well as between matrices and reinforcing fillers [3, 4]. If recycled fibers are used as a reinforcement, the availability of composite wastes with different compositions and geometries, and the resulting variability of the filler properties, introduce other variables to be governed into the process. Therefore, it is necessary to validate the opportunity to develop an additive remanufacturing process of composites reinforced with recycled glass fibers (GFs) and carbon fibers (CFs). This requires the development of new 3D printable composite inks together with a new custom-made 3D printing system to obtain a good flowability of inks during the extrusion and high shape retention after the extrusion through the nozzle.

For this purpose, the aim of this study was firstly to characterize recycled fibers in terms of diameter sizes and dimension distributions. Secondly, different formulations of 3D printable inks were investigated by starting from a UV-curable acrylic resin combined with different amounts of recycled glass fiber reinforced polymers (rGFRP) and carbon fiber reinforced polymer (rCFRP). Rheological properties of inks were studied and optimized considering various recycled fiber percentages and mixing technologies. This led to the definition of a few requirements for 3D printable ink compositions. Moreover, a set-up for the 3D printing apparatus coupled with a UV-light to crosslink UV-curable ink was built up for these new 3D printable inks. Post-curing treatments after the 3D printing process enabled to maximize the crosslinking degree of 3D printed materials and their final mechanical properties. Various 3D structures and components were also 3D printed to validate the feasibility of the additive remanufacturing process with the reuse of end-of-life (EoL) composites.

2   Additive Manufacturing of Composites State of the Art

2.1   Commercial Solutions

3D printing of polymers is spreading fast in the field of automotive industries, particularly in many different brands of sports and standard cars. Sports car brands are currently developing the AM technology to improve their production process since it allows the fabrication of very complex geometries and, in some cases, reduces sensibly the production time and costs.

In 2014, Local Motors presented, for the first time, a car completely 3D printed by fused deposition modeling (FDM) technology and composed of acrylonitrile butadiene styrene polymer (ABS) reinforced with carbon fibers [5]. Since 2017, McLaren Formula 1 team has been using AM to produce race-ready parts of the car and tooling for immediate evaluation during practical sessions that need specific features not achievable with traditional manufacturing methods [6]. For instance, the material used to produce a structural bracket was a thermoplastic polyamide reinforced with chopped carbon fibers (FDM® Nylon 12CF), and the component was obtained in just 4 h instead of 2 weeks, which were necessary for standard processes.

Very recently, Moi composites s.r.l produced the world's first 3D printed real boat in continuous fiberglass thermoset material called MAMBO (Motor Additive Manufacturing Boat) with a patented additive manufacturing process [7]. Another example of AM technology currently used for the 3D printing of long fiber reinforced composite 3D printing is the continuous fiber printer developed by MarkForged© [8]. For instance, this printer and technology were deployed for the prototyping of fins for surfboards, showing a high level of mechanical strength when they were 3D printed with a thermoplastic filament of polyetherimide (ULTEM) reinforced with carbon fibers [9].

2.2 Academic Advancements

Over the last decade, many academic researchers have investigated and employed the FDM technology for the additive manufacturing of short fiber reinforced composites consisted of thermoplastic matrices and different types of reinforcements [10, 11]. The most common thermoplastic materials used as matrices are polylactic acid (PLA), ABS, polycarbonate (PC), nylon, and polyether ether ketone (PEEK), which seemed to provide some advantages such as a higher tensile strength compared to the other 3D printed thermoplastic polymers [12–14]. Regarding the reinforcement, natural fibers, which are for example made of hemp and wood have been applied and tested for the AM of polymer composites [15, 16]. However, synthetic fibers, i.e. short GFs as well as short and long CFs, are the most widespread reinforcing fillers added in the 3D printing of thermoplastics polymers [4, 17].

Another 3D printing technology used for the additive manufacturing of composites is liquid deposition modeling (LDM), based on the extrusion of liquid or viscous inks that can solidify just after the extrusion through the nozzle. A strategy commonly employed to induce a fast solidification is the application of heat and/or UV light that can trigger the polymerization of ink components using a UV and thermal initiators. In 2014, Compton and Lewis developed the LDM process, also called direct ink writing (DIW), capable of extruding short carbon fiber epoxy-based composites [18]. After the 3D printing of the composite structures, a thermal curing cycle was needed to completely solidify the resin. This study evidenced how the high aspect ratio fibers were aligned in the direction of the print during the process, conferring to the material the behavior of a classical composite. In 2017, the same authors prosecuted their previous study focusing on the orientation of the fibers in the desired orientation [19]. They developed and used a rotating printing head capable of imparting a particular "rope-like" orientation to the printed fibers, leading to a material with great performances in both compression and impact. The production of 3D printed carbon fiber composites was also attempted by Mahajan and Cormier with the aim of fabricating an oriented short carbon fiber epoxy composite [20]. A similar approach was used to 3D print CFRPs with orthotropic physical properties [21]. Even in this study, the ink material used in the DIW process exhibited a shear-thinning behavior with a moderate yield stress, which was necessary to allow the 3D printing and show excellent shape retention after the deposition to maintain the printed geometry. A thermally cured aromatic thermoset resin system with latent thermal cure catalysis was developed to induce the generation of a strong Lewis acid above ~70 °C and initiate a thermally activated cationic self-crosslinking among the epoxide function-alities of the resin. The formulated resin can remain stable at 20 °C for over 5 days at catalyst concentrations below 2% by weight and the gelation of the system and curing to full network density can be achieved in less than 5 s when the applied temperatures reach 150–200 °C.

Another study was also focused on the 3D printing of carbon fiber and glass fiber reinforced thermosetting resins with a standard desktop 3D printer modified for the LDM deposition [22]. In this study, a novel setup of an extrusion-based printer coupled with UV-light sources was employed to print a batch of specimens reinforced both with glass fibers and carbon fibers (up to 30wt%). Moreover, an interpenetrating polymer network (IPN) resin system was used in the UV-assisted 3D printing of composites. The formulated IPN system was composed of a thermal activated epoxy resin and a UV-curable acrylic resin. The resin system undergoes two sequential curing mechanisms: the former was due to the UV-curing of the acrylic resin during the extrusion process and the latter increased the crosslinking degree of the 3D printed enhanced mechanical properties. The enhanced stiffness and mechanical resistance of the glass/carbon fiber reinforced IPN resin systems were also demonstrated in another paper [23]. The mechanical properties of GFRP and CFRP systems exhibit an average increase in elastic modulus of 37% and 44%, respectively, compared to unfilled blends and a tensile strength increase of 19% and 91%.

Recently, 3D printed structures obtained with continuous carbon fiber reinforced PLA filaments were recycled and reused to produce the raw material for a subsequent 3D printing process [24]. The reuse of mechanically recycled EoL GFRPs as a reinforcement for the 3D printing of composites was very recently achieved [25]. The 3D printed composites with mechanically recycled GFs showed results comparable to 3D printed composites filled with virgin GFs, thus fostering the implementation of a circular economy case for thermoset composites [26].

3   UV-Assisted Additive Manufacturing

3.1   Working Principle

As depicted in Fig. 1 the process starts with the 3D printable ink preparation, then the UV curable ink is extruded from a syringe and a UV light-emitting diodes (LEDs) apparatus is switched on during ink deposition. As soon as the ink flows out from

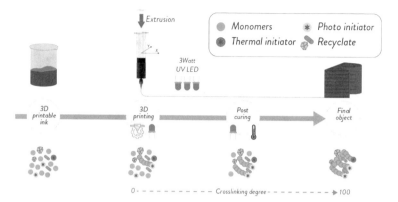

Fig. 1 UV-assisted liquid deposition modeling process description

the syringe nozzle it starts to cure. The free-radical photo-crosslinking process is normally fast enough to assure the layer-by-layer build-up process.

3.2 Machine Modifications

At the time of this work, a commercial UV-assisted 3D printer capable of processing composite inks was not available. For this reason, an open-source 3D printer was modified as follows.

The first modifications of the 3Drag printer (version 1.2, Futura Elettronica, Italy) were developed in previous work [22]. It is an open-source cartesian desktop-size 3D printer with a Z-head mechanical arrangement, meaning that the X and Y axis movements are delegated to the plate, while the tiny Z-axis movements are left to the head. This arrangement is very favorable in the case of bigger and heavier extruders since the limited accelerations do not cause any inertia-related issues. The main changes were done by modifying the extruder as shown in Fig. 2: the standard system was completely removed, and a custom extruder was installed in its place [27]. The custom system transfers the rotation of a stepper motor to the linear motion of a syringe piston to extrude the inks. Due to the difference in the material feed system, the motor steps per mm of feedstock had to be modified in the firmware matching the actual value. This and other modifications to the firmware were needed to allow the extrusion at ambient temperature [28].

Considering the UV source, three UV-LEDs were used (395 nm, 3 W) powered by a dimmable electronic power supply [29]. Three fans cool the LEDs, extending their lifetime, maintaining constant power, controlling the heat-up, and preventing damage to their supports. A relay, connected to the fan port of the 3D printer, could be used to controls the UV source. Therefore, by switching on and off the fan port through the Gcode it is possible to directly control the LEDs. Figure 3 shows two

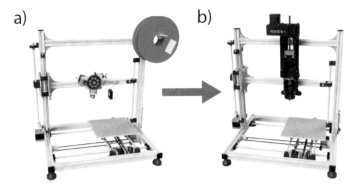

Fig. 2 Modified 3Drag with custom extrusion head: **a** commercial 3Drag version 1.2, **b** modified 3Drag with custom extrusion head

different configurations of the UV-LEDs apparatus. The first one has the three LEDs arranged in a circle around the nozzle, pointing at its tip. The second layout has the LEDs positioned alongside the extrusion head. In this way, the polymerization of the resin can be activated after the deposition of the material. Different examples of this approach can be found in the literature [30, 31]. A physical division of the extrusion step from the resin activation step is crucial to prevent nozzle clogging or for particular situations in which the polymerization may take a longer time than the layer deposition time.

Finally, the nozzle plays a vital role in the material deposition process for LDM systems. Generally speaking, conical nozzles proved to be a better alternative to cylindrical ones, due to their geometry that reduces backpressure build-up. So, this allows for a more constant flow, less clogging, and a generally more precise extrusion [33–35]. A fundamental constraint on nozzle selection was linked to UV-shielding. To prevent the polymerization of the ink inside the nozzle, metal nozzles have to be preferred.

Fig. 3 UV source apparatus: **a** standard configuration with LEDs pointing at the nozzle tip, **b** side configuration with LEDs pointing to the build plate alongside the extruder head [32]

3.3 In-Situ Polymerization Challenges and Opportunities

The challenges related to the material crosslink itself during the manufacturing process are mainly related to the rapidity of the process and its activation. Among the processes suitable for rapid crosslinking, radical photo-crosslinking is the most promising. These systems can provide a set time of seconds, or fractions of seconds, allowing to build the 3D object layer by layer. Examples can be found in literature where a self-standing ink is crosslinked after the 3D object manufacturing [18, 36]. The opportunities of the in-situ polymerization are both to use simpler ink systems that do not need to be self-sustaining or to have the possibility of crosslinking part of the 3D object to build bigger and more complex structures [25].

4 Experimental Methods, Materials Description, and Requirements

To better understand and follow the results that will be presented in the next chapter, the experimental methods, and the materials used will be described. The workflow followed in the development of the research can be divided into five interconnected steps: recyclate characterization, ink formulation, curing cycle, process development, material characterization. Figure 4 shows a flow chart that depicts the connection between the five steps and the characterization methods.

Optical microscopy and scanning electron microscopy (SEM) were performed respectively with Olympus BX60 (Olympus Corp., Japan) and Cambridge Stereoscan 360 (Cambridge Instrument Company Ltd., UK). Recyclate fibers were analyzed with a particle analyzer software [37]. Thermogravimetric Analyses (TGA) were

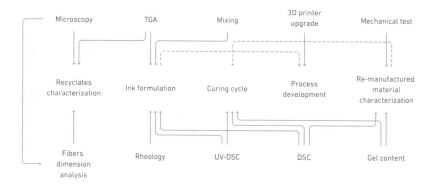

Fig. 4 A Flow chart representing the tests performed for the development of the remanufacturing process and the 3D printable inks

performed on TA INSTRUMENTS Q500 TGA (TA Instruments, Inc, US). Rheological tests were performed with Kinexus DSR (Malvern Panalytical Ltd., UK)
and Discovery HR-2 (TA Instruments, Inc, US). Differential scanning calorimetry
(DSC) was performed with a Mettler–Toledo DSC/823e (Mettler Toledo, USA)
equipped with Lightningcure LC8 (Hamamatsu Photonics, Japan) for UV-DSC.
Tensile tests were performed with Zwick Roell Z010 (ZwickRoell GmbH & Co. KG,
Germany). Detailed descriptions of the methodologies were described in previous
works [25, 26].

4.1 Materials Description

3D Printable inks were mainly composed of a photo and thermo-curable acrylic-based
resin matrix and recycled GFRP or CF reinforcement, mixed in different percentages. The matrix consisted of the ethoxylate bisphenol A diacrylate resin (hereinafter
named SR349), purchased from Arkema and locally distributed by Came S.r.l., Italy.
Butanediol dimethacrylate (hereinafter called BDDMA) and BYK-7411 ES (hereinafter called BYK) were added in different proportion as a reactive diluent and
as a rheological modifier respectively, both purchased from Sigma-Aldrich, Italy;
Dicumyl peroxide (Sigma-Aldrich, Italy) and Ethyl phenyl(2,4,6-trimethylbenzoyl)
phosphinate (hereinafter named TPO-L, purchased from Lambson Limited, UK)
were added at the fixed proportion of 0.3%wt and 3%wt as thermal activator and as
photo activator respectively.

Recycled GF powders can be divided into two main groups: Rivierasca GFRPs
(RIV) and Gamesa GFRPs (GAM). RIV GFRPs originally consisted of scraps of a
styrene-based unsaturated polyester resin reinforced with 5 cm long glass fibers with
a diameter of 13 μm (supplied by Rivierasca S.p.A., Italy). GAM GFRPs derive from
Siemens Gamesa Renewable Energy S.A. End-of-Life (EoL) wind turbines, and they
were made of an epoxy resin reinforced by continuous GF. Successively, they were
shredded and supplied from Consiglio Nazionale di Ricerca—Sistemi e Tecnologie
Industriali Intelligenti per il Manifatturiero Avanzato (Stiima-CNR), Italy, please
refers to Chap. 4 of the book for detail of the shredding process. Recycled CFs were
provided by Stiima-CNR after shredding with a cutting mill by Retsch GmbH (model
SM-300, Haan, Germany): they were previously pyrolyzed by Tecnalia (TEC) from
Aernnova (AER) expired prepreg. After the pyrolysis, a sizing material was applied
by Tampere University of Technology (TUT) ad described in Chap. 7 of the book.
Finally, a cryogenic shredding was performed with a quad blade chopper CH580
(Kenwood Limited, Havant, United Kingdom) and the powder was sieved manually
with a 100 μm sieve.

4.2 Additive Remanufacturing Requirements

Recyclate particle dimensions. Many works related to clogging mechanisms of spherical particle-filled liquids have been published [38–40]. In particular, the nozzle and particle diameter can be related to defining a maximum particle diameter to prevent clogging [41]:

$$D/d > 6.2 \tag{1}$$

where D is the nozzle diameter and d the particle diameter. Dealing with fibers literature shows their alignment to the flow under extrusion [18]. Nevertheless, sudden restriction of the channel and small nozzle diameter can cause their blocking and, consequently, the clogging of the nozzle. Tentatively, Eq. (1) can be used to define a maximum length of the fibers, substituting d with l: considering a nozzle of 1 mm in diameter, the corresponding value is 160 μm.

Ink rheology. This material property is limited by the process and the maximum pressure supported by the extrusion system. The extrusion head described before works by applying a force to the syringe's piston which generates pressure inside the syringe. Overloading a stepper motor will not compromise the motor necessarily. However, the motor will not rotate. The most used model in literature links the power law rheological model with the maximum pressure or force exerted by the extrusion system. Starting from the derivation made for a commercial FDM extruder [41], it is possible to express the maximum viscosity in function of the maximum pressure exerted by the extrusion head:

$$\eta < \frac{\pi \cdot c \cdot P_0 \cdot d^4 \cdot n}{32 Q \cdot l \cdot (3n + 1)} \tag{2}$$

where c is the ratio between the pressure drop inside the last section of the nozzle and the total pressure drop inside the syringe, P_0 is the maximum pressure sustained by the extrusion system, d is the nozzle diameter, n is the flow behavior index, Q is the ink volumetric flow and l is the length of the last section of the nozzle. Different from the derivations found in a previous work [41], Eq. (2) does not make the approximation of Newtonian flow and express the viscosity in function of the volumetric flow rate instead of the printing velocity, which is more suitable for the LDM process. More information about the derivation of Eq. (2) has been extensively described in previous work [42].

Ink reactivity. The importance of the crosslinking kinetics has been yet described in a previous work [25]. The deposition of a layer of material over the previous one relies on the stability of the deposited material.

Rheological modifiers are eventually used to produce inks with suitable yield stress [18, 36]. A drawback of this solution is the height limit of the 3D printed object since the weight of the object itself can exceed the yield stress, causing the structure to collapse [43]. Photo-crosslinking was set as a requirement for 3D printable inks,

Table 1 Recyclate properties

Recyclate name	RIV ultrafine	GAM fine	AER fine
Short name	RIVUF	GAMF	AERF
Recycling method	Mechanical	Mechanical	Thermal/Mechanical
Nominal granulometry [μm]	3.5 ± 4.2	80	100
rGF/rCF content [%wt.] by TGA	12 ± 1	70 ± 0.5	100
rGF/rCF average length [μm]	102.8 ± 33.5[1]	34.4 ± 47.7	74.6 ± 48.3
rGF/rCF average diameter [μm]	9.4 ± 4.31	13.5 ± 6.0	6.5
rGF/rCF average aspect ratio (l/d)	11.4 ± 1.71	2.3 ± 3.2	11.5 ± 7.4

to take advantage of 3D printing freedom. Contemporarily, big volume models can be fabricated without maximum dimensions limits, because the material will rapidly crosslink and the shape set.

5 Results and Validation: Material Characterization and Remanufacturing Description

As previously described the material development is crucial to fully exploit the potentiality of 3D printing technology. This section will be focused on the development of the 3D printable inks, the process optimization, and the material properties.

5.1 Recyclate Characterization

Recyclates' characteristics are intrinsically variable and their characterization is important in order to control the development of 3D printable inks.

Considering mechanically recycled fiber-reinforced composites, both granulometry of the recyclate and length and diameters of the fibers have been evaluated. Table 1 shows the morphological, compositional, and physical characteristics of the recyclates. RIV recyclate granulometry was measured by optical image analysis as described in a previous work [25]. For the other recyclate, the sieve size was considered. The average length and diameter of the recycled fibers were evaluated by SEM micrographs analysis performed with a specifically designed MATLAB application [37]. Figure 5. shows a sample micrograph of each recyclate.

---

[1] A low number of fibers were detected during the microscopy analysis.

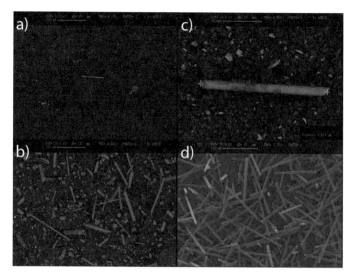

Fig. 5 SEM micrographs of recyclates: **a** RIFUF recyclate, **b** rGF inside RIVUF recyclate, **c** GAMF recyclate, **d** AERF recyclate

5.2 3D Printable Inks Formulation and Mixing

To better understand and simplify the dissertation, inks containing RIVUF and GAMF recyclates will be named with the following nomenclature: XDYR, where X is the wt% of reactive diluent BDDMA (D), Y is the wt% of recyclate, and R will be substituted by the short name of the recyclates as reported in Table 1. Inks containing AERF recyclate will be named with the following nomenclature: XBYAERF, where X is the wt% of BYK (B) and Y is the wt% of AERF recyclate.

Considering RIVUF and GAMF recyclates, their formulation was extensively studied, by changing the proportion between BDDMA, SR349, and recyclates, in order to define boundaries in terms of mixing technology and recyclate content. As depicted in Fig. 6 the use of a reactive diluent does not have a significant influence on the recyclate content limit. Concerning mixing technology, high-velocity impeller mixers work best with material with a viscosity lower than 10 Pa·s. With high viscosity material, it is possible the creation of isolated cavities in the vessel where the material is not mixed. Double arm kneading mixers use viscous forces instead of turbulence to mix, and no isolated cavities are created [44]. Experimentally a limit in the use of a high-velocity impeller mixer was found as shown in Fig. 6.

A second rheological modifier was used in combination with the rCF. As will be discussed later, the UV conversion of AERF inks is low. It was proven that the addition of 6 wt% of BYK to the resin was enough to create hydrogen bonding structures as confirmed by rheological analysis.

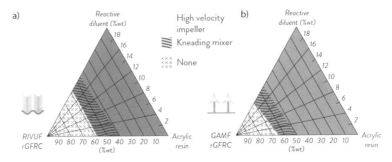

Fig. 6 Compositional graph for RIVUF and GAMF recyclate ink formulation

5.3 Reactivity Measurements on the 3D Printable Inks

The reactivity of the 3D printable inks has been studied through the measurements of the enthalpy of the reaction. More in detail, the enthalpy was measured employing both DSC and photo-DSC. The latter method was used to investigate the UV conversion of the 3D printable inks and 3D printed samples. DSC was used to verify the thermal reactivity of the 3D printed samples and to measure the glass transition temperatures of the different materials at the end of the curing cycle.

Firstly, UV reactivity was studied. Figure 7 shows that increasing the recyclate content a decrease in UV conversion compared to the neat resin was obtained. Moreover, Fig. 7b shows the presence of a residual UV conversion after the manufacturing process. As a consequence, UV post-curing was performed to complete the photo-crosslinking process. As expected AERF ink shows a low UV conversion. The presence of black carbon fibers competes in the absorption of the UV radiation. Consequently, a lower amount of photo-initiator is activated. The problem was solved by adding a rheological modifier to the ink as described in Sect. 5.4 of this chapter.

After the thermal post-curing, the glass transition temperature (T_g) was measured and gel content evaluation was performed. As reported in Table 2 AERF sample

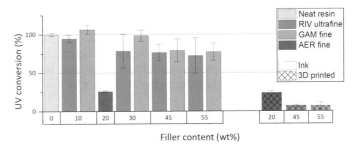

Fig. 7 UV conversion of 3D printable inks (**a**) compared to the neat resin and UV conversion of 3D printed samples (**b**)

Table 2 Glass transition temperature and gel content for three remanufactured materials

Material	Glass transition temperature [°C]	Gel content [%]
20D45RIVUF	110	94
0D55GAMF	100	99.4
6B20AERF	82	99

exhibit a lower T_g compared to all other remanufactured samples. Nevertheless, the gel content confirmed that a high level of crosslinking was achieved. The lower value obtained for the RIVUF sample can be related to the presence of unreacted monomers of the recyclate material. More information about the post-curing process and monitoring of crosslinking were described in previous works [25, 26].

5.4 Printability and Rheological Modification

A systematic rheological characterization has been performed for the different recyclate formulations. The rheological behavior can be used to predict which formulation can be suitable for the particular 3D printing process used in this work. Nevertheless, the following study can be used both as a benchmark for future works with different 3D printing processes and as a guideline for determining the printability of inks via LDM process.

RIVUF inks. Independently from the recyclate and reactive diluent content, the behavior of the material is pseudoplastic. As shown in Fig. 8, a quasi-Newtonian behavior was shown by all the formulations at a lower shear rate. By increasing the content of recyclates, the range in which the materials show quasi-Newtonian behavior shifts towards higher viscosity and lower shear rate. On the other side, by increasing the reactive diluent content, the quasi-Newtonian behavior shifts towards lower viscosity and higher shear rate. The effect of shelf-time on the viscosity of already prepared formulations was investigated. As shown in Fig. 8d, the ink has a thixotropic characteristic. Therefore, its behavior change through time. Nevertheless, after 72 h the remixed ink recovers the original rheological behavior and could be processed in the same condition as after the ink preparation. Comparing the rheological behavior of Fig. 8 to the maximum viscosity model of Eq. (2), inks containing 55 wt% of RIVUF recyclate cannot be extruded by the extruder used in this work.

GAMF inks. As done for RIVUF inks, the flow ramp test on GAMF inks has shown a pseudoplastic behavior with a flow behavior index change (Fig. 9). The behavior at the higher shear rate is constant for inks with up to 60 wt% of recyclate. The addition of the reactive diluent decreases the overall viscosity, which is similar to what was expected from the RIVUF rheology study. However, the lower viscosity and the overall bigger particle of GAMF recyclate, respect to RIVUF recyclate, evidenced a problem of separation of the recyclate particles from the resin. The

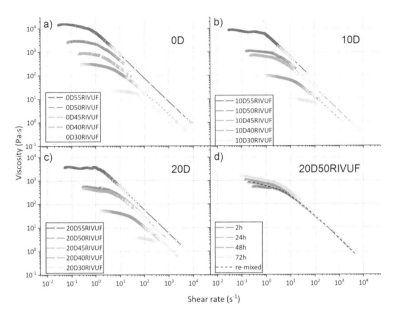

Fig. 8 Flow ramp test for RIVUF recyclate inks: **a** tests on RIVUF inks without reactive diluent, **b** tests on RIVUF inks containing 10 wt% of reactive diluent, **c** tests on RIVUF inks containing 20 wt% of reactive diluent, **d** effect of shelf-time and re-mixing on the viscosity of 20D50RIVUF ink

behavior of GAMF inks without the presence of BDDMA was furtherly studied. As shown in Fig. 9c–d, the viscosity at lower shear rates is different if the pre-shear was not applied or was applied. Consequently, the ink behavior before the loading into the syringe will be different from that after the loading. Moreover, this difference highlights the thixotropic characteristic of the inks' rheological behavior. Considering Fig. 9a–b and calculation performed with Eq. (2), only the inks with 65 wt% of GAMF recyclate will not be suitable for the extrusion process considered.

AERF inks. The addition of BYK additive can introduce in the material a rapid recovery of the deformation, and it induces a rapid fluid/solid-like transition thanks to the formation of hydrogen bonding structure [45]. A three-step strain test has been developed to simulate the 3D printing process steps and measure the response of the material before, during, and after the extrusion. The graph in Fig. 10 shows the strain stimulus applied and the responses of different inks. The rest properties of the material are recorded in the first and third step where a strain very close to zero was applied. In the second step, a strain is applied to simulate the extrusion of the material.

The addition of BYK additive is not enough to assure a rapid fluid/solid-like behavior transition. In order to exhibit a rapid recovery, a minimum of 20 wt% of rCF has to be added to the formulation.

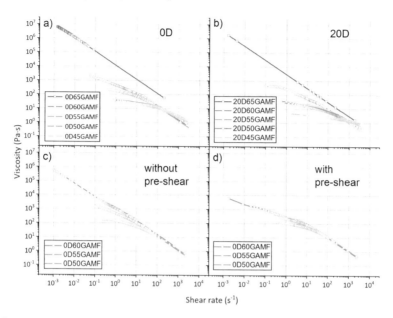

Fig. 9 Flow ramp test for GAMF recyclate inks: **a** tests on GAMF ink without reactive diluent, **b** tests on GAMF ink containing 20 wt% of reactive diluent, **c** tests on GAMF ink performed without the pre-shear step, **d** tests on GAMF ink performed with the pre-shear step

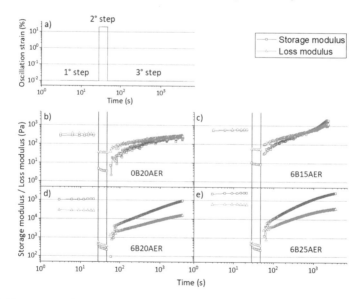

Fig. 10 Three-step strain tests on different AERF ink formulations

5.5 Additive Remanufacturing Process and Surface Finishing

As discussed before, 3D printable ink could be improved by modifying recyclate content and/or by adding rheological modifiers. For these reasons, 3D printing parameters optimization is connected to the development of the inks. Starting from the consideration about reactivity and rheological behavior discussed before, one ink for each recyclate has been selected and their process parameters are reported in Table 3.

Differences between inks developed from different recyclates are not only related to process parameters and composition but also on appearance: for instance, RIVUF 3D printed objects have a more visible layer-by-layer regular appearance (Fig. 11), while random textured appearance is exhibited by GAMF 3D printed samples (Fig. 12). Furthermore, AERF pieces show similarities with GAM ones, since layer-by-layer surface finishing is less noticeable (Fig. 13). These big differences can affect 3D printable ink applications and product perception for users.

As shown in Fig. 12e, f the arch model 3D printing exploits the use of UV-assisted LDM process by enabling the deposition of overhang structures. Promising results have been achieved with AERF ink by exploiting the use of the LED apparatus positioned alongside the extruder head. By lowering the layer height and splitting the polymerization step it was possible to irradiate for a longer period the deposited ink achieving better object stability for the following post-curing. However, further developments are needed for the manufacturing of complex 3D models with the rCF inks.

Table 3 Process parameters for 20D0B45RIVUF, 0D0B55GAMF and 0D6B20AERF inks

Material	Nozzle diameter [mm]	Layer height [mm]	Speed [mm/s]	Flow [%]
20D45RIVUF	1.6	0.4–0.5	10	130
0D55GAMF	1–1.5	0.25–0.75	8–12	100–105
6B20AERF	1	0.2	12–15	100

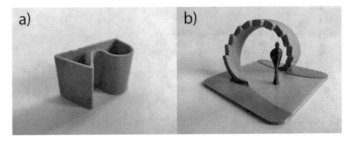

Fig. 11 3D printed structures with 20D45RIVUF ink: **a** test sample, **b** scale model of an arch

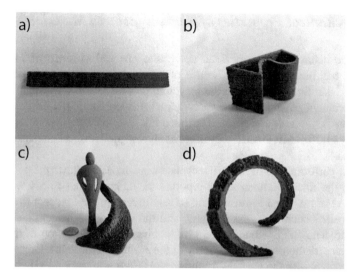

Fig. 12 3D printed structures with 0D55GAMF ink: **a** tensile specimen, **b** test sample, **c** scale model of an arch, **d** scale model of a fountain

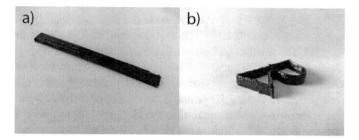

Fig. 13 3D printed structures with 6B20AERF ink: **a** tensile specimen, **b** test sample

Finally, some guidelines for UV-assisted LDM additive remanufacturing processes can be defined: crossing tool-paths are to be avoided using for example hexagonal infill; tilted surface up to 30° could be easily realized if the ink provides rapid enough photo-crosslinking. Spiral and Shell 3D Printing modes can be achieved easily, according to different geometrical shapes; layer height should be about ¼ of the nozzle diameter: thicker layer height could result in less UV penetration and crosslinking. Printing speed is affected not only by viscosity but also by UV-conversion and cross-section geometries, for bigger areas it is necessary to increase the UV-irradiation time, decreasing the printing speed or splitting the deposition from the crosslinking step.

5.6 Mechanical Properties of Remanufactured Materials

Finally, the behavior of reprocessed composite material in terms of mechanical properties has been evaluated. The use of a recyclate material as reinforcement has an intrinsic limitation due to the recycling process which intrinsically reduces its reinforcement potential. The overall behavior is similar, and a brittle failure of the samples is observed for every 3D printed ink. Besides the outstanding stiffness reached by 6B20AERF, GAMF recyclate has shown good results compared to virgin GFs (Fig. 14).

The low performances of RIVUF recyclate were expected due to the poor amount of rGF in the final formulation, as reported in Table 4 equal to 5.4 wt%. Even though GAMF ink has a lower amount of reinforcement with respect to the virgin GF composite benchmark, the stiffness, and the overall mechanical behavior are comparable. The Halpin–Tsai model [46], shows a good prediction of the experimental values, demonstrating the quality of the evaluation through a large dataset of fibers measured with the Particle Analyzer MATLAB application [37]. A two-fold increase in modulus with respect to neat resin was measured for the 0D55GAMF formulation, and a three-fold increase was measured for 6B20AERF one as reported in Table 4.

Even though a lower elongation at break for the composite samples could be expected, SEM micrographs analysis evidenced the poor adhesion between fibers and matrix. For what concerns RIVUF ink, it was hard to find rGF at the fracture surface (Fig. 15a). The fracture surface of a 0D55GAMF specimen (Fig. 15b) shows indirectly the preferential alignment of the fibers after 3D printing. Most of the fibers are perpendicular to the fracture surface, consequently, the majority is aligned with the tensile test direction. During 3D printing, the strands of material were deposited parallelly to the tensile test direction, forcing the alignment of the fibers. This effect has been extensively studied in literature [18, 47]. Moreover, the detachment of the

Fig. 14 Average mechanical behavior of additive remanufactured materials compared to neat resin behavior and virgin glass fiber 3D printed specimens

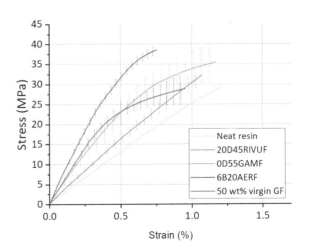

Table 4 Average mechanical properties of additive remanufactured materials compared to neat resin behavior and virgin glass fiber 3D printed specimens

Material	Content of reinforcement [wt%]	Halpin–tsai elastic modulus [GPa]	Elastic modulus [GPa]	Tensile strength [MPa]	Elongation at break [%]
Neat resin [26]	0	n.d	2.6 ± 0.3	30.0 ± 5.0	1.2 ± 0.3
50 wt% GF [26]	50	8.9 ± 2.2	6.1 ± 0.4	26.3 ± 1.7	0.8 ± 0.2
20D45RIVUF	5.4	3.4 ± 0.2	3.6 ± 0.3	26.4 ± 5.2	0.9 ± 0.2
0D55GAMF	38.5	5.5 ± 1.8	5.5 ± 1.0	33.6 ± 5.6	0.9 ± 0.2
6B20AERF	20	8.1 ± 2.0	8.1 ± 0.5	39.4 ± 2.1	0.9 ± 0.1

fibers from the matrix is shown in Fig. 15c. Finally the fracture surfaces of 6B20AERF specimen exhibit similar conditions to GAMF specimen. As shown in Fig. 15d the detachment of the fibers from the matrix is not clear. Nevertheless, the tensile test results indicate a low adhesion between them.

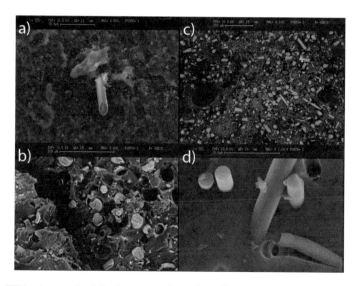

Fig. 15 SEM micrograph of the fracture surface of tensile test specimens: **a** 20D45RIVUF, **b** and **c** 0D55GAMF, **d** 6B20AERF. Fibers aligned perpendicularly to the section are visible in the micrographs

6 Towards Urban Furniture and Large-Format Additive Manufacturing Concepts

6.1 Urban Furniture from Recycled Plastics: A Case Study

As previously discussed, an increasing quantity of wind blade turbines is going to be dismissed in the next few years. A noticeable amount of rGFRPs will be therefore available as a new raw reinforcement for remanufacturing processes. Considering also the properties of the above-mentioned 3D printable inks, other 3D printing apparatus should be taken into account, since "desktop-size" setups may limit the range of the potential applications.

Recently, the average size of the building volumes is increasing for industrial level 3D printers. This is primarily related to the rising demand for these printing volumes from the manufacturers, which are relying ever more on additive manufacturing processes. When the building volumes are equal or higher than 1 m^3, the technology is better known as Large-Format Additive Manufacturing (LFAM). These technologies allow the processability of a wide range of materials, such as thermoplastics, composites, and ceramics [48, 49]. In its turn, extrusion-based LFAM is also known as Big Area Additive Manufacturing (BAAM) thanks to the gantry-based 3D printer developed at the Oak Ridge National Laboratory [50]. As an alternative, LFAM apparatus can also be based on other movement systems, such as 6-axis robotic arms.

Due to the possibility to directly construct big parts from a 3D model, the use of LFAM technology is spreading in those fields of applications with a high presence of handcrafted manufacturing processes and artisanal experiences, such as naval industry, construction, and furniture [51, 52]. A catamaran boat hull with an overall length of 10.36 m was successfully manufactured from a 3D printed mold made with CF-ABS and, very recently, a GF boat was entirely 3D printed with a 6-axis robot arm [53, 54]. As a matter of fact, the use of LFAM allows the reduction of times and costs for small batches of complex products.

Especially for the construction sector, the main innovation from LFAM is the chance to use new sustainable and efficient materials with a higher level of automation through the process [55]. From literature, many case studies range from high-performance concrete structures for construction to real buildings and homes [56, 57]. Furthermore, 3D printer manufacturers are focusing their attention on the development of new setups that allow the in situ production of architectural structures with sustainable raw materials, i.e. "Gaia" project from Wasp [58].

Similarly, furniture products were directly manufactured by using LFAM. Even if it is possible to produce artifacts for indoor spaces with these technologies [59], urban furniture represents a more interesting field of application, especially considering composite materials. The first 3D printed footpath bridge was opened to the public in 2016 (Madrid). Its main structure was made of a concrete powder reinforced with polypropylene fibers manufactured by means of a D-Shape printer developed by Enrico Dini [60]. Thanks to a 6-axis robot arm, a metal footpath bridge was developed and 3D printed in Amsterdam (Fig. 16). Its overall length is 12.5 m, and the structure was inaugurated during the Dutch Design Week 2018 [61]. In both cases, the automation offered from the digitalization simplified the manufacturing process for the production of complex shapes. Other projects demonstrate the feasibility of urban furniture production from recycled materials, i.e. plastic. The custom-made chairs from "Moro Collection" (Studio Achoo) were 3D printed with plastic wastes from Belgium employing a large-scale 3D printer. They were developed for local-scale production in order to generate circular economy cases at the local level. A similar approach was followed by the "Print your City" project (The New Raw). In this case, the main goal was to raise awareness among citizens about the reuse of plastics by transforming wastes in street furniture. The project, firstly implemented only in Amsterdam (2016), was also carried out in Thessaloniki. The product collection is mainly composed of 3D printed tree pots and benches [62, 63]. Similar considerations could be made for theater and amusement park scenery parts. Their realization is limited to unique complex pieces, and the process is often material and time-consuming. In 2017, the first scenery for the Opera Theater in Rome was 3D printed with a DeltaWASP 3MT industrial 3D printer. Particularly, this 3D printed scenery could be easily recycled by shredding it and reusing the material for a new work [64].

In light of the above, urban furniture and amusement park elements seem the most suitable design application for LFAM with the 3D printable inks shown in this chapter. Accordingly, the products would be manufactured and post-processed in situ, reducing the environmental footprint related to supply chains [65]. Moreover, the final product could be produced avoiding molds and reducing the number of parts in assemblies. Two concepts were finally designed and 3D printed with the 3Drag LDM apparatus as a preliminary feasibility test [25, 66].

Fig. 16 Metal footpath bridge from the Dutch Design Week 2018 [61]. Reprinted from Journal of Constructional Steel Research. 172, Gardner, L., Kyvelou, P., Herbert, G., Buchanan, C.: Testing and initial veri-fication of the world's first metal 3D printed bridge, 106,233 (2020), with permission from Elsevier

6.2 Scaling-Up Challenges and Perspectives

In order to develop the before-mentioned concepts, a scaling-up of the 3Drag LDM process should be carried out. Further work would be finalized not only to increase the building volumes but also to the improvement of the LDM process. Delta bot systems, due to the fixed build plate which transfers lower vibration to the 3D printed structure and their expandable construction, seem to be the best solution for the development of a large-format LDM 3D printer. Moreover, the moving structure of these systems can carry heavier tools and a bigger reservoir could be assembled to the extruder. Besides, the use of a screw extruder could be beneficial for the processing of a wider range of inks, the reduction of air bubbles entrapped in the material, and the prevention of the formation of aggregates.

7 Conclusions and Future Research Perspectives

In this work, an additive manufacturing process for rGFRP and rCF inks was developed at a desktop-scale according to LDM principles. At first, three essential remanufacturing requirements were detected to define the optimal formulations of the 3D printable inks.

RGFRPs and rCF recyclates were characterized to define their fiber content, granulometry, and morphology. Afterward, the ink formulations and the mixing technologies were determined to maximize the recyclate content. Reactive diluent or rheological modifiers were added to the inks to increase the filler content or to better control the rheological behavior of the formulation. Furthermore, a proper curing cycle was defined to assure a good cross-linking degree of the inks.

The behavior of the 3D printable inks during the extrusion was predicted thanks to a systematic rheological characterization and the definition of an analysis that emulates the 3D printing process. To sum up, inks with a pseudoplastic behavior, rapid recovery of the deformation after the extrusion, and a quick fluid/solid-like transition are the most suitable materials, especially in case of incomplete UV cross-linking like for rCF.

Some representative geometries and design concepts were designed and 3D printed with the most performing ink formulation for each kind of recyclate. In this way, structures with overhangs or complex geometries were successfully achieved. Despite its low reactivity, promising results were obtained with rCF ink thanks to a further modification of the 3D printer setup. Afterward, a shortlist of guidelines for the use of rGFRP and rCF inks with the developed 3D printing system was provided.

The mechanical behavior of these remanufactured materials was also investigated, and remarkable results were achieved. In particular, AER Fine samples demonstrated the high added value of rCF reprocessing even at low filler percentages.

At last, a focused analysis of Large-scale Additive Manufacturing was carried out to better define a possible field of application. Some considerations related to the scale-up of the remanufacturing process were then mentioned identifying alternatives to the desktop system developed yet. Future research efforts should be targeted not only to the process scale-up but also on the exploration of different fields of applications as well as new matrix and filler systems.

References

1. Zindani, D., Kumar, K.: An insight into additive manufacturing of fiber reinforced polymer composite. Int. J. Lightweight Mater. Manuf. **2**, 267–278 (2019). https://doi.org/10.1016/j.ijlmm.2019.08.004
2. Jiang, Z., Diggle, B., Tan, M.L., Viktorova, J., Bennett, C.W., Connal, L.A.: Extrusion 3D printing of polymeric materials with advanced properties. Adv. Sci. **7**, 2001379 (2020). https://doi.org/10.1002/advs.202001379
3. Ali, Md.H., Batai, S., Sarbassov, D.: 3D printing: a critical review of current development and future prospects. RPJ. **25**, 1108–1126 (2019). https://doi.org/10.1108/RPJ-11-2018-0293
4. Wang, Y., Zhou, Y., Lin, L., Corker, J., Fan, M.: Overview of 3D additive manufacturing (AM) and corresponding AM composites. Compos. Part A: Appl. Sci. Manuf. **139**, 106114 (2020). https://doi.org/10.1016/j.compositesa.2020.106114
5. Kerns, J.: How 3D printing is changing auto manufacturing. https://www.machinedesign.com/3d-printing-cad/article/21834982/how-3d-printing-is-changing-auto-manufacturing
6. McLaren website. https://www.mclaren.com/racing/partners/stratasys/mclaren-deploys-stratasys-additive-manufacturing-improve-2017-car-performance/. Last Accessed 28 Nov 2020
7. Moi Composites homepage. https://www.moi.am/. Last Accessed 28 Nov 2020
8. Markforged, Composite 3D Printing. https://markforged.com/materials. Last Accessed 28 Nov 2020
9. Gately, R.D., Beirne, S., Latimer, G., Shirlaw, M., Kosasih, B., Warren, A., Steele, J.R., in het Panhuis, M.: Additive manufacturing, modeling and performance evaluation of 3D printed fins for surfboards. MRS Adv. **2**, 913–920 (2017). https://doi.org/10.1557/adv.2017.107
10. Parandoush, P., Lin, D.: A review on additive manufacturing of polymer-fiber composites. Compos. Struct. **182**, 36–53 (2017). https://doi.org/10.1016/j.compstruct.2017.08.088
11. Brenken, B., Barocio, E., Favaloro, A., Kunc, V., Pipes, R.B.: Fused filament fabrication of fiber-reinforced polymers: a review. Addit. Manuf. **21**, 1–16 (2018). https://doi.org/10.1016/j.addma.2018.01.002
12. Deng, X., Zeng, Z., Peng, B., Yan, S., Ke, W.: Mechanical properties optimization of poly-ether-ether-ketone via fused deposition modeling. Materials. **11**, 216 (2018). https://doi.org/10.3390/ma11020216
13. Cicala, G., Latteri, A., Del Curto, B., Lo Russo, A., Recca, G., Farè, S.: Engineering thermoplastics for additive manufacturing: a critical perspective with experimental evidence to support functional applications. J. Appl. Biomater. Funct. Mater. **15**, 10–18 (2017). https://doi.org/10.5301/jabfm.5000343
14. Rahman, K.M., Letcher, T., Reese, R.: Mechanical properties of additively manufactured peek components using fused filament fabrication. In: Volume 2A: Advanced Manufacturing. p. V02AT02A009. American Society of Mechanical Engineers, Houston, Texas, USA (2015)
15. Stoof, D., Pickering, K.: Sustainable composite fused deposition modelling filament using recycled pre-consumer polypropylene. Compos. B Eng. **135**, 110–118 (2018). https://doi.org/10.1016/j.compositesb.2017.10.005
16. Le Duigou, A., Castro, M., Bevan, R., Martin, N.: 3D printing of wood fibre biocomposites: from mechanical to actuation functionality. Mater. Des. **96**, 106–114 (2016). https://doi.org/10.1016/j.matdes.2016.02.018

17. Blok, L.G., Longana, M.L., Yu, H., Woods, B.K.S.: An investigation into 3D printing of fibre reinforced thermoplastic composites. Addit. Manuf. **22**, 176–186 (2018). https://doi.org/10. 1016/j.addma.2018.04.039
18. Compton, B.G., Lewis, J.A.: 3D-printing of lightweight cellular composites. Adv. Mater. **26**, 5930–5935 (2014). https://doi.org/10.1002/adma.201401804
19. Raney, J.R., Compton, B.G., Mueller, J., Ober, T.J., Shea, K., Lewis, J.A.: Rotational 3D printing of damage-tolerant composites with programmable mechanics. Proc. Natl. Acad. Sci. USA **115**, 1198–1203 (2018). https://doi.org/10.1073/pnas.1715157115
20. Mahajan, C., Cormier, D.: 3D printing of carbon fiber composites with preferentially aligned fibers. In: Cetinkaya, S., Ryan, J.K. (eds.) Proceeding of 2015 Industrial and Systems Engineering Research Conference. Nashville, Tennessee (2015)
21. Lewicki, J.P., Rodriguez, J.N., Zhu, C., Worsley, M.A., Wu, A.S., Kanarska, Y., Horn, J.D., Duoss, E.B., Ortega, J.M., Elmer, W., Hensleigh, R., Fellini, R.A., King, M.J.: 3D-printing of meso-structurally ordered carbon fiber/polymer composites with unprecedented orthotropic physical properties. Sci. Rep. **7**, 43401 (2017). https://doi.org/10.1038/srep43401
22. Griffini, G., Invernizzi, M., Levi, M., Natale, G., Postiglione, G., Turri, S.: 3D-printable CFR polymer composites with dual-cure sequential IPNs. Polymer **91**, 174–179 (2016). https://doi. org/10.1016/j.polymer.2016.03.048
23. Invernizzi, M., Natale, G., Levi, M., Turri, S., Griffini, G.: UV-assisted 3D printing of glass and carbon fiber-reinforced dual-cure polymer composites. Materials. **9**, 583 (2016). https:// doi.org/10.3390/ma9070583
24. Tian, X., Liu, T., Wang, Q., Dilmurat, A., Li, D., Ziegmann, G.: Recycling and remanufacturing of 3D printed continuous carbon fiber reinforced PLA composites. J. Clean. Prod. **142**, 1609–1618 (2017). https://doi.org/10.1016/j.jclepro.2016.11.139
25. Mantelli, A., Levi, M., Turri, S., Suriano, R.: Remanufacturing of end-of-life glass-fiber reinforced composites via UV-assisted 3D printing. RPJ. **26**, 981–992 (2019). https://doi.org/10. 1108/RPJ-01-2019-0011
26. Romani, A., Mantelli, A., Suriano, R., Levi, M., Turri, S.: Additive re-manufacturing of mechanically recycled end-of-life glass fiber-reinforced polymers for value-added circular design. Materials. **13**, 3545 (2020). https://doi.org/10.3390/ma13163545
27. Mantelli, A.: Syringe extruder for 3Drag printer. (2020). https://doi.org/10.5281/ZENODO. 4283444
28. Mantelli, A.: 3Drag printer custom firmware for Liquid Deposition Modeling mode. Zenodo (2020)
29. DI001LE dimmable electronic power supply, https://tecnoswitch.com/prodotti/alimentatori-ali mentatore-dimmerabile-e-dimmer-per-led/. Last Accessed 28 Nov 2020
30. Chen, K., Kuang, X., Li, V., Kang, G., Qi, H.J.: Fabrication of tough epoxy with shape memory effects by UV-assisted direct-ink write printing. Soft. Matter. **14**, 1879–1886 (2018). https:// doi.org/10.1039/c7sm02362f
31. Scott, P.J., Rau, D.A., Wen, J., Nguyen, M., Kasprzak, C.R., Williams, C.B., Long, T.E.: Polymer-inorganic hybrid colloids for ultraviolet-assisted direct ink write of polymer nanocomposites. Addit. Manuf. **35**, 101393 (2020). https://doi.org/10.1016/j.addma.2020.101393
32. Mantelli, A.: 3Drag UV curing addon. (2020). https://doi.org/10.5281/ZENODO.4298907
33. Martanto, W., Baisch, S.M., Costner, E.A., Prausnitz, M.R., Smith, M.K.: Fluid dynamics in conically tapered microneedles. AIChE J. **51**, 1599–1607 (2005). https://doi.org/10.1002/aic. 10424
34. Li, M., Tian, X., Schreyer, D.J., Chen, X.: Effect of needle geometry on flow rate and cell damage in the dispensing-based biofabrication process. Biotechnol Progress. **27**, 1777–1784 (2011). https://doi.org/10.1002/btpr.679
35. Udofia, E.N., Zhou, W.: A guiding framework for microextrusion additive manufacturing. J. Manuf. Sci. Eng. **141**, 050801 (2019). https://doi.org/10.1115/1.4042607
36. Sun, H., Kim, Y., Kim, Y.C., Park, I.K., Suhr, J., Byun, D., Choi, H.R., Kuk, K., Baek, O.H., Jung, Y.K., Choi, H.J., Kim, K.J., Nam, J.D.: Self-standing and shape-memorable UV-curing epoxy polymers for three-dimensional (3D) continuous-filament printing. J. Mater. Chem. C. **6**, 2996–3003 (2018). https://doi.org/10.1039/C7TC04873D

37. Mantelli, A.: Particle analyzer. Zenodo (2020)
38. Sharp, K.V., Adrian, R.J.: On flow-blocking particle structures in microtubes. Microfluid. Nanofluid. **1**, 376–380 (2005). https://doi.org/10.1007/s10404-005-0043-x
39. Agbangla, G.C., Climent, É., Bacchin, P.: Numerical investigation of channel blockage by flowing microparticles. Comput. Fluids **94**, 69–83 (2014). https://doi.org/10.1016/j.compfluid.2014.01.018
40. Shahzad, K., D'Avino, G., Greco, F., Guido, S., Maffettone, P.L.: Numerical investigation of hard-gel microparticle suspension dynamics in microfluidic channels: aggregation/fragmentation phenomena, and incipient clogging. Chem. Eng. J. **303**, 202–216 (2016). https://doi.org/10.1016/j.cej.2016.05.134
41. Beran, T., Mulholland, T., Henning, F., Rudolph, N., Osswald, T.A.: Nozzle clogging factors during fused filament fabrication of spherical particle filled polymers. Addit. Manuf. **23**, 206–214 (2018). https://doi.org/10.1016/j.addma.2018.08.009
42. Mantelli, A., Romani, A., Suriano, R., Levi, M., Turri, S.: Direct ink writing of recycled composites with complex shapes: process parameters and ink optimization. Adv. Eng. Mater. **23**, 2100116 (2021). https://doi.org/10.1002/adem.202100116
43. Agnoli, E., Ciapponi, R., Levi, M., Turri, S.: Additive manufacturing of geopolymers modified with microalgal biomass biofiller from wastewater treatment plants. Materials. **12**, 1004 (2019). https://doi.org/10.3390/ma12071004
44. Paul, E.L., Atiemo-Obeng, V.A., Kresta, S.M.: Handbook of Industrial Mixing. John Wiley & Sons Inc., Hoboken, NJ, USA (2003)
45. De Capua, V.: LDM—3D printable polymer ink filled with EoL glass fibre reinforced composites, http://hdl.handle.net/10589/148913. (2019)
46. Halpin, J.C., Kardos, J.L.: The Halpin-Tsai equations: a review. Polym. Eng. Sci. **16**, 344–352 (1976). https://doi.org/10.1002/pen.760160512
47. Shofner, M.L., Lozano, K., Rodríguez-Macías, F.J., Barrera, E.V.: Nanofiber-reinforced polymers prepared by fused deposition modeling. J. Appl. Polym. Sci. **89**, 3081–3090 (2003). https://doi.org/10.1002/app.12496
48. Nieto, D.M., Molina, S.I.: Large-format fused deposition additive manufacturing: a review. Rapid Prototyping J. (2019). https://doi.org/10.1108/RPJ-05-2018-0126
49. Al Jassmi, H., Al Najjar, F., Mourad, A.-H.I.: Large-scale 3D printing: the way forward. IOP Conf. Ser.: Mater. Sci. Eng. **324**, 012088 (2018). https://doi.org/10.1088/1757-899X/324/1/012088
50. Roschli, A., Gaul, K.T., Boulger, A.M., Post, B.K., Chesser, P.C., Love, L.J., Blue, F., Borish, M.: Designing for big area additive manufacturing. Addit. Manuf. **25**, 275–285 (2019). https://doi.org/10.1016/j.addma.2018.11.006
51. Moreno Nieto, D., Casal López, V., Molina, S.I.: Large-format polymeric pellet-based additive manufacturing for the naval industry. Addit. Manuf. **23**, 79–85 (2018). https://doi.org/10.1016/j.addma.2018.07.012
52. Naboni, R., Kunic, A.: Bone-inspired 3D printed structures for construction applications. GT Projetos. **14**, 111–124 (2019). https://doi.org/10.11606/gtp.v14i1.148496
53. Post, B.K., Chesser, P.C., Lind, R.F., Roschli, A., Love, L.J., Gaul, K.T., Sallas, M., Blue, F., Wu, S.: Using big area additive manufacturing to directly manufacture a boat hull mould using big area additive manufacturing to directly manufacture a boat hull mould. Virtual Phys. Prototyping. 2759, (2019). https://doi.org/10.1080/17452759.2018.1532798
54. Mambo Project—MOI Composites, https://www.moi.am/projects/mambo. Accessed 19 Nov 2020, Last Accessed 19 Nov 2020
55. Ghaffar, S.H.: Additive manufacturing technology and its implementation in construction as an eco-innovative solution. Autom. Constr. **11** (2018)
56. Gosselin, C., Duballet, R., Roux, P., Gaudillière, N., Dirrenberger, J., Morel, P.: Large-scale 3D printing of ultra-high performance concrete—a new processing route for architects and builders. Mater. Des. **100**, 102–109 (2016). https://doi.org/10.1016/j.matdes.2016.03.097
57. Paolini, A., Kollmannsberger, S., Rank, E.: Additive manufacturing in construction: a review on processes, applications, and digital planning methods. Addit. Manuf. **30**, 100894 (2019). https://doi.org/10.1016/j.addma.2019.100894

58. 3D Printing for sustainable living. https://www.3dwasp.com/en/3d-printing-for-sustainable-living/. Accessed 19 Nov 2020
59. Novak, J.I., O'Neill, J.: A design for additive manufacturing case study: fingerprint stool on a BigRep ONE. RPJ. **25**, 1069–1079 (2019). https://doi.org/10.1108/RPJ-10-2018-0278
60. Lowke, D., Dini, E., Perrot, A., Weger, D., Gehlen, C., Dillenburger, B.: Particle-bed 3D printing in concrete construction—possibilities and challenges. Cem. Concr. Res. **112**, 50–65 (2018). https://doi.org/10.1016/j.cemconres.2018.05.018
61. Gardner, L., Kyvelou, P., Herbert, G., Buchanan, C.: Testing and initial verification of the world's first metal 3D printed bridge. J. Constr. Steel Res. **172**, 106233 (2020). https://doi.org/10.1016/j.jcsr.2020.106233
62. Moro Collection. https://www.achoo.be/blackbaboon?lang=en. Accessed 19 Nov 2020
63. Print your City. https://www.printyour.city/. Accessed 19 Nov 2020
64. The First 3D Printed Scenography. https://www.3dwasp.com/en/the-first-3d-printed-theatre-scenic-design/. Accessed 19 Nov 2020
65. Despeisse, M., Baumers, M., Brown, P., Charnley, F., Ford, S.J., Garmulewicz, A., Knowles, S., Minshall, T.H.W., Mortara, L., Reed-Tsochas, F.P., Rowley, J.: Unlocking value for a circular economy through 3D printing: a research agenda. Technol. Forecast. Soc. Chang. **115**, 75–84 (2017). https://doi.org/10.1016/j.techfore.2016.09.021
66. Romani, A., Mantelli, A., Levi, M.: Circular design for value-added remanufactured end-of-life composite material via additive manufacturing technology. In: Segalàs, J., Lazzarini, B. (eds.) Proceeding of 19th European Roundtable for Sustainable Consumption and Production. Circular Europe for Sustainability: Design, Production and Consumption. pp. 491–512. Institute for Sustainability Science and Technology, Universitat Politècnica de Catalunya, Barcelona (2019)

Composite Finishing for Reuse

Alessia Romaniⓘ, **Raffaella Suriano**ⓘ, **Andrea Mantelli**ⓘ,
Marinella Leviⓘ, **Paolo Tralli, Jussi Laurila**ⓘ, **and Petri Vuoristo**ⓘ

Abstract Coating processes are emerging for new applications related to remanufactured products from End-of-Life materials. In this perspective, their employment
can generate interesting scenarios for the design of products and solutions in circular
economy frameworks, especially for composite materials. This chapter would give an
overview of coating design and application for recycled glass fiber reinforced polymers on the base of the experimentation made within the FiberEUse project. New
cosmetic and functional coatings were developed and tested on different polymer
composite substrates filled with mechanically recycled End-of-Life glass fibers.
Afterwards, recycled glass fiber reinforced polymer samples from water-solvable 3D
printed molds were successfully coated. Finally, new industrial applications for the
developed coatings and general guidelines for the coating of recycled glass fiber reinforced polymers were proposed by using the FiberEUse Demo Cases as a theoretical
proof-of-concept.

Keywords Glass fibers · Polymer composites · Surface finishing · Physical vapor
deposition · Thermal spray coatings · Indirect 3D printing

A. Romani (✉) · R. Suriano · A. Mantelli · M. Levi
Department of Chemistry, Materials and Chemical Engineering "Giulio Natta", Politecnico Di
Milano, Piazza Leonardo da Vinci 32, 20133 Milano, Italy
e-mail: alessia.romani@polimi.it

A. Romani
Department of Design, Politecnico Di Milano, Via Durando 38/A, 20158 Milano, Italy

P. Tralli
Green Coat S.R.L., Strada Romana Nord 1, 46027 San Benedetto Po, Italy

J. Laurila · P. Vuoristo
Department of Material Science, Tampere University of Technology, 33720 Tampere, Finland

M. Colledani and S. Turri (eds.), *Systemic Circular Economy Solutions for Fiber
Reinforced Composites*, Digital Innovations in Architecture, Engineering
and Construction, https://doi.org/10.1007/978-3-031-22352-5_9

1 Introduction, Motivation and Objectives

1.1 Re-processing and Post-processing for Recycled Composite Materials

Composites, as a whole, are in a vivid and challenging market. A big portion of the market is composed by established and standardized production and processes which are responsible for the bigger production volume. In Europe, glass fibers reinforced plastics (GFRPs) accounts for over the 90% of reinforced composites production, and the 85% of the GFRPs market is covered by construction, transport and electronic sectors. Over the past 20 years, resin transfer molding (RTM) and reinforced thermoplastic composites represent the two most significant segments of GFRPs market for the flexibility of the process and for the particular interest gained for large-scales series production, respectively [1].

In particular, boat hulls and wind turbine rotor blades account for most of the End-of-Life (EoL) GFRPs products. Because of their thermosetting matrix, the recovery of the original components is impossible, and the amount of GFRPs wastes is expected to raise significantly considering the increased use of wind energy. Very recently, a work shows the possibility to exploit the high strength, high water resistance of EoL GFRPs products by keeping the composite structure intact but sectioning the structure in smaller parts, strips or flakes suitable for the production of simple shape profile, beams or plates. Thanks to this methods, structural reuse of EoL thermoset composite has been successfully proven with infra-structural demonstrator such as retaining walls and crane-mats or bridge decks [2].

The composite substrates, in general, can be coated for two main reason: the reduction of their surface roughness, or the protection against corrosion and erosion. Gelcoats are used for the roughness reduction particularly for hand-made components, which are quite frequent in composite industry. At the same time, gelcoats assume a key role for wind blades. As a matter of fact, their protection against corrosion and erosion is essential, given their exposure to harsh environment [3]. However, literature does not provide specific information for the coatings for re-processed composites.

Using a coating is not only advantageous from the mechanical and durability point of view, but also for the possibility to cover imperfection and in homogeneities of recycled material. As a consequence, coatings could produce a beneficial effect on the perception of the recycled material, giving more confidence on the properties of the material.

1.2 Research Goal: Coatings for Reprocessed Composite Materials

The development of a new circular economy model is becoming not only a significant objective in global economy, but also an ethical duty. In fact, the growing costs linked to the disposal of EoL parts are no longer sustainable, as well as the environmental footprint of wastes. Nevertheless, EoL composite products are still landfilled. Considering the heterogeneity of these kinds of materials, this method is considered as the fastest and cheapest solution [4].

The attention on the remanufacturing and reforming of new products starting from EoL composites has grown significantly. In this scenario, new products obtained from recycled GFRP can be remanufactured relying on circular economy models. But their surface finishing may not satisfy the aesthetic standards of potential end-users. Functionality and aesthetic could be extremely affected by the finishing of these reformed products [5]. Therefore, the use of rGFRP for real industrial applications could be more challenging.

Accordingly, the main research goal was to investigate methods and solutions to provide aesthetic (i.e. bright chrome finishing) or functional improvements (i.e. anti-bacterial, anti-scratch, self-cleaning properties) to the final reformed product in order to increase its perceived quality. For this reason, the work was mainly focused on the development of cosmetic and functional coatings for rGFRP.

Subsequently, the aim focused on the implementation of some of the developed coatings on 3D objects with a particular shape and geometry, to observe the advantages and the limitations of the processes under investigation on more complex surface and realistic context. At the end, some industrial applications and theoretical links with the FiberEUse demo-cases are given, in order to close the gap related to the development of real products made with rGFRP.

2 Positioning of the Solution

2.1 Composite Coatings

Surface finishing of GFRPs parts or products obtained by using recycled glass fibers are often of low quality [6]. However, a surface coating can be desirable not only for aesthetic purposes but also for protection against environmental erosion. An example of this protective effect against rain erosion is the use of polyurethane coatings applied on GFRP airfoils to damp the stress waves induced by the impact of water droplets and therefore to reduce the stress transmitted to the substrate [7].

An enhanced level of resistance to water droplet erosion was also studied for GFRPs coated with various electro-deposited metal layers, such as chrome and nickel with an intermediate Cu layer [8]. The impact resistance of GFRPs with metallic coatings was found to increase by more than 20 times when compared to the resistance of the uncoated material. Taking into account that composite materials obtained with recycled glass fibers can find probably applications in the building construction and sanitary ware industry, erosion protection coatings against damage caused by liquid impact can be advantageous even for this case. Moreover, an epoxy-based coating was recently developed and studied to enhance the erosion resistance of GFRPs in the marine environment. This erosion-resistant polymeric coating demonstrated to reduce the surface abrasion of epoxy glass laminate composites induced by the impact with sand particles in marine simulated conditions [9].

As already mentioned before, the recycling of EoL GFPRs is becoming a big environmental issue. Only a better understanding of the relationships between the processing, the formulations and the properties of GFRPs filled with recycled composites can promote the actual reuse of composite wastes. For example, a lower level of hardness and therefore a higher wear rate was observed for composites only filled with mechanically recycled GFRPs and an increase of wear resistance was obtained with the introduction of virgin glass fibers together with recycled GFRPs [10].

In light of the above, the development of functional and protective coatings can be very helpful for fostering the recycling and the reuse of GFRPs. A thorough literature search highlighted that there are no examples of coatings on GFRP filled with mechanical recycled GFRPs. Only one paper about the development of a surface layer on a structural composite reinforced with recycled glass fibers was reported very recently [11]. This surface layer was composed of a polyester matrix filled with small particles of calcium carbonate ($CaCO_3$, size up to 2 μm). The thickness of the surface layers seems quite variable. However, this surface showed a homogeneous morphology without cracks, contributing to the protection of the underneath porous composite.

2.2 Limitations and Needs

As shown in the previous paragraph, only one paper showing the study of surface layers applied on composite substrates reinforced with recycled GFRPs was found. This clearly suggests the need to investigate new polymeric and metallic coatings for composite substrates obtained reusing GFRPs. A finishing material or coating can boost the aesthetic properties of GFRPs obtained by recycling EoL products. Moreover a protective function can be surely identified for coatings applied on composites, which find outdoor applications and therefore the study of coating resistance against atmospheric agents by hardness and abrasion tests emerges as a crucial factor. Few examples of works regarding the characterization of hardness for composite coatings are present in literature, but no previous studies like these were found for coatings applied on recycled GFRPs. Moreover, the level of adhesion between the substrates and coatings can be of a fundamental importance for a functional coating and this is another missing point in literature regarding the coatings of GFRPs.

Excellent mechanical and physical properties can be achieved when using reinforced polymer material. Among these low manufacturing costs, excellent resistance to corrosion, and low weight together with relatively high strength can be found. Wear resistance can be good enough, but it can be significantly improved by coating the surface with a wear resistant coating material. Moreover, thermal and electrical conductivities can be improved by coating polymer-based composite. In addition to this, the surface coatings can aim at imparting new added-value functionalities to rGFRPs. These added-value functionalities can be, for instance, anti-scratch resistance, antibacterial properties and self-cleaning abilities.

When thermal spray coating methods are used to produce coatings on the polymer-based composite surfaces, anti-scratch resistance can be introduced for example by thermally sprayed hard metal coatings containing tungsten and chromium carbides. Antibacterial properties can be obtained with thermally sprayed thick metallic coatings (e.g. silver, copper, zinc, nickel–chromium, bronze), and self/easy-cleaning surfaces with titanium oxide coatings.

The composite surfaces can be therefore modified by coatings to have desired properties. It is evident that these type of coating procedures can be also applied on the surfaces of recycled composites as it will be presented below in this chapter.

3 Methods and Workflow

3.1 Approach and Workflow

The main workflow of the experimentation can be resumed in Fig. 1. Two different kinds of mechanically recycled glass fibers (rGF) were used as a filler for the reformed composite substrates. The first one derived from EoL wind blade, while the second one was mainly composed by internal waste scraps of corrugated GFRP sheet production. Starting from them and optimizing the process parameters, two different kinds of samples were obtained: epoxy-based substrates with EoL wind blade rGF, and unsaturated polyester-based (UP-based) substrates filled with internal waste rGF. After that, two different approaches were defined for the surface optimization of the developed substrates:

1. appearance improvement aiming at the increase of the aesthetic appeal for design products (cosmetic coatings);
2. functional improvement aiming at the increase of the value-added features of the reprocessed composite products (functional coatings).

The first approach provided for the liquid deposition of UV- or thermally-curable primers on the substrates, followed by the coating deposition of a thin chrome layer by means of PVD (Physical Vapor Deposition) sputtering process. Methods and parameters were optimized specifically for rGFRP materials.

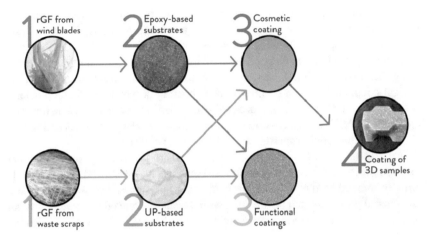

Fig. 1 Workflow of the experimentation from the development of new coatings for rGFRP composites to their preliminary implementation on 3D samples with complex surfaces

In the second approach, new functionalities were implemented on the substrates through the deposition of thick metal coatings with thermal spray processes for new anti-bacterial, anti-scratch and self-cleaning properties. Finally, samples obtained from water-solvable 3D printed molds were coated with PVD, implementing this surface finishing on components with a complex shape.

3.2 Substrates

Mechanically rGFs were used as a reinforcement for the reprocessed GFRP samples, employed as substrates for coating tests. These rGFs were provided both by Rivieresca (RIV), recovering RIV's ground GFRP internal waste, and by STIIMA-CNR (Istituto di Tecnologie Industriali e Automazione-Consiglio Nazionale delle Ricerche), grinding EoL wind turbine blades (Siemens Gamesa Renewable Energy S.A.).

UP-based samples with rGFs provided by RIV were composed by a mixture of virgin glass fibers and rGFs combined with an orthophthalic unsaturated polyester resin. These substrates contain approximately 44 wt% of recycled waste, and they were coated with a chrome layer deposited by PVD, using either a UV-curable solvent-less acrylic primer or a two-component acrylic primer with solvent.

As for rGFs from wind blades, the resulting fibers had a maximum size of around 4 mm, after the processing through grid holes with this dimension. An epoxy-based resin (commercial resin name: Araldite BY158, Huntsman, BY158 for brevity) was used for substrates filled with the rGFs obtained from EoL wind blades. BY158, which was composed of an epoxy-terminated bisphenol-A-based oligomer and 1,4-butanediol diglycidyl ether, was crosslinked with a curing agent (Aradur 21, Huntsman) using a weight ratio of 100/28 between BY158 and Aradur 21. The rGFs were added to the resin and mixed for 15 min by a mechanical stirrer (50 rpm). A nominal percentage of rGFs ranging from 40 to 60% wt. with respect to the mixture of BY158 and Aradur21 was used. The real content of glass fibers in samples prepared with a nominal 60wt% of rGFs was found to be 51.3 ± 5.9wt% by thermogravimetric analysis (TGA). The resulting mixture was subjected to a low vacuum level for 15 min and processed by compression molding for 24 h at room temperature with a post-curing at 100 °C for 1 h. These substrates were employed to develop a UV-curable or thermo-curable epoxy-based primer for a consequent chrome deposition.

3.3 Cosmetic Coatings

A PVD process was employed to obtain a decorative bright chrome finishing on rGFRP substrates. Prior to PVD sputtering deposition, the application of a primer was necessary to provide a smooth surface and improve the hardness and brightness of the coating. The absence of Cr^{6+} and Cr^{3+} compounds and the possibility to use a solvent-less primer make this process a more sustainable option to electroplating.

Two different commercial primers were initially used in the experimentation. The first one is a UV-curable solvent-less primer commonly used in PVD coating of injection molded parts (Cromogenia Units SA., Spain). It is composed by a blend of different acrylates and a mixture of two different photoinitiators with range of absorption of 255–380 nm. The second one is a two-component solvent-based acrylic primer (Lechler S.p.A., Italy), suitable for composite substrates. For PVD, chrome target with purity 99.95% was used.

Araldite BY158 was employed for the development of the thermally-curable epoxy-based primer. Aradur 21 was added as curing agents to increase the crosslinking with a weight ratio of 100/28. Epoxy-resin primer formulations were cured following the same procedure used for the epoxy-resin substrates.

Four different methods for surface preparation were tested. Degreasing with acid solution took place in pre-treatment spray tunnel, and the parts were sprayed with a phosphoric acid based solution at 60 °C for 4 min. Samples were then rinsed using demineralized water of 5 μS/cm of conductivity. Finally, a drying phase was carried out in oven at 60 °C for 30 min. Degreasing was carried out manually with an isopropanol moistened cloth rubbing the surfaces, and followed by 10 min of drying at room temperature. Surface sanding was carried out manually or with a sanding machine, using P100 and P120 grit sandpaper.

Plasma surface activation was carried out in a vacuum chamber equipped with two electrodes, a power source up to 6 kW, a turbo-pump allowing 10^{-3} mbar vacuum and a flow-controlled gas inlet. The treatment was performed for 40 s with a power of 4 kW. The processing gas used in the experimentation was oxygen.

Primer was applied both with manual and automatic equipment. UV-curable solvent-less primer was applied in the painting booth by anthropomorphic robot equipped with rotary bell atomizers. UV-curing was carried out in a station equipped with 26 microwave-powered bulbs, delivering a peak irradiance up to 500 mW/cm^2 and an energy density up to 2600 mJ/cm^2.

The steps of the PVD sputtering and its parameters were set according to the standard process, which is already consolidated for injection molded plastic parts. The chrome layer was deposited using four chrome target at 10 kW of power and 120 s of takt time (240 s of total deposition time). The tests were mainly focused on the optimization of the methods and the process parameters in order to improve the process ability of rGFRPs surfaces.

Adhesion properties of the chrome coating were tested performing manual cross-cut test according to ISO2409:2013 [12]. A linear Taber equipment was used to test abrasion resistance with a H1 felt (total weight of 1 kg and 60 cycles/min). Abrasion test results are expressed in 5 levels depending on the surface area of coating removed by abrasion. In detail, 0 means no abrasion, and level 5 means more than 50% of area was affected by removal. Chrome layer thickness was measured using a Helmut Fischer Fischerscope X-RAY XDL spectrometer.

3.4 Functional Coatings

Thermal spray coating onto polymer-based base materials are defined as the application of thermal spray processes to deposit metallic, ceramic or polymeric feedstock materials (powder or wire) onto the surface of a polymeric composite base material.

According to the primary energy source used for particle acceleration and heating, thermal spray coating processes can be categorized into four groups: cold spraying, flame spraying, electric arc wire spray and plasma spraying [13, 14]. In this experimentation, all these processes were used for the deposition of functional coatings. However, the most suitable thermal spray processes to deposit heat sensitive substrates, such as polymer composites, are electric arc spraying and wire flame spraying. Other processes are possible, but they often require the preparation of the surfaces to make the substrates more heat resistant.

To improve the adhesion of thermal spray coatings, different kinds of primers were studied, such as low melting polymer (melting region of 80–120 °C), metallic layer (aluminium, nickel–chromium and WC–CoCr powders, chromium or aluminium oxides). The thickness of the primer layer was controlled by adding some quartz particles with a size of 200–600 μm, when the primer layer was applied between the base material and the PTFE plate.

Experiments were carried out with different materials and equipment, according to the specific thermal spray technology. For low-pressure cold spray (LPCS) trials, a Dymet cold spray unit was used to produce copper and aluminium coatings, that were sprayed using a thin low melting polymer layer on the rGFRP surface.

Table 1 List of thermal spray treatment for the characterization of the rGFRP samples

Coating method	Substrates	Primer	Coatings
Cold spray (Dymet)	Composite	80 °C melting polymer	Cu, Al, or AlZn
Flame spray (Castodyn DS 8000)	Composite	80 °C melting polymer	Diam 4008, Amp. 481.002, Amp 481.003, or LDPE
Flame spray (Master Jet)	Epoxy + Cr_2O_3, Epoxy + WC–CoCr, Epoxy + Al, Epoxy + NiCrSib	RODCUR 6740, NiChrome, Chroma Supra, or TI-Elite	NiCrBSi + 40%WC, 80%Ni + 20%Cr, Cr_2O_3, or Al_2O_3 + TiO_2
Plasma spray (Technik A3000 S)	Composite	None	Diam 2001
Arc spray (Sulzer Metco Smart Arc)	Composite, Epoxy	None, or Zn	Zn wire, or Cu + 10Al + 1Fe

Powder flame spraying equipment Castodyn DS 8000 was used with LDPE polymer powder. The coating was sprayed directly on slightly sand blasted surfaces. Wire flame spraying equipment Saint-Gobain Master Jet was used for metallic coatings with RODCUR 6740 (60% NiCrBSi with 40% WC) and NiChrome (NiCr) wires. Ceramic TI-Elite (Al_2O_3 + TiO_2) coatings were also made using this method, and prepared on WC–CoCr, Cr_2O_3 or NiCrBSi filled epoxy based precoated surfaces.

Electric arc spraying equipment Sulzer Metco Smart Arc was used for arc spraying. Zinc wire was sprayed directly on slightly sand blast composite surfaces, whereas aluminium bronze coatings were sprayed on the composites and on the zinc precoat surface. Plasma Technik A3000 S unit was used for atmospheric plasma spraying with Diamalloy 2001 (NiCrSiB) as feed material.

To better understand the quality and real thicknesses of the coatings, polished cross-section specimens cast were prepared and studied using scanning electron microscopy (SEM) equipped with an Energy-dispersive X-ray spectroscopy (EDX) analyser. The list of the thermal spray treatments and the kinds of substrates, primers and coatings used during the experimentation are shown in Table 1.

3.5 Surface Finishing Implementation on 3D Complex Shapes

To evaluate the surface finishing on more complex substrates, two different 3D samples were designed. The samples were obtained by casting the epoxy-based rGFRP (40wt% of wastes) in water-solvable 3D printed molds. The final pieces were then treated with the PVD sputtering process.

In this work, Fusion 360 (Autodesk, San Rafael, California, US) and Rhinoceros (Robert McNeel & Associates, Seattle, WA, USA) were the CAD software used for the design of the samples and molds. Afterwards, the water-solvable molds were produced with a BVOH (butane diol vinyl alcohol copolymer) filament (Verbatim Italia S.p.A., Cassina de Pecchi, Italy) through a Prusa i3 MK3S Fused Deposition Modeling (FDM) 3D printer (Prusa Research, Prague, Czech Republic). The Gcode files were created with the open source software Prusa Slicer from the same manufacturer.

The epoxy-based composite was prepared following the procedure previously described, and poured into the molds. After the first curing phase (24 h at room temperature), the molds were immersed in a beaker with water. To easily dissolve BVOH, the beaker was heated at 60 °C for 2 h on a magnetic stirrer. The 3D samples were manually washed to remove the residues of BVOH, and a layer of the thermally-curable epoxy-based primer was applied on the surface with a soft-bristled brush. The curing treatment was accomplished following the same steps for the epoxy-based flat substrates.

Finally, the PVD treatment was performed according to the process parameters set for the flat samples. In Table 2, the main features of the two 3D samples are shown, whereas the appearance of the casted pieces and of the molds is visible in Fig. 2.

Table 2 Main parameters for the brick, surface tile samples for the PVD sputtering process

3D sample shape	Dimensions (mm)	Mold layer height (mm)	Filler content (% wt.)	Number of samples (−)
Brick	58 × 25 × 30	0.15	40	6
Surface tile	40 × 40 × 18	0.10	40	6

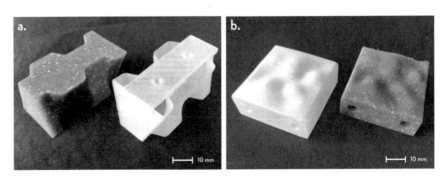

Fig. 2 3D Samples for the PVD cosmetic coating with the corresponding water-solvable 3D printed mold: brick shape (**a**), surface tile shape (**b**)

4 Coating Results and Validation

4.1 Cosmetic Coatings

In first test, the UP-based samples were divided into two groups, on which two different surface preparation methods were applied: acid degreasing on the former group, isopropanol cleaning on the latter. Both groups were coated with a UV-curable solvent-less acrylic primer applied by robots equipped with rotary atomizers. For levelling the rough surface of the samples, the primer flow-rate was increased from the standard values of 15–35 μm to 60–80 μm. As a consequence, the samples were cured with 100% of UV power.

Since it was not possible to further increase the primer thickness, the final result was highly dependent on the substrate roughness, which showed an uneven surface with several cavities as visible in Fig. 3. In this case, the adhesion level between the primer and the substrate was found to be equal to 5, which means that more than the 65% of the cross-cut area was affected by detachment. This represented a primary issue to be solved, since the standard level for furniture and creative products must be at least equal to 1 (less than 5% of cross-cut area affected).

Poor adhesion can be linked to the chemical incompatibility between the primer and the matrix of the substrate. Therefore, further rGFRP samples made with different resins were tested, and other methods to improve adhesion were investigated.

According to Table 3, machine sanding significantly improved the adhesion levels of the coating. In addition, the abrasion level was satisfactory for machine-sanded UP-based substrates, since less than 1% of the coating was removed at 1000 strokes. Nevertheless, the adhesion level should be furtherly increased to fulfill the minimum requirements of the industrial sectors.

Fig. 3 UP-based sample before (on the left) and after the chrome coating using (on the right) UV-curable primer

Table 3 Adhesion and abrasion level tests of the UP-based and vinylester coated samples

Samples	Adhesion level test		Abrasion test
	Manual sanding	Machine sanding	Machine sanding
UP-based	5	3	1
Vinylester	N.A	2	2

Two-component acrylic primer. Since an alternative solvent-less UV-curable primer was not available, a two-component acrylic primer for PVD process on composite surfaces was then chosen, and manually applied with an handgun.

As visible in Fig. 4, the use of this alternative primer did not change the aesthetical appearance of the chrome coating with respect to the previous tests on the UP-based substrates. On the contrary, the adhesion level was equal to 1, which is generally enough for most of the technical applications for indoor furniture and home appliances. However, the abrasion level was inadequate if compared to the substrates treated with the UV-curable primer, since the chrome layer was fully removed at 500 strokes. For this reason, this primer could be used for surfaces that are not subjected to frequent mechanical stresses (i.e. reflective surfaces of lightning appliances).

Primer development for epoxy-based composites. Considering the low adhesion level observed with a standard UV-curable acrylic primer on epoxy-based matrix, a new primer should be developed for this kind of rGFRP. To minimize the surface tension differences, a thermally-curable primer formulation with the same composition of the substrate matrix (BY158 and Aradur 21) was tested. As observed after the deposition, the overall quality of the primer coating was better with respect to the UV-curable commercial solutions, and the cratering effect almost disappeared as shown in Fig. 5a. Accordingly, the samples were coated with PVD sputtering deposition process. The result after the chrome deposition is visible in Fig. 5b.

Fig. 4 UP-based sample after chrome coating using solvent-based acrylic primer

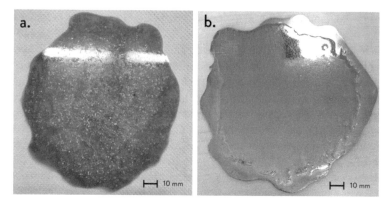

Fig. 5 Epoxy-based sample before (on the left) and after (on the right) the chrome deposition on the thermally-curable primer formulation

Since no detachment occurred, the adhesion level was equal to 0. On the contrary, the abrasion resistance was low. As a matter of fact, the chrome layer was almost entirely removed at 100 strokes.

4.2 Functional Coatings

At first stage, the quality of the formed coatings were visually evaluated. As mentioned before, the adhesion of thermal spray coatings was improved by developing several kinds of pre-coatings. Initially, the layer consisted in a low melting polymer layer, but it was too soft especially for cold sprayed coatings. Consequently, the next step was to use a metallic or ceramic powder primer (aluminium, nickel–chromium and WC–CoCr or chromium and aluminium oxides) as pre-layers under the thermal sprayed coating. Their use seemed promising, since grow and adhesion of the coating was found to be improved especially for flame spraying methods.

Cold spraying. To test cold spray technology to form coatings on polymer composite surface low pressure cold spray Dymet cold spraying equipment was used. Cold spraying method was also tried directly on the composite surface without any precoat layer. However, the high-speed coating material particles hitting the surface were only removing the composite material, and the coating was not growing on the base material surface. Accordingly, a pre-coat layer is needed.

Coating materials used for these experiments were Cu and Al–Zn powders sprayed on low melting polymer primer layer. Mainly, it was found that the primer layer had too low melting point, and was too soft. As a result, formed coatings were not uniform and dense, even though individual feed particles were stick on the surface forming metallic colored layer.

Flame spraying. Two flame spraying methods were used within the experiments. At first Castodyn DS 8000 flame spray gun was used. Two Ni–Al coating materials were utilized: Diamalloy 4008 and Amperit 281.002 on low melting polymer primer layer.

Also in this case, no uniform metallic coating layer was formed on the composite surface. It is obvious that the primer used is too soft to form dense thermal sprayed coating layer. Therefore, the process with this kind of soft bonding layer was not suitable for the coating of rGFRPs. This primer layer was used only because the metallic coating did not deposit directly on the surface with this flame spray technology.

Recycled GFRPs can also be coated with polymer layers. Of course there are several suitable polymers to choose as the coating material, but in this case low melting LDPE powder was used. The adhesion of the LDPE coating on the sand blasted composite surface was good. On the interface between the coating and base material, some porosity was observed in the cross-sectional specimens. Thickness of the coating layer was measured to be between 250 and 300 μm.

Using the Saint-Gobain Master Jet unit, metallic and ceramic coatings were made on the recycled composite surface. For the first trials, the selected coating material was RODCUR 6740. The coating was mainly sprayed on slightly sand blasted composite surface and thin coating layer formed. Metallic NiChrome (Ni + Cr) and ceramic Ti-Elite (Al_2O_3 + TiO_2) were successfully sprayed on recycled composite surfaces, as can be seen in Fig. 6. Also epoxy-based precoat layers filled with metallic or ceramic powder were obtained on the rGFRP surface. Thanks to this kind of precoat layer, the coating can easily grow on the surface, improving the adhesion. According to the EDX analysis map measured from the cross-sectional RODCUR 6740 presented in Fig. 7, the coating is well built on the rGFRP surface.

To sum up, Powder flame spraying coating was sprayed directly on slightly sand blasted composite surface with great success. Different kinds of coating materials could be used with this technology, and there are several filler materials for the precoat to be tested. In this work, only some were tested with promising results.

Plasma spraying. For the plasma spraying experiments, Diamalloy 2001 powder was used as coating material. The coating was sprayed directly on the rGFRP surface or on thermally sprayed HDPE precoat. A uniform layer was not formed with this latter method. The polymer precoat layer should be more heat resistant and harder to

Fig. 6 Cross-section of ceramic Al_2O_3 + TiO_2 coating on the rGFRP surface with the NiChrome precoat (light) under the ceramic coating sprayed using Saint-Gobain Master Jet unit

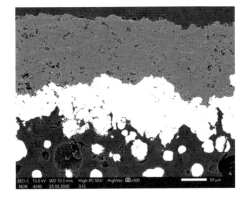

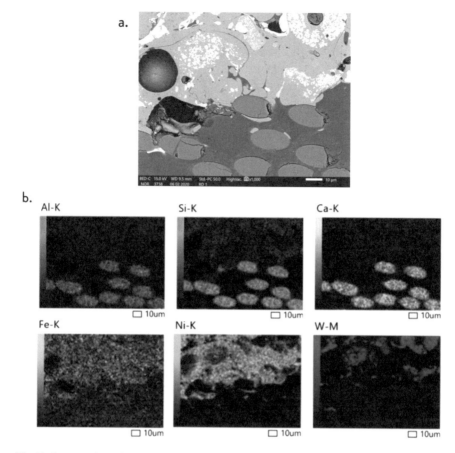

Fig. 7 Cross-section of Saint-Gobain MasterJet flame spray coating on the recycled composite surface with RODCUR 6740 layer (**a**) and EDX elemental maps (**b**)

last to the melt coating particles hitting the surface. As shown in Fig. 8, dense coating was formed on sand blasted rGFRP surface. As a conclusion, plasma spraying has potential for composites. Material processed with plasma spraying has good wear resistance and it was sprayed successfully directly on softly sand blasted recycled composite surface. Moreover, other suitable coating material could be used.

Arc spray coatings. Coating materials chosen for these experiments were zinc and bronze wires sprayed. Thick and dense coatings were achieved especially with zinc as coating material. Zinc was used also as a precoat under the bronze layer to improve the adhesion. The coating surfaces are visible in Fig. 9a and b, whereas cross-sectional image and elemental maps measured from the zinc-bronze construction cross-section are presented in Fig. 10. According to the results, aluminum was oxidized during the process, forming an antibacterial coating primarily made of copper, and the most promising results were obtained with the zinc as a precoat. Similarly, other coating materials could be potentially obtained with this technology.

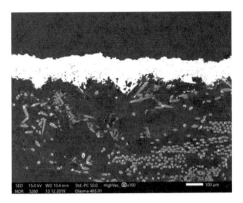

Fig. 8 Cross-section of the plasma sprayed Diamalloy 2001 coating on the rGFRP surface

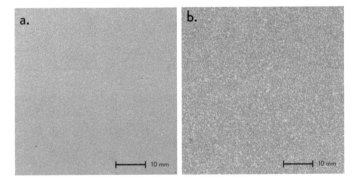

Fig. 9 Arc sprayed zinc coating on the sand blast epoxy based rGFRP surface (**a**), and antibacterial bronze coating sprayed on zinc precoat (**b**)

4.3 Surface Finishing Implementation for New Applications

As described in the previous paragraphs, the surface finishing for reformed composite substrates with rGF was deeply investigated in the FiberEUse context. However, the application of these coatings should be done at larger scales in order to promote their real application. From literature, the importance of the material surface finishing for the definition of new products is a well-established concept. This is especially true in the field of industrial design, since surface finishing strongly affects the technical properties of a new product and its expressive-sensorial qualities [15, 16].

For these reasons, after the development on new cosmetic and functional metal coatings for reprocessed GFRPs on flat substrates, an ad-hoc experimentation on 3D samples was carried out for PVD processes to give the possibility not only to validate the surface finishing process but also directly experience expressive-sensorial qualities. As seen in a previous experimentation with PVD sputtering onto 3D printed surfaces, the choice of a proper shape of the samples was crucial. On the

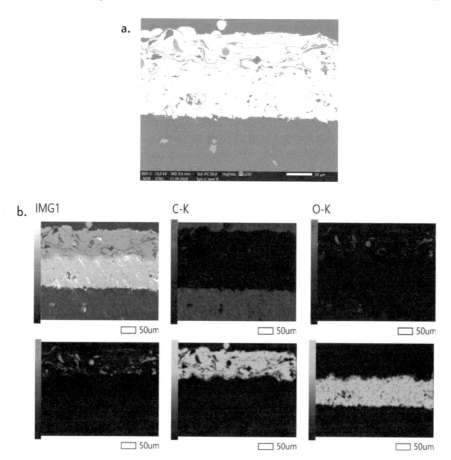

Fig. 10 Cross-section of arc sprayed antibacterial coating of zinc and bronze on the rGFRP epoxy based composite surface (**a**) and EDX elemental maps (**b**)

one hand, the shape itself could affect the surface perception. On the other hand, a more complex surface could simulate the coating application on new reprocessed products, showcasing the achievable results from the designer and manufacturer perspectives [17].

As mentioned, the samples were produced by pouring the epoxy-resin based composite in water-solvable 3D printed molds. From literature, this process is known as "Indirect 3D Printing". Accordingly, the final piece is not directly 3D printed, but its shape is defined by means of 3D printed tooling such as molds or inserts. A wider range of materials can be processed, including thermosetting resins and composites. Moreover, solvable molds allow the production of parts with complex overhangs. At the moment, this approach is mainly adopted in electronics, bioengineering or medical fields [18, 19], but it could be suitable for all those applications that need high-performance composite pieces, complex shapes or heat resistance properties [20].

Two different shapes were designed to evaluate the PVD coating on a planar ("Brick" sample) and on a complex curved surface ("Surface Tile" sample). Mainly, molds were designed to avoid resin leakages after casting. At the same time, one of the goal was to minimize the amount of 3D printed solvable material for each mold.

First, a batch of six brick samples was successfully produced. After the application of the thermally-curable epoxy-based primer, the surface of the bricks did not show cratering effect except for the area near to the sharp angles, as shown in Fig. 11a. Moreover, no marks on the primer surface related to the FDM process are visible. After the PVD, the appearance of the chrome layer is similar to the previous planar substrates made with the epoxy-based rGF composite. The final appearance of the bricks is shown in Fig. 11b.

Afterwards, the PVD coating was performed on six Surface Tile samples following the above-mentioned steps. In this case, the primer tended to accumulate in the surface lows of the shape, as visible in Fig. 12a. This is mainly due to the polymerization time needed for the primer curing, that does not allow to obtain a uniform thickness. Nevertheless, no cratering effects was visible on the surface.

The chrome coating emphasized the inhomogeneities related to the thickness of the primer. According to Fig. 12b, the 3D printing extrusion marks were clearly visible in all those points where the primer layer was thinner. As expected, no signs were found in the surface lows. However, some cracks on the chrome surface were detected in correspondence of the highest thicknesses of the primer layer. Even though the chrome layer was successfully deposited, other efforts should be done for the optimization of the primer application, since it negatively affects not only the performances of this coating, but also its expressive-sensorial qualities.

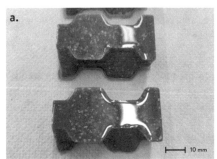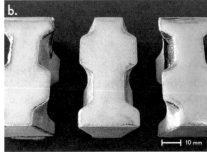

Fig. 11 Brick samples for surface finishing evaluation: samples after the primer application (**a**), and after the PVD cosmetic coating (**b**)

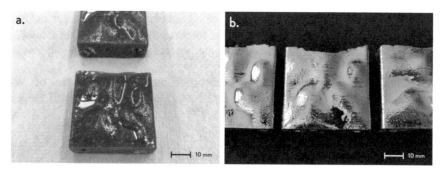

Fig. 12 Surface Tile samples for surface finishing evaluation: samples after the primer application (**a**), and after the PVD cosmetic coating (**b**)

5 Fields of Application

5.1 Surface Coatings for Industrial Application

Cosmetic Coatings. The most potential industrial applications of chrome coating on rGFRPs are represented from all those fields with a high relevance of the expressive-sensorial qualities. Mainly, these coatings are able to enhance end user perceived quality. Moreover, they can give unique finishing characteristics to rGFRPs that are not commonly linked to these kinds of recycled composites. As a consequence, their use perfectly fits for the design of new products in the furniture and interior design sectors. More in detail, UP-based and epoxy-based substrates could be used at this purpose, since high levels of abrasion resistance are not required. For these reasons, design-driven products from the FiberEUse Demo Cases could extremely benefit from the application of this kind of coating.

Other fields of application could considered once the functional features will be optimized, and the current limits will be overcome (i.e. outdoor products, automotive). Thanks to the promising results of the adhesion test, UP-based rGFRPs could also be suitable for technical applications for indoor furniture and home appliances. Similarly, epoxy-resin based rGFRPs could be considered for the same sectors after fixing the current issues related to the primer development and deposition. Further opportunities could be then explored by including additional properties in the surface treatment (i.e. antibacterial power or wear-resistant features), or taking into account rCFs products. Therefore, Automotive Demo Cases as well as products from Additive Remanufacturing could be potentially considered for future developments.

Functional Coatings. Improvement of wear resistance and better visual appearance can be achieved using thermal spraying methods for rGFRP surfaces. Moreover, improving the thermal and electrical conductivities of the coated rGFRPs could lead to new technical high-performance application. Accordingly, Automotive Demo Cases seem to be the most promising application within the FiberEUse framework.

Protective and antibacterial functions could also be implemented for outdoor applications (i.e. against the weathering). Moreover, antibacterial coating should be considered in those contexts where a specific product gets in contact with several users.

Further novel applications could be detected after the development of new functional coatings, since their coating can give new properties to the rGFRP surface.

5.2 Guidelines for Reformed Composites Coating

Cosmetic Coatings. The results of the experimental phase underline the importance, for the coating of recycled composites, of surface pre-treatment of the substrate before proceeding with PVD deposition. When using UV primers, best results have been achieved by machine sanding the surface with sandpaper 100–120 grit. It is therefore necessary that all surfaces to be coated are easily accessible to sanding operations.

For solvent-based acrylic primer instead, accurate surface cleaning with isopropanol is sufficient. After the application, it is important to follow the curing times indicated on the technical datasheet in order to assure a complete cross-linking before proceeding with PVD deposition. Otherwise, defects like cracking lines may occur on the surface after the chrome layer deposition.

PVD deposition is carried out in vacuum environment. This requires that all parts are completely humidity-free and used resins are fully polymerized in order to avoid degassing during the process. This could lead to defects such as darker color, iridescent shades and thickness of chrome layer lower than expected.

Since temperature during PVD process can reach 70 °C, the parts need to be sized to avoid deformation. Deposition power and time can be set to reduce the process temperature. However, this would negatively impact layer thickness and productivity.

Since the PVD process takes place in a closed chamber, parts must be designed taking into account the maximum size which can be processed by the plant.

Functional Coatings. There are several thermal spray technologies, which can be utilized when making functional coatings on the recycled composite surface. Each of them is giving different qualities to the formed coating, since feed materials are different. Therefore, several coating methods can be used for different needs. Coatings can be sprayed successfully on the rGFRP surface using several technologies. In general, the sand blasting of the rGFRP surface can significantly improve the adhesion and the quality of the coating. This is also true for the deposition of the precoat.

For powder flame spray with Ni–Al, the polymer primer was too soft to form dense coating layer. Accordingly, the process with this kind of soft bonding layer is not suitable for coating of polymer composites, even though that layer was used only as a precoat layer for the metallic coating on the rGFRP surface. Nevertheless, polymer coating could be possible, as demonstrated thanks to the good adhesion of LDPE coating on composite surface.

By changing the flame spray technology, metallic and ceramic coatings are possible on the rGFRP surface. In this case, an epoxy based precoat filled with metallic or ceramic powders is recommended, since it makes the coating to grow easier on the surface. Moreover, it could improve the adhesion on the composite surface.

Through plasma spraying, the coating can be sprayed directly on the sand blasted composite surface without the need of a precoat. Therefore, this technology has potential for coatings on rGFRP surfaces, also considering the suitable coating powders.

Similarly to plasma spraying, it is possible to directly spray Zn and aluminium bronze wires sprayed on the sand blasted rGFRP surface with arc spraying. Zinc can be use both as coating material and as precoat layer under the bronze layer. In the first case, thick and dense coatings can be achieved, whereas the adhesion of the bronze layer improves in the latter one. Consequently, it is worthful to use more demanding coating materials with this technology.

6   Conclusions and Future Research Perspectives

6.1   Cosmetic Coatings

Best results with solvent-less UV-curing primer were obtained on UP-based and vinyl-ester substrates treated with sandpaper machine. The chrome coating showed a good abrasion resistance on UP-based, fulfilling the requirement from the industrial sectors of 1000 linear Taber strokes without detachment. However, the adhesion requirement was not met in this work.

In terms of adhesion, the most promising results were achieved with the alternative solvent-based primer. Anyway, the use of solvent-based primer limits the environmental benefits prospected at the beginning of the experimentation. If compared to typical requirements for furniture and design products, the adhesion of solvent-based primer proves to be satisfactory on UP-based. The abrasion resistance of the top chrome layer is not satisfactory if compared to the requirements of the industrial sectors. Nevertheless, results may be adequate for lighting design.

Regarding the developed epoxy-based primer, very high level of adhesion between the primer and the rGFRP substrates was achieved. However, the adhesion between the chrome layer and the epoxy-based primer seems not enough satisfactory according to the abrasion tests. Therefore, this coating can be successfully employed for design-driven applications, which do not require high levels of abrasion resistance.

6.2   Functional Coatings

Most promising results during the studies were obtained with plasma spraying, arc spraying and flame spraying methods. With these technologies tailored coatings was formed also directly on the slightly sand blast composite surface. However, deposition of composites often requires some sort of pre-treatment, or actually a precoat layer, in order to prepare the coatings successfully with adequate coating adhesion to the composite substrate and proper microstructure. Thin epoxy-based coatings mixed with fine metallic or ceramic particles were used successfully in this purpose.

To highlight some results, it can be concluded that wear resistant tungsten carbide containing coatings can be deposited on the recycled composite surface. Also, other metallic coatings such as copper, bronze or zinc can be made on the composite surface giving metallic properties and appearance for the recycled composite surface.

Ceramic coatings sprayed on precoat layer were also made on the recycled composite surface using flame spaying technology, e.g. for providing wear and scratch protection and for visual appearance. Also, LDPE polymer coatings were sprayed successfully directly on the slightly sand blast recycled polymer composite surface using conventional flame spray technology.

For the first time in literature, chrome coating from PVD was successfully deposited on rGFRP pieces obtained from water-solvable 3D printed molds. Two different kind of samples were manufactured in order to test the primer application and the PVD coating both on planar and curved surfaces. Even though cratering effect was not detected, some issues occurred, mainly related to the primer thickness. As a matter of fact, its inhomogeneity negatively affected the overall aspect of the chrome layer.

In the light of the above, further work would be done to improve the quality of the coated surface and its aesthetic properties. At this purpose, different application methods for the epoxy-based primer should be investigated, such as airbrush deposition or dip coating. Alternatively, other primer formulation could be developed to reduce the polymerization time. These solutions may lead to a more accurate control of the primer layer thickness, improving the final outcome. Furthermore, more complex 3D samples could be developed in order to test this coating onto different surface features, i.e. embosses or engraves, with a view to new design-driven applications.

References

1. Witten, D.E., Mathes, V.: The Market for Glass Fibre Reinforced Plastics (GRP) in 2019, https://www.avk-tv.de/files/20190911_avk_market_report_e_2019_final.pdf, (2019)
2. ten Busschen, A.: Industrial re-use of composites. Reinf. Plast. **64**, 155–160 (2020). https://doi.org/10.1016/j.repl.2020.04.073
3. Kjærside Storm, B.: Surface protection and coatings for wind turbine rotor blades. In: Advances in Wind Turbine Blade Design and Materials, pp. 387–412. Elsevier (2013)

4. Yang, Y., Boom, R., Irion, B., van Heerden, D.J., Kuiper, P., de Wit, H.: Recycling of composite materials. Chem. Eng. Process. Process Intensif. **51**, 53–68 (2012). https://doi.org/10.1016/j. cep.2011.09.007

5. Karlsson, M., Velasco, A.V.: Designing for the tactile sense: investigating the relation between surface properties, perceptions and preferences. CoDesign **3**, 123–133 (2007). https://doi.org/ 10.1080/15710880701356192

6. Final Report Summary—EURECOMP (Recycling Thermoset Composites of the SST), https:// cordis.europa.eu/project/id/218609/reporting/it. Accessed 16 Dec 2020

7. Zahavi, J., Nadiv, S., Schmitt, G.F.: Indirect damage in composite materials due to raindrop impact. Wear **72**, 305–313 (1981). https://doi.org/10.1016/0043-1648(81)90257-X

8. Lammel, P., Whitehead, A.H., Simunkova, H., Rohr, O., Gollas, B.: Droplet erosion performance of composite materials electroplated with a hard metal layer. Wear **271**, 1341–1348 (2011). https://doi.org/10.1016/j.wear.2010.12.034

9. Rasool, G., Stack, M.M.: Some views on the mapping of erosion of coated composites in tidal turbine simulated conditions. Tribol. Trans. **62**, 512–523 (2019). https://doi.org/10.1080/104 02004.2019.1581313

10. Souza, J.R., Silva, R.C., Silva, L.V., Medeiros, J.T., Amico, S.C., Brostow, W.: Tribology of composites produced with recycled GFRP waste. J. Compos. Mat. **49**, 2849–2858 (2015). https://doi.org/10.1177/0021998314557296

11. Sabău, E., Udroiu, R., Bere, P., Buranský, I., Miron-Borzan, C.-Ş: A novel polymer concrete composite with GFRP waste: applications, morphology, and porosity characterization. Appl. Sci. **10**, 2060 (2020). https://doi.org/10.3390/app10062060

12. EN ISO 2409:2013: Paints and varnishes—Cross-cut test. Available at: https://www.iso.org/ obp/ui/#iso:std:iso:2409:ed-4:v1:en, (2013)

13. Davis, J.R.: Handbook of Thermal Spray Technology. ASM International, Materials Park, OH (2004)

14. Gonzalez, R., Ashrafizadeh, H., Lopera, A., Mertiny, P., McDonald, A.: A review of thermal spray metallization of polymer-based structures. J. Therm. Spray Technol. **25**, 897–919 (2016). https://doi.org/10.1007/s11666-016-0415-7

15. Chen, X., Shao, F., Barnes, C., Childs, T., Henson, B.: Exploring relationships between touch perception and surface physical properties. Int. J. Des. **3**, 67–76 (2009)

16. Zuo, H.: The selection of materials to match human sensory adaptation and aesthetic expectation in industrial design. Metu J. Fac. Archit. **27**, 301–319 (2010). https://doi.org/10.4305/METU. JFA.2010.2.17

17. Romani, A., Mantelli, A., Tralli, P., Turri, S., Levi, M., Suriano, R.: Metallization of thermoplastic polymers and composites 3D printed by fused filament fabrication. Technologies. **9**, 49 (2021). https://doi.org/10.3390/technologies9030049

18. He, S., Feng, S., Nag, A., Afsarimanesh, N., Han, T., Mukhopadhyay, S.C.: Recent progress in 3D printed mold-based sensors. Sensors. **20**, 28 (2020). https://doi.org/10.3390/s20030703

19. Mohanty, S., Bashir, L., Trifol, J., Szabo, P., Vardhan, H., Burri, R., Canali, C., Dufva, M., Emnéus, J., Wolff, A.: Fabrication of scalable and structured tissue engineering scaffolds using water dissolvable sacrificial 3D printed moulds. Mat. Sci. Eng. C. **55**, 569–578 (2015). https:// doi.org/10.1016/j.msec.2015.06.002

20. Montero, J., Vitale, P., Weber, S., Bleckmann, M., Paetzold, K.: Indirect additive manufacturing of resin components using polyvinyl alcohol sacrificial moulds. Procedia CIRP. **91**, 388–395 (2020). https://doi.org/10.1016/j.procir.2020.02.191

Composite Repair and Remanufacturing

Justus von Freeden[iD]**, Jesper de Wit**[iD]**, Stefan Caba**[iD]**, Susanne Kroll**[iD]**,
Huan Zhao**[iD]**, Jinchang Ren**[iD]**, Yijun Yan**[iD]**, Farhan Arshed**[iD]**,
Abdul Ahmad**[iD]**, and Paul Xirouchakis**[iD]

Abstract For the reuse of components and structures made of fiber composite materials, a complete remanufacturing process chain is necessary to prepare the parts for a further life cycle. The first step is to dismantle the parts to be reused. Fiber composite components are mostly joined using adhesive technology, so that solution techniques are required for adhesive connections. One possibility is the separation of the adhesive layer by means of thermally expanding particles. Adhesive residues are removed by laser so that the components can be glued again after reprocessing. The decisive factor for which process is used for the remanufacturing of the components is the state at the end of the life cycle. Non-destructive testing methods offer a very good option for detecting damage, planning necessary repairs and direct reuse of damage-free components. Repairs to fiber composite structures have been carried out in aviation for a long time and are accordingly established. These processes can be transferred to the repair of automotive fiber composite components. Many technical solutions were developed and tested as part of the project. Future research work is aimed at further development, particularly with regard to the automation of the technologies in order to enable an industrial application of the recycling of automobile components made of fiber composites.

Keywords Fiber reinforced plastics · Reuse of composite components ·
Non-destructive testing methods · Composite repair · Detachable adhesive ·
Laser-based remanufacturing processes

J. von Freeden (✉) · S. Kroll
Fraunhofer Institute for Machine Tools and Forming Technology, Chemnitz, Germany
e-mail: justus.freeden@iwu.fraunhofer.de

J. de Wit
INVENT Innovative Verbundwerkstoffe Realisation Und Vermarktung Neuer Technologien
GmbH, Braunschweig, Germany

S. Caba
EDAG Engineering GmbH, Fulda, Germany

H. Zhao · J. Ren · Y. Yan · F. Arshed · A. Ahmad · P. Xirouchakis
University of Strathclyde, Glasgow, UK

© The Author(s) 2022
M. Colledani and S. Turri (eds.), *Systemic Circular Economy Solutions for Fiber
Reinforced Composites*, Digital Innovations in Architecture, Engineering
and Construction, https://doi.org/10.1007/978-3-031-22352-5_10

1 Introduction: Remanufacturing and Repair for Composite Part Re-Use

Fiber-plastic composites are difficult to recycle due to the combination of two different materials (fiber and polymer matrix) and the resulting variety of materials and production processes. Nevertheless, due to the rising amount of end-of-life (EOL) products and components of fiber-plastic composites recycling solutions and reuse options are necessary. Due to the variety of materials mentioned, no single solution is possible. Therefore, many different individual and application-adapted recycling and reuse processes are required. In the previous chapters, solutions regarding thermal and mechanical processes were presented and discussed. Thermal processes are based on a separation of the composite through the pyrolysis of the polymer matrix, whereas mechanical processes do not separate the fiber and matrix. Instead the fiber reinforced plastic (FRP) structures and components are shredded into scraps, granulate and even powder. As already outlined in the previous chapters, these processes offer good opportunities for the reuse of fiber plastic composites in different sectors and applications. Nevertheless, due to the existing weaknesses (mechanical recycling: reduction in properties by shortening the fibers, thermal recycling: high costs, fiber damage due to high temperature), further supplementary solutions are necessary to reuse more FRP material.

While the thermal and mechanical processes consider recycling and reuse at the material level, the approach in this chapter is based on reusing at component level. This innovative approach is based on the concept that EOL products are dismantled into the individual components and these are reused in new products afterwards (Fig. 1). In this context, various new technologies are necessary as the adaptation of existing technologies in order to enable the reuse of components and structures. The overall objective in this topic is therefore the development of a complete process chain starting with the dismantling of the EOL product through the remanufacturing of the components and structures up to the reuse in a new product. Corresponding specific goals for different processes along the entire process chain are derived from this. To reuse components or structures, EOL products must first be dismantled. Therefore, detachable joints and corresponding dismantling processes must be developed. The condition of EOL components is decisive for their reuse. Therefore, before re-using, a condition analysis of the EOL component to determine whether there is (irreparable) damage is essential. This means, the process chain for reuse requires a technology that determines the condition of the components and structures. For example, non-destructive testing methods or structural health monitoring are very suitable for this. Components that are badly damaged cannot be reused. However, these can possibly be fed to the thermal or mechanical recycling processes at the material level. Components with minor damage can be repaired. For other parts, for example, only an aesthetic refinishing in the form of a new paint or coating is necessary. Another goal is to find process engineering solutions for these requirements.

The products and components must be designed accordingly for the implementation of a corresponding circular economy at component level. In general, the term

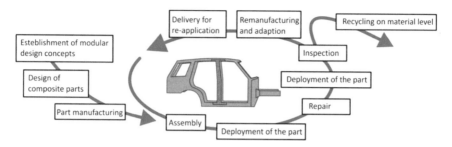

Fig. 1 Scenario of the reuse chain in the automotive industry

"design for reuse" means that the reuse of components and structures is taken into account in the context of product design. Therefore, there are several aspects to consider when designing structures and components for reuse, including geometry, material selection and joining techniques. Important points are simple and quick dismantling options as well as a modular design. In addition, the components should be designed for a longer period of use over several life cycles. The selection of FRP composites as a material for reusable structures and components offers several advantages. In addition to high fatigue strength and corrosion resistance, the use of FRP also enables lightweight construction concepts to be implemented, especially in the mobility sector.

A modular design also opens further possibilities. Depending on the future development of customer requirements for the product, this can be adapted regarding these requirements. By inserting longer or wider components, the product can completely change its external shape. In the exemplary case of the automotive industry, this could for example mean that the length of the body could change from a small car to a family van.

2 State of the Art

The reuse of components and structures made of FRP is a new and innovative approach, especially in the chosen automotive application area. FRP with continuous fibers are only used in the automotive industry in the area of specific vehicles with small quantities and in the motorsport sector. For cost reasons, mainly short fiber reinforced materials are used in large-scale production. In order to describe the state of the art, it is therefore not expedient to use the automotive industry. For this reason, a comparison with aviation is more useful. In the aviation sector it is common to carry out a complete check of the aircraft, remanufacture and repair parts after a certain lifetime and then use them for a further life cycle. Due to the use of continuous FRP in aircraft construction, these materials are also reused at component level. Accordingly, certain technologies and processes for reuse from aviation can be transferred to a process chain for the reuse of automotive components. These include in particular the methods of repairing FRP and non-destructive testing methods.

2.1 Non-destructive Testing Methods

Many non-destructive testing techniques have been investigated for composite inspection in the past decades including ultrasound, active thermography, shearography, acoustic emission, etc. [1–3]. Ultrasound is one of the most common non-destructive testing techniques in composite inspection. It covers a range of transducer types including single element, phased array [4], air-coupled [5] and immersed transducers. Combining with flexible data acquisition methods such as pulse-echo, pitch-catch and Full Matrix Capture, ultrasonic testing is able to detect diverse defect types with position, size and depth information in reused composite [6, 7] as well as new composite material [8, 9]. Active thermography is also a broadly used technique for composite inspection. It applies an infrared camera to record the heat distribution in a test object. Consequently, the damage on the composite surface and sub-surface can be detected [6, 10]. Due to the distance between camera and test object, active thermography is a contactless method, which allows to analyse complex structures. Shearography is a laser based non-destructive testing method to create an interference pattern, therefore, the defect as a nonuniform part can be presented in the shearing image [11]. It is able to identify surface deformation and defects in composite. X-ray radiography usually displays an image with different densities to indicate material and defects inside, while computed tomography can provide a 3D image to present detailed structure information [7]. However, X-ray is time and cost consuming and requires huge image processing work. Vibration analysis is usually employed to determine the resonance frequency, damping and mode shapes of the materials following the analysis of measured signals in the frequency domains [12]. Hyperspectral imaging has been successfully applied in many emerging applications such as food and drinks [13], remote sensing [14], arts verification [15], and other. Through capturing spectral data from a wide frequency range along with the spatial information, hyperspectral imaging can detect minor differences in terms of temperature, moisture and chemical composition, etc. Therefore, it can deliver a special solution to address the challenge of material composition.

Through the inspection of FRP components by suitable non-destructive testing methods, a decision can be made on whether the components are recycled on material level or reused on a component level.

2.2 Repair of Fiber-Reinforced Plastics

FRP components are widely used in the aerospace industry, making it a benchmark for repair processes currently in use. In the products of this industry, components damage is also regular, which is why their repair processes are available to be used as new methods for automotive products, and therefore are described below. The most serious fault with FRP components is damage from impacts. The fibers are damaged directly and delaminations occur, which is why the defect must be separated from the component over a large area (Fig. 2).

These exposed areas are closed again with adapted patches. A mixture of riveting and gluing is mainly used in aircraft construction. Since pre-drilled holes are necessary for riveting, although these mean a weakening of the material for FRP, the primary goal of current research is to develop certifiable and approvable repair adhesives to avoid riveting. Repair by means of patches is the topic where the greatest challenge but also potential savings in terms of processing costs are to be found. Usually, these patches are applied manually, but research is currently underway to find ways to automate them. The greatest potential is offered by laser processing, which enables layer-precise removal, and the automated insertion of new fiber layers (Fig. 3).

A repair in the broadest sense is necessary, if separated components are to be glued together again for a re-use. Here the automatable technologies milling- and laser-processing are to be emphasized. These techniques allow the separated components to be prepared by removing the adhesive from the initial bond without damaging the underlying fiber layers. Studies have shown that repairing with a surface preparation brings the same strength as original bonding. The same results were obtained when a bonded joint was destructively separated, i.e. a fiber layer of one component was separated. So, a repair of failed components is possible without problems.

These studies have also shown that repairing with only cleaned parts and without surface-preparation causes a loss of strength, which is expected. Due to the lack of preparation and the unevenness of the surface, a sufficiently good bond between the sample and the adhesive cannot be achieved in order to reach similar high strengths.

Fig. 2 Removed defect caused by an impact

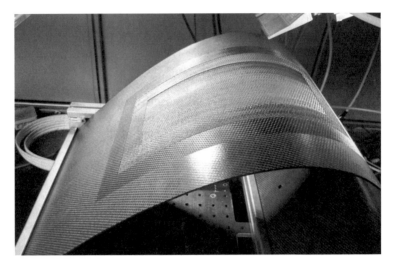

Fig. 3 Laser processing of a complex curved CFRP component. *Source* Laser-Zentrum Hannover e.V

But still, high values can be reached, if the components are designed accordingly, and these reduced strengths may be enough.

2.3   Reversible Joints

For the reuse of individual components or structures, separation based on detachable connections is an important aspect. Screws, rivets and bolts are among the classic detachable joints. With this method, the joining partners are connected via an auxiliary joining part (screw, rivet). As this part is removed, the connection between the joining partners is also separated. However, with fiber-reinforced plastics this type of joint connection is rarely used, as it is necessary to drill holes in the joint partners. In the process, the continuous fibers are separated at the holes, resulting in reduced mechanical properties and stress peaks. Therefore, the preferred joining method for FRP or FRP/metal structures is an adhesive connection. Adhesive joints are normally not separable, which is why new solutions are necessary here.

3 Thermal Aging as Proof of the Suitability of Fiber-Reinforced Plastics for Durable and Reusable Components

The selection of fiber-reinforced plastics as a material for durable and reusable components and structures is based on their high resistance to corrosion and fatigue. Nevertheless, there are aspects that need to be examined and analysed in order to reuse these materials. This includes the aging of plastics, especially FRP. Plastics are generally subject to various aging mechanisms that lead to reduced properties over time. Influencing factors are, among other things, different temperatures or the influence of UV radiation. In the automotive application examples examined in the FiberEUse project, the structures to be reused are inside the vehicle and are therefore not exposed to UV rays. For this reason, the focus of the project was on thermally induced aging effects. For this purpose, tests were carried out on accelerated aging at higher temperatures and temperature changes.

Due to the accelerated aging at higher temperatures, which is based on the Arrhenius equation, aging processes can be simulated in shorter periods of time. The aim of the tests carried out on accelerated aging was to simulate the aging of various FRP materials and adhesive bonds by 5, 10 and 20 years and to determine the influence of accelerated aging on mechanical properties. The specimens were stored at various temperatures (80 °C, 100 °C) for different lengths of time (29, 58 days) and then tested in mechanical tests. The test results showed no significant reductions in the mechanical parameters compared to unaged samples.

A test cycle from the automotive industry is used for the temperature change test. The test specimens are alternately cooled to −40 °C and heated to 80 °C. The heating or cooling time is always 2 h. this cycle is repeated a total of 54 times. The test procedure is graphically presented in Fig. 4.

Following the temperature change test program, the test specimens were subjected to mechanical tests. Analogous to accelerated aging, no significant reductions in the mechanical parameters compared to unaged test specimens could be observed.

Examples of the tests carried out are the results of the additional test on specimens that have been thermally aged and exposed to different degrees in comparison to untreated specimens. These results are shown in Fig. 5.

In summary, the results of the tests have shown that both accelerated aging and temperature changes have a very small influence on the mechanical properties of the examined FRP. Since accelerated aging was used to gain knowledge about the

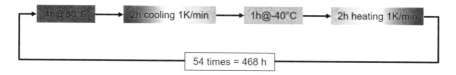

Fig. 4 Temperature change test program

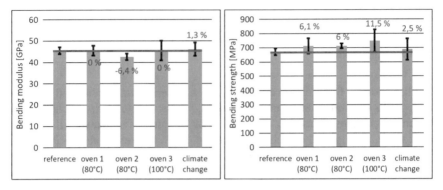

Fig. 5 Bending test results for thermal aged RTM samples

condition of the materials after 5, 10 and 20 years of use, it can be concluded that the reusable components made of FRP are only subject to very little aging effects even after 20 years. It should be noted, however, that this conclusion only relates to thermally induced aging processes. Long-term mechanical loads or other influences such as the UV radiation mentioned above can of course cause aging effects that prevent components from being reused. In the case of mechanical long-term loads, procedures such as non-destructive testing (NDT) or structural health monitoring are necessary again to assess the overall condition of a component that is to be reused.

4   New Circular Economy Processes for Composite Parts

The reuse of carbon fiber reinforced plastic (CFRP) structures and components is a relatively new approach. Accordingly, new processes have to be developed, existing processes have to be adapted, and the entire the process chain have to be implemented. The process chain can be divided into the following areas.

4.1   De-Manufacturing, Inspection Phase

At the end of the life cycle of a system, which uses components or structures made of FRP the system must be dismantled first in order to be able to separate the corresponding structures and components. As already mentioned, when dismantling the parts and structures made of FRP, the focus is on the separation of adhesive bonds. For this purpose, the research results on a separation process using a saw and separation using thermally expanding particles are presented later in this chapter.

The condition of the structures and components is decisive for reuse. The EOL structures and components can be roughly divided into three categories:

- Undamaged structures and components

  These are not damaged by the previous life cycle, so that they can be reused directly. Before this, at most minor visual work-ups, such as a coating application, should be carried out.
- Damaged components and structures with the possibility of repair

  These parts or structures have damage from the previous product life, but which can be repaired. The components will be forwarded accordingly for repair.
- Damaged components and structures that cannot be repaired and reused

  Components and structures that are so badly damaged that they cannot be repaired for a further life cycle can be transferred to appropriate thermal or mechanical recycling processes and to a circular economy at the material level.

In order to classify the structures, it is necessary to determine the condition of the components. This can initially be done optically. Major damage, such as in the event of a crash on automobile components, is immediately visible and the components can be categorized accordingly. In contrast, internal damage that for example have arisen from long-term use of the structures and components are not visible from the outside. However, components and structures for reuse must still be examined for this kind of damage. Primarily NDT methods are suitable for this. Structural health monitoring (SHM) allows components to be checked during usage and can be used in addition to NDT. Ultrasonic-based SHM systems are used to detect and localize changes and damages in fiber composite structures with high precision. These SHM systems are based on a piezoelectric sensor network in which the wave propagation of the ultrasonic waves within the component between piezoelectric transducers is evaluated. The sound waves are reflected differently when obstacles are encountered, thus enabling the detection of these damages. The development and production of an SHM network is of course associated with increased costs. However, the fundamental advantage of using an SHM system is that new damage is detectable very early. In the past, initial damage was compensated by an increased wall thickness. By using an SHM system, an increase in wall thickness can be avoided and the lightweight concept of a FRP can be fully exploited. In the past, research projects as well as industrial applications have proven the basic suitability of SHM systems at laboratory scale as well as on complex structures. That means, components or structures could be monitored over their entire service life using suitable sensors. Using sensor data such as stresses or deformations, the component condition can be determined immediately after the end of a life cycle during usage on a special point or areas. In the case of NDT methods, the determination of the condition only happens before and after usage, but can cover the whole part. Both NDT methods and SHM causes costs for determining the condition of components. Within SHM, the costs result from the sensors and their integration into the components. For NDT methods, investments in the corresponding equipment are usually necessary. The tests are usually time-consuming and therefore costly. As part of the FiberEUse project, the focus is on various NDT methods to get an optimal decision within the circular economy about the material flow. Furthermore, automation of the NDT processes allows NDT cost reductions. The research on these techniques is presented later in this chapter.

4.2   Re-Manufacturing, Repair Phase

Following the de-manufacturing and inspection phase, the components that are suitable for reuse are sent for repair and re-manufacturing. In addition to the repair of damage or optical refurbishment, it is particularly necessary to prepare the joints and joint location. As described in this chapter, adhesive bonds are one of the main structural connections for FRP. After disassembly, adhesive residues will still adhere to the components and structures. Therefore, the adhesive residues must be removed and the surfaces prepared before a new bond. Possible solutions for the removal of adhesive residues are the use of milling machines or lasers. Lasers provide high quality processing of CFRP materials and using optics some shapes on the car structure can be processed where milling machines fail. Laser equipment costs are high, but savings could be made from an overall point of view since the laser can process the components faster than milling.

As already pointed out in this chapter, a method of repair is inevitable in order to reuse composite components. In order to ensure a significant saving of resources, the re-used components should survive at least three life cycles depending on the size and complexity of the structure. It is therefore an important issue to be able to repair any damage that occurs in order to maintain the structure through this period. In order to define the correct repair method, it is important to know and evaluate sensitive damage cases. In the exemplary automotive application, FRP materials have not yet been fully used and have only recently been used in current series. This means that other industries must be used as benchmarks for these new remanufacturing processes. Repairing CFRP in case of damage still requires manual work. However, since components made of this material are becoming increasingly widespread—especially in lightweight construction—greater automation in the repair of CFRP would be desirable. In order to integrate the FRP material class more strongly into automotive construction, it is therefore necessary to establish automation techniques.

If used components are to be reused, they must be prepared for re-joining after the described repair. FRPs are largely bonded, as other joining technologies such as screws or rivets result in a fiber-destroying intervention of the structure. Pre-treatment methods for bonded joints are usually cleaning of the affected area and roughening of the surface. Depending on the application, this can be done manually by grinding or automatically by milling or lasers. After application of the adhesive, it must be cured according to specific conditions. The reused component can then be used again without any problems.

5 Development and Implementation of New Technologies for EoL Composite Parts

In order to enable the reuse of components various processes are necessary, as described above. The technical solutions and technologies developed as part of the FiberEUse project are explained below.

5.1 *Detachable Adhesive Bonds*

As composite structures are used for lightweight design, the composite parts are usually thin and require a distributed load application in order to avoid stress concentrations in a joint. This issue can be overcome using adhesive bonding. Adhesive bonding is an appropriate composite-based joining method. This method enables a uniform non-concentrated stress distribution in the joint. Structural adhesives provide a high strength bond between the parts.

However this method has a significant disadvantage, when considering the reuse of composite parts: in contrast to mechanical fasteners such as screws and rivets, it is difficult to separate adhesive bonds. After gluing, the parts cannot longer easily be separated without damaging them. There are various ways to bring about separation [16], but in the course of FiberEUse a novel approach was taken.

To solve this problem, different solutions have been investigated. One of these solutions is based on a mechanical process by using a computer-controlled saw. In addition, there is another innovative solution of separating adhesive joints. This is how the target of developing reusable designs and remanufacturing technologies can be achieved. Thermally expandable particles (TEPs) are mixed into the adhesive. The adhesive bonding process including TEPs is nearly the same as using a conventional adhesive. It is only important to know what quantity of TEPs is necessary for an expansion at a certain temperature. Then the appropriate quantity of TEPs is mixed into the prepared adhesive. The other steps of manufacturing remain the same: applying the adhesive, pressing the parts together and finally curing. The TEPs are thermoplastic particles with encapsulated gas [17]. The diameter of these particles is 10–15 μm. At the activation temperature the shell softens, and the gas expands drastically. The shell expands to more than 35 μm, whereby the adhesive bond will be destroyed. Using this technology, the mechanical requirements when in service, are still maintained and there will be no accidental debonding of the joint. A fast and simple damage-free debonding is activated when required. With this technology it is possible to separate both metallic and FRP parts.

For the separation of the samples, they can be simply put in an oven. After setting the correct temperature, the samples are exposed to the heat. After a few minutes, the adhesive is completely heated, and the samples are removed from the oven. Now, only a small force is required to separate the composite parts from each other. In Fig. 6 (left) separated parts using TEPs are visible.

Fig. 6 Separated parts using of TEPs (left) and using electric energy (right)

However heating by an oven leads to several disadvantages. For the exemplary case of the automotive sector, this would mean that complete components or systems (e.g. the entire car) would have to be transported and heated in an oven. This means a lot of effort, since large ovens are required. These have to be heated accordingly, which leads to a high energy consumption. In addition, it is not possible to separate individual parts in a targeted manner, since all adhesive connections are separated in an oven with TEPs.

The solution might be to bring in the energy directly into the adhesive layer, for example by using a Joule heater. A wire with high ohmic resistance can be actually integrated into the adhesive layer. To activate the separation the wire is contacted with a voltage source and a voltage is applied. This results in heating the wire and the expanding of the particles starts. In Fig. 6 separated parts using a heating wire are visible.

The advantage of this method based on TEPs compared to the sawing process is that the separation can be carried out regardless of the geometry and adhesive layer thickness. The sawing process, on the other hand, is particularly suitable for straight bonds with a large adhesive layer. An example of a joint, where cutting by using a saw is appropriate, is the bond between the CFRP pultrusion profiles of the reusable vehicle platform and the outer aluminium crash absorbers.

5.2 Inspection Due to Non-destructive Testing for Composite Parts

In order to ensure the safety of EoL composite parts or to determine the re-use, repair and recycling, it is significant to analyse the condition of the composite structures.

Non-Destructive Testing (NDT) is to use different technologies to detect and characterize damages present in composite parts. It can detect damage without changing the material properties and can display hardly visible defects inside the material. NDT can provide useful information for composite repair and remanufacturing such as damage location, size and characterization. Many NDT techniques have been applied for composite inspection in the past decades and three methods will be introduced in this chapter: ultrasonic testing, hyperspectral imaging and active thermography.

Ultrasonic testing

Ultrasonic testing uses a transducer to transmit ultrasonic waves into a test sample until they interact with a feature or the boundary of the sample. The waves are reflected and scattered by such features and these echo waves are subsequently received by the same transducer [18]. Consequently, the size and position of the feature, which is typically a defect, are related to the amplitude of the echo signal and the time when the signal arrives at the transducer, now working as a receiver (Fig. 7). Advanced ultrasonic phased array technology is also implemented in composite testing, which contains multiple elements in one transducer [19]. It can cover a large range during testing and enable different focus laws and sophisticated post-processing methods to improve testing accuracy. However, this requires a more complex data acquisition system. Figure 7 demonstrates an example of ultrasonic testing results when detecting delaminations inside of a composite plate. It can be found that ultrasonic testing has a high accuracy for defect size measurement. An automatic scanning system can be built to make the testing process automatically that can speed up the testing and avoid human operator error [20].

Hyperspectral imaging (HSI)

A standard RGB image, includes three bands which are red, green and blue (Approx. 650, 550 and 450 nm respectively). For a hyperspectral image, there are hundreds of bands and each pixel is made by hundreds of spectral values (Fig. 8) [21]. Through capturing spectral data from a wide frequency range along with the spatial information, hyperspectral imaging can detect minor differences in terms of physical and chemical characteristics on the surface of materials [22]. In addition, hyperspectral imaging has capabilities including fast data acquisition and non-destructive inspection, such that it will cause no damage to the materials and make defect

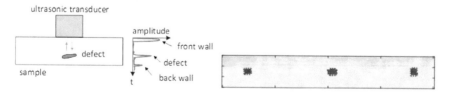

Fig. 7 Basic principle of ultrasonic testing (left) and an example of ultrasonic C-scan image of a composite plate with delamination defects (right). The blue areas are delaminations detected by ultrasonic testing, the red dotted rectangles indicate the real defect size

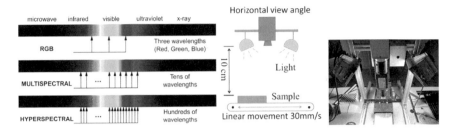

Fig. 8 Band description (left), simulated experimental setup (middle) and real experimental setup (right)

detection and analysis more effective and efficient. As a result, it provides a unique way to characterize material composition, which makes it extremely useful in EOL composite re-manufacturing in terms of characterising and grading of materials and contaminants/defects detection.

In the FiberEUse project, HSI has been successfully used to detect the adhesives on the bonded joints of aluminium and CFRP, the wearing and crack on the surface of aluminium, and the damage caused by pointed body on RTM, BMC and rBMC materials [23]. The damage caused by a wedge-shaped body can be partially detected depending on the illumination and background conditions (Fig. 9). Subsequently, an automated solution is implemented using robotic-arm-based scanning for complex and difficult to reach geometries, and utilising data fusion methods to provide a full, automatic analysis of the component being inspected and characterised.

Active thermography

Active thermography is an NDT method, which allows analysis of defects like contaminations, porosity, delamination, material breakage, scratches, debonding,

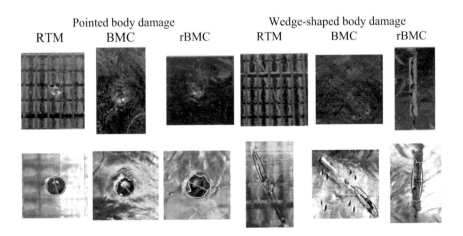

Fig. 9 Example of damage detection for pointed body and wedge-shaped body on three materials

Fig. 10 Infrared camera with excitation unit and sample carrier (left); adhesively bonded joint samples (centre); phase picture of the joint samples made by active thermography (right)

contact corrosion or lack of painting. The process is based on different thermal conductivities such as the ones from CFRP, metal and air. During active thermography, the part is thermally excited by an energy source (Excitation system). The thermal wave flows through the part, which depends on the material, and is partly reflected on defects like delamination's or part-borders [24, 25]. An infrared (IR) camera records the thermal diffusion of infrared radiation on the surface.

There are different excitation systems available, based on electrical, inductive, laser or optical excitation. For contactless excitation in a building environment, an optical excitation with halogen lamps were used (see Fig. 10, left). Depending on the part size and accessibility, a reflection or transmission arrangement is possible. In addition, there the different techniques of Lock-In Thermography and Pulse-Phase Thermography. During the NDT, different modulation frequencies were used to detect defects in different layers of the part. Furthermore, the variation of different process parameters (like distance of excitation unit, distance of camera, number of lights, period number) took place to optimize the resolution and results. The phase picture describes temporal irregularities in the heat flow and was mainly used for damage detection, e.g. Figure 10 (right) showing an unequal adhesive thickness of bonded CFRP-CFRP-specimen.

The tests exhibited that near surface defects can be properly analysed using a cooled IR camera. Uncooled IR cameras can also be used, but the resolution of the defects is not as good as that when using cooled IR cameras. Defects in parts made by fabrics or unidirectional material can be easily detected. The examined RTM structures with epoxy resin and the adhesive joints particularly allow good detection. The depth of the detected defects strongly depends on the structure (thickness of the single layer, layup, material etc.) and the process parameters used in the active thermography process. The material and its heat conductivity have a big influence on the results. Furthermore, if the Lock-In frequencies are reduced, the heat transfer allows deeper defects to be reached. Active thermography thus provides essential information to decide on the re-use, repair or recycling of components and can be automated for reducing NDT costs.

The mentioned NDT techniques play a significant role for validation of the composite parts within the circular economy. On the other hand, the various NDT techniques complement each other in the testing of composite structures. For example, different defects can be detected at different component depths and non-contact methods allow the analysis of complex geometries. Therefore, these techniques provide important information for the EOL composites in order to check whether they can go through a further life cycle.

5.3   Laser-Based Composite Remanufacturing Processes

Processing and preparing CFRP components for re-use either by cutting the material or removing remnants of adhesive that was used to bond parts together has become a topic of major interest from many different types of industry. High quality processing allows components to be re-used effectively. Conventional methods such as milling and waterjet cutting for processing CFRP are not only likely to internally rupture the matrix [24, 25], but also might make it impossible to process some complex 3D shapes, such as those associated with automotive structures.

A single mode fiber laser with average power of 1500 W, wavelength of 1080 μm and spot size of 28 μm was used for laser cutting CFRP material. The process window was defined in terms of cutting speed, number of passes, nozzle distance from surface and nozzle diameter. The focus was on reducing the fiber damage extent (μm) by varying these parameters. The fiber was attached in a collimation and refocusing optical system with 200 mm focal length and can be adjusted manually. An argon jet gas was also delivered coaxially with the beam through exchangeable gas jet nozzles of different aperture diameters.

The investigation carried out concluded that the behaviour of the material during cutting is attributed to the wide weaving, specifically on the top and bottom layers, where the material suffers significant amounts of fraying when processed. It is possible to compensate multiple passes with processing each pass at an equivalent higher speed. In fact, an overall speed advantage can be gained when taking such an approach and making larger than proportionate speed increments. Multi-pass with 3 passes was found to significantly suppress top and bottom fraying and improved the cut quality, whether executed with large, small or double apertures. Laser cutting speeds of 2.5 m/min and over were achieved with fiber damage below 100 μm. The SEM micrograph of the laser cut surface with a single pass using 1.0 mm gas nozzle diameter, cutting speed of 3.5 m/min and fiber damage extent of 55 μm is shown in Fig. 11.

For the removal of adhesive, a novel approach was developed to use a laser to clean the remnants adhesive after the separation of the CFRP parts for future re-use. The laser used was a flash pump Nd:YAG laser with average power of 25 W, wavelength of 1064 nm, maximum pulse energy of 2.5 J and pulse duration of 10 ns. Two types of adhesives were removed with a laser and they were Polyurethane- and Epoxy-based adhesives, and they both were different in chemical composition with maximum thicknesses of around 2 and 0.3 mm, respectively. The beam was focused via a focusing lens on the sample. The samples were moved across the focused beam by an automated XYZ platform, while held vertical to the direction of the incoming

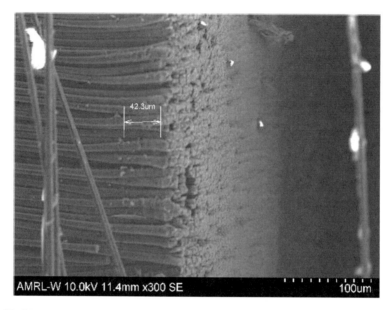

Fig. 11 SEM micrograph of laser cut surface using single pass technique

beam. The spot size and thus laser fluence per pulse was adjusted by changing the focus position along the beam propagation axis.

Furthermore, the results showed that it is possible to laser clean adhesives from CFRP substrates using a pulsed Near Infra-Red laser. There is a natural process window that limits damage on the substrate but does not shield the substrate completely. This effect is best exploited at this NIR wavelength of around 1 μm by releasing high energy pulses over a large spot size. The EP-based material debonds well from the fibers and binder as well as sealant if the layer thickness is consistent over the irradiated area. For non-consistent thickness, this is not possible as irradiation reaches the surface and damages surface fibers. The PU-based adhesive removed in an ablation manner. A purely ablative process would not differentiate between the materials, especially considering the very high optical absorption of both materials. Surface topography is more of a concern for PU-based adhesive due to its nature and large changes in topography. The damage occurred on the CFRP from removing the PU-based adhesive was 60% greater than removing the EP-based adhesive since the latter required less energy to be removed. But for both adhesives, the damage on the substrate was less than the maximum surface abrasion of 0.1 mm allowed by industry. The process that was developed also managed to control the temperature at any time below 76 °C for the adhesives, thus maintaining the integrity of the polymers involved in the bulk composite structure.

5.4 Damage to Fiber Composite Material and Repair Methods

The aim of the new innovative approach is to reuse end of life parts at component and structure level. It is therefore an important issue to be able to repair any damage that occurs in order to maintain the structure regarding this period of time. Before developing new, or improving and testing existing, repair strategies an analysis of possible damage types for these components is necessary. Therefore, a collection of possible damage types was created, and the damage types divided into four different categories: damage through manufacturing, damage during the lifetime or through misuse, temporary damage and damage to joint parts. After the collection of possible damage types an evaluation regarding relevance for reuse and probability was done. Damage related to manufacturing problems are not addressed because damage-free production is assumed for durable components and structures. Only damage types related to lifetime will be investigated. This resulted in the following damage types, which were investigated:

- Damage by impact
- Damage by fault separation of bonded parts
- Damage through aging effects.

Damage by impact

The aim of the investigations in this area was to determine the influence of various impact damages on the properties of glass fiber reinforced plastic (GRP) material and to carry out a repair. In this context, various impact bodies damaged test specimens made of continuous fiber reinforced epoxy resin with carbon fibers. The experimental setup consists of a down pipe with a height of 2 m. Each falling body has a weight of 607 g, which results in an impact energy of 11.9 J. The results of the impact tests are shown in Fig. 12. Almost no external damage is visible for the flat-rounded falling body. A minimal dent has only been detected on the upper side. No damage is visible on the underside. Regarding the pointed falling body, a clear impact can be seen on the top. A minimal dent can be seen on the underside. This means that the damage has clearly gone through the laminate. With the wedge-shaped falling body, a clear damage is also visible on the top. As in the first case, there is no visible damage on the underside.

To ensure that the damage was actually implemented in the laminate, ultrasonic (US) measurements were carried out on the relevant zone of each sample. These measurements also made it possible to determine the extent of the damage zone. An exemplary representation of a sample damaged with the wedge-shaped falling body can be seen in Fig. 13. Here, in the middle and the right results in various US-output options of the sample on the left are visible.

After the damage in the laminate has been detected in depth and extent, the repair can be started. In this area, no innovative processes were developed, but consolidated methods from aviation, as described in the state of the art, were used. To do this, the

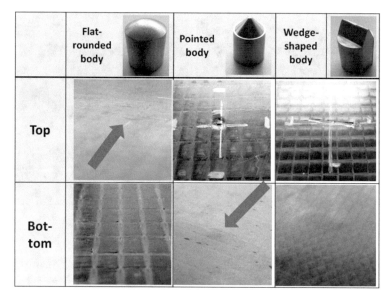

Fig. 12 Results investigating damage by impact

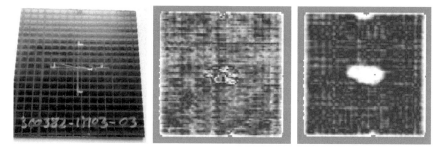

Fig. 13 Sharp-wedged impact (left) and corresponding results of US-measurement

relevant damage is first removed by milling and then a patch is applied. The repaired laminate is shown in Fig. 14.

Damage by fault separation of bonded parts

Regarding the damage by faulty separation of adhesive-bonded parts, the investigation was carried out using flat samples. The samples were manufactured from carbon fiber fabric. In the first step, the simulation of the damage was done by separating the parts in a universal test machine. In Fig. 15 the results of this destructive testing are visible. The failure of the first layer can be clearly seen in the right picture.

To re-glue, the samples must first be repaired or prepared respectively in a second step. This was done in two ways:

Fig. 14 Repaired laminate. Viewed from above(left) and below (right)

Fig. 15 Failure mode of
these samples

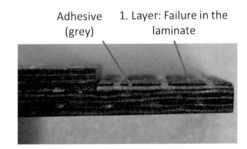

- Removing the adhesive inclusive failed first layer and cleaning
- Without removing, only cleaning the parts.

In the first case the adhesive including the faulty first layer was removed by milling. The sample was then only cleaned because the milling machine had pretreated and activated the surface of the sample. Afterwards the bonding was carried out. The second case (without removing) was much simpler. Here the sample was only cleaned and then glued. No additional pretreatment was carried out. Both batches of samples were then tested in the universal test machine. In Fig. 16 the exemplary failure modes of both cases are visible.

To sum up with the results of the mechanical testing, it can be said that the repair with a surface preparation brings same the strength than original bonding, even though a layer of fiber has been removed. A repair of failed components is possible without problems, which is one of the targets of the FiberEUse project. A repair without preparation causes a loss of strength, which was expected. Due to the lack of preparation and the unevenness of the surface, a sufficiently good bond between the sample and the adhesive cannot be achieved in order to reach similar high strengths. But still high values can be reached, if the components are designed accordingly, such that these reduced strengths may be sufficient.

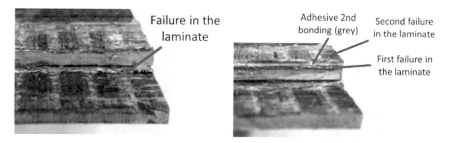

Fig. 16 Failure modes: with removing the adhesive (left) and only cleaning (right)

6   Conclusions

The concept of reusing components and structures made of fiber-reinforced plastics in the automotive sector is a completely innovative approach. On the one hand, this approach offers many advantages; on the other hand, there are still some challenges that have to be solved for an industrial implementation. As part of the FiberEUse project, many technologies for a remanufacturing process chain have already been developed and tested. Based on this, however, further research and development work is necessary.

6.1   Advantages of Composite Part Re-Use

The main advantage of reusing components and structures made of fiber-reinforced plastics is that the composite of fiber and matrix and the continuous fiber are retained and thus the best possible mechanical properties. In this context, the lifetime of the components and structures is also increased by enabling several life cycles. This can enable an increased use of fiber-reinforced plastics. Due to the very high cost of the fiber and the complex manufacturing processes, carbon fiber-reinforced plastics in particular are used very little in large-scale automotive production. Increased use would enable a further reduction in vehicle mass and thus more energy-efficient vehicles. The high material and production costs can be offset by reusing and extending the useful life of the components and structures made of fiber-reinforced plastics. This enables an increased use of these materials in large-scale automotive production.

6.2 Perspective

As part of the project, various technologies and solutions for a remanufacturing process chain for the reuse of composite components and structures are being developed and tested. There is still a need for research here with regard to the industrialization of the processes, but the processes show great potential. The implementation of the circular economy based on durable and reusable composite components and structures in the automotive sector is still a long way off. The reason for this is that vehicles will have to be designed and built completely differently in the future. The majority of vehicles are currently made up of steel bodies and the plants and production processes are optimized for these materials. A complete change in the design of vehicles would require enormous investments in new production facilities and factories and the old machines could no longer be used. For this reason, the established automotive OEMs in particular will not change their basic vehicle design. However, the automotive industry is currently changing. More and more small start-ups are emerging and bringing their own vehicles onto the market. They are much more flexible in the design of their vehicles than the large OEMs and thus have the opportunity to implement new, innovative technologies and vehicles. There is also a great opportunity here to reuse automotive composite components and structures.

6.3 Future Research Questions

The overarching focus of future research work will particularly be on the further industrialization of the developed processes. Many of the technologies are still carried out by hand. At this point, an automation of processes or process steps must be developed. In particular, further developments are necessary for the individual methods presented in this chapter.

A preselection is necessary in the area of recording the condition of EoL components, as not all components have to be examined using complex, non-destructive testing procedures. For example, components can be equipped with sensors and, based on the life history, they can be categorized into "reusable", "not reusable" and "to be examined more closely" at the end of a life cycle. Only the components of the last category would be examined using NDT. In the area of detachable connections based on TEPs, it will be important to find solutions for how the TEPs are activated or how the heat can be introduced directly into the adhesive layer. Automated processes leading to this would allow even faster and easier dismantling. For repairs to automotive structures, it will be important to investigate how existing technologies can be transferred even better and eventually automated.

References

1. David, K.: Nondestructive inspection of composite structures: methods and practice. In: 17th World Conference On Nondestructive Testing, pp. 1689–1699 (2008). https://doi.org/10.1017/CBO9781107415324.004

2. Vaara, P., Leinonen, J.: Technology Survey on NDT of Carbon-Fiber Composites (2012). [Online]. Available: http://www3.tokem.fi/kirjasto/tiedostot/Vaara_Leinonen_B_8_2012.pdf

3. Gholizadeh, S.: A review of non-destructive testing methods of composite materials. Procedia Struct. Integr. **1**, 50–57 (2016). https://doi.org/10.1016/j.prostr.2016.02.008

4. Habermehl, J., Lamarre, A.: Ultrasonic phased array tools for composite inspection during maintenance and manufacturing. In: 17th World Conference on Nondestructive Testing, pp. 25–28 (2008). https://doi.org/10.1063/1.3114343

5. Guo, C., Xu, C., Xiao, D., Hao, J., Zhang, H.: Trajectory planning method for improving alignment accuracy of probes for dual-robot air-coupled ultrasonic testing system. Int. J. Adv. Robot. Syst. **16**(2), 1–11 (2019). https://doi.org/10.1177/1729881419842713

6. Ocal, F., Xu, Y.: Using NDT techniques to detect and characterise the damage of end-of-life components in remanufacturing. J. Innov. Impact 136–147 (2014)

7. Jin, B.C., Li, X., Jain, A., González, C., LLorca, J., Nutt, S.: Optimization of microstructures and mechanical properties of composite oriented strand board from reused prepreg. Compos. Struct. **174**, 389–398 (2017). https://doi.org/10.1016/j.compstruct.2017.05.002

8. Ibrahim, M.E.: Nondestructive evaluation of thick-section composites and sandwich structures: a review. Compos. Part A Appl. Sci. Manuf. **64**, 36–48 (2014). https://doi.org/10.1016/j.compositesa.2014.04.010

9. Katunin, A., Dragan, K., Dziendzikowski, M.: Damage identification in aircraft composite structures: a case study using various non-destructive testing techniques. Compos. Struct. **127**, 1–9 (2015). https://doi.org/10.1016/j.compstruct.2015.02.080

10. Dutta, S., Huber, A., Schuster, A., Kupke, M., Drechsler, K.: Automated NDT inspection based on high precision 3-D thermo-tomography model combined with engineering and manufacturing data. In: 2nd CIRP Conference on Composite Material Parts Manufacturing, pp. 321–328 (2019)

11. Santhanakrishnan Balakrishnan, V., Seidlitz, H.: Potential repair techniques for automotive composites: a review. Compos. Part B Eng. **145**(March), 28–38 (2018). https://doi.org/10.1016/j.compositesb.2018.03.016

12. Budhwani, K.I., Janney, M., Vaidya, U.K.: Vibration-based nondestructive testing to determine viability of parts produced with recycled thermoplastic composites. Mater. Eval. **74**(2), 181–193 (2016)

13. Tschannerl, J. et al.: Potential of UV and SWIR hyperspectral imaging for determination of levels of phenolic flavour compounds in peated barley malt. Food Chem. **270**(January 2018), 105–112 (2019). https://doi.org/10.1016/j.foodchem.2018.07.089

14. Sun, H., Ren, J., Zhao, H., Yan, Y., Zabalza, J., Marshall, S.: Superpixel based feature specific sparse representation for spectral-spatial classification of hyperspectral images. Remote Sens. **11**(5) (2019). https://doi.org/10.3390/rs11050536

15. Sun, M., Zhang, D., Wang, Z., Ren, J., Chai, B., Sun, J.: What's wrong with the murals at the Mogao Grottoes: a near-infrared hyperspectral imaging method. Nat. Publ. Group (2015). https://doi.org/10.1038/srep14371

16. Broughton, J.G., Hutchinson, A.R., Winfield, P.: Dismantlable adhesive joints for decommissioning, repair and upgrade. S&T Organization

17. https://www.nouryon.com/products/expancel-microspheres/. Accessed on 12 Aug 2021

18. Song, S.: Ultrasonic nondestructive evaluation systems—models and measurements. Springer Science+Business Media, LLC (2007)

19. Drinkwater, B.W., Wilcox, P.D.: Ultrasonic arrays for non-destructive evaluation: a review. NDT E Int. **39**(7), 525–541 (2006). https://doi.org/10.1016/j.ndteint.2006.03.006

20. Mineo, C. et al.: Flexible integration of robotics, ultrasonics and metrology for the inspection of aerospace components. In: AIP Conference Proceedings, vol. 1806 (2017). https://doi.org/10.1063/1.4974567
21. Yan, Y., Ren, J., Liu, Q., Zhao, H., Sun, H., Zabalza, J.: PCA-domain fused singular spectral analysis for fast and noise-robust spectral-spatial feature mining in hyperspectral classification. IEEE Geosci. Remote Sens. Lett. (2021)
22. Yan, Y., Ren, J., Tschannerl, J., Zhao, H., Harrison, B., Jack, F.: Nondestructive phenolic compounds measurement and origin discrimination of peated barley malt using near-infrared hyperspectral imagery and machine learning. IEEE Trans. Instrum. Meas. **70**, 1–15 (2021)
23. Yan, Y., Ren, J., Zhao, H., Windmill, J.F.C., Ijomah, W., de Wit, J., von Freeden, J.: Nondestructive testing of composite fibre materials with hyperspectral imaging: evaluative studies in the EU H2020 FibreEUse project. arXiv preprint arXiv:2111.03443
24. Prinzip der optische angeregten Lockin-Thermografie (n.d.) https://www.edevis.com/content/de/optical_lockin_thermography.php
25. Deutsches Institut für Normung. Zerstörungsfreie Prüfung—Aktive Thermografie. DIN Standard Nr. 54192 n.d.

Co-Design of Creative Products Embedding Recycled Fibers

Sarah Behnam⏾, **Giacomo Bonaiti**⏾, **Severin Filek, and Tamara König**

Abstract The involvement of designers in the sustainable transition from linear to circular economy is crucial since they significantly contribute to the realization of products and services. In the FiberEUse project, a multiple-step approach to co-design was employed, starting with the definition of a first and second design brief in order to clarify the task objectives for designers. This was followed by the description of the co-design process, which aims to engage designers to contribute innovative design concepts for recycled composites. By publishing design concepts in the feedback collection software module Idea Manager, designers and users were able to exchange information, insights, visions, and thoughts digitally. The Idea Manager comprises a feedback collection tool that supports a first assessment of design concepts. Depending on the design briefing and/or confidentiality agreements, the feedback collection and the assessment can either be done (stakeholder-)internally or publicly. A flowchart illustrates the multi-step approach of co-design within the FiberEUse project. The feedback collection process was aided by a progress analysis to detect new value chains for business cases. For the selection of product design concepts, a progress analysis partitioned into four main criteria, the following aspects are drawn on for assessment: (i) quantitative and qualitative production feasibility, (ii) closeness to market introduction, (iii) potential volume of the market, (iv) circularity, (v) type of market, (vi) service opportunities, and (vii) take-back/deposit systems. Aside from bringing out the advantages of co-design for consumers as well as production companies, this chapter also discusses general challenges of co-design and co-creation in a broader sense when intellectual property rights (IPR) are not respected appropriately. The participation in a publicly accessible co-design of

S. Behnam
STIIMA-CNR, Milan, Italy

G. Bonaiti
Rivierasca S.P.A, Via Strasburgo 7, 24040 Bottanuco, BG, Italy

S. Filek · T. König (✉)
Designaustria—Knowledge Centre and Interest Organization, Designforum Wien, Vienna, Austria
e-mail: tamara.koenig@designaustria.at; koenig@designaustria.at

© The Author(s) 2022
M. Colledani and S. Turri (eds.), *Systemic Circular Economy Solutions for Fiber Reinforced Composites*, Digital Innovations in Architecture, Engineering and Construction, https://doi.org/10.1007/978-3-031-22352-5_11

concepts must be clearly communicated and accepted by each participant by agreeing to intelligible terms and conditions.

Keywords Co-design · Creative products · Re-use · Glass fibers · Recycling

1 Introduction

'Design is a crucial element for European competitiveness', said Günter Verheugen, former Vice-President of the European Commission [1]. Design is often misinterpreted as sole aesthetical form-giving for decorative purposes, or as a 'finishing touch'. However, design is far more complex and diverse. Design is a strategic instrument with which entrepreneurs can define how their products and services are perceived on the market [2].

Designers are key for a sustainable transition from linear to circular economy as they contribute products and services within a co-design approach for circular design. The transformation of our current linear economy into a circular economy, which is driven by 'smart consumption' and 'smart products' [3] appears most promising. However, the shift to a circular economy is gradual and not radical, it is a transition and not a switch. Creators and decision makers, such as designers, engineers, and industrial managers, play a crucial role for a successful change. Multi-step approaches in defining targets, analyzing markets and data need to involve players for co-creating product and service solutions that are long-lasting and of high added value, yet flexible in adaption for optimization.

The objective of the co-design methodology for FiberEUse is the development and improvement of circular economy-oriented products via close cooperation with users and/or (end-)customers as well as design professionals to unlock potential profitable circular economy solutions for the re-use of composite material. Open innovation activities in the development phase of co-design projects, for instance collaborative IT-supported feedback collection, is assessed as crucial [4].

The feedback collection accelerates the:

- development of product design concepts based on the FiberEUse circular economy paradigm for composites.
- improvement of these circular design concepts in a collaborative co-design process and by integrating market data.

The following chapter describes the methodology of co-design and feedback collection in detail.

2 Methodology of Co-Design Approach and Customer Feedback Collection

2.1 Exposé and Selection of Co-Design Partners

An exposé on the FiberEUse paradigm for communicating the overall scope of the FiberEUse project is a recommended starting point for each co-design process and its corresponding design briefings. The aim of the exposé is both to raise interest in the industrial, and product design community, and to collect questions and insights into the status quo. A 'Call for Interest' through various media channels supports the promotion of knowledge transfer with regard to experience in designing products made of recycled glass-fiber reinforced plastics (rGFRP) in line with circular-economy principles.

In order to find a cooperation partner, the following criteria are recommended to be met:

- Design and material expertise (especially with composites such as rGFRP).
- Access to research and workshop facilities.
- Knowledge of incorporating circular co-design methods.
- High commitment to co-design for circularity.
- Interdisciplinary working attitude as a common prerequisite for professionals.

The communication phase between co-designing partners helps reveal the level of engagement and intensify the flow of information. Keeping up a high-quality level of communication is key for success of the co-design process and its outcomes. As soon as the co-design partners were defined, a document (the design brief) was drafted to outline the scope of the cooperation project. The following design process flow chart illustrates the co-design process from the first design brief to the actual results, namely circular-designed product proposals (Fig. 1).

2.2 The Design Briefings

The first design brief of the FiberEUse project included the following five activities for co-design to be performed:

- market research and analysis on the status quo of designs complying with circular economy and circular co-design practices in special regard to end users and/or end consumers needs;
- material research on recycled composites, including physical testing and development of a comprehensive material characteristics information sheet;
- examination of current waste streams in Europe to understand the transition points from linear to circular economy;

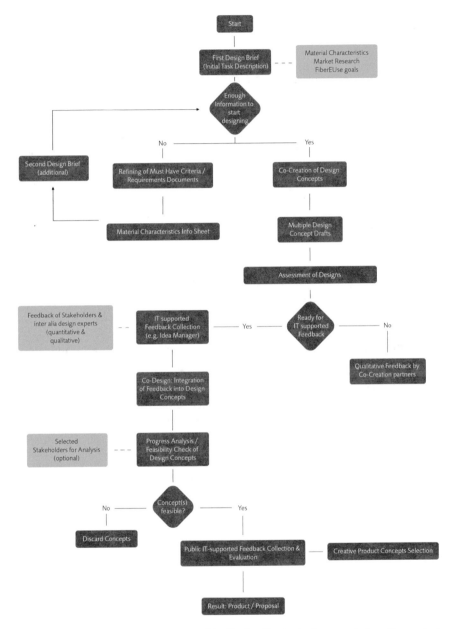

Fig. 1 Design process flow of co-design and feedback collection in the context of creative products embedding recycled fibers

- creation of a design brief to provide description on tasks/objectives and to support the individual task descriptions for designers from various expert backgrounds, e.g. designers providing specific chemical and technical expertise or designers specialized in interior design (furniture), exterior design (construction/architecture);
- Multi-step co-design processes through personal meetings and IT-support.

This first briefing helps to clarify issues—most commonly regarding functions and aesthetics of design concepts, timing, and budget. Furthermore, it simplifies eventually demanded cooperation agreement contracts. Those first considerations lead the way to a second, more detailed, design brief.

The second design brief encompasses processes and contents to determine basic objectives and scopes of designs that are going to be optimized in the co-design procedure.

- Activities/content chapters include purpose, function, deadlines;
- Designs should have high social, emotional/aesthetical, and functional utility;
- Marketing objectives are determined, e.g. strengthening the unique selling proposition (USP) or market position through product/material innovation; quantity/quality requirements foresee high-volume and medium-quality products and high-quality and medium-quantity products respectively;
- Material analysis and development of an extensive document that contains relevant material characteristics data for the design project to gain the best results in the conceptualization.

Furthermore, written information on the recycled material to be used was collected and shared. This material is the recycled fiber reinforced plastic material called 'Glebanite' [5]. It is grinded scraps of glass reinforced polyester with unsaturated polyester resins. In addition to the written documents, examples of early stage use-cases of Glebanite, such as prototypes of home décor pieces, as well as handy-size samples of Glebanite were provided to convey both written data and a haptic impression of qualities and characteristic.

Since Glebanite is a material for manifold purposes, a catalogue of criteria of 'must-haves' provides for the continuous alignment with the respective project's objectives and to create a framework for the product design development.

'Must-haves'—design criteria for creative products embedding recycled fibers:

- Design for circularity: First of all, designs for potentially slowing the resource loops by creating designs for reliability and durability are demanded. Research results of potential reuse, repair, redistribution, remanufacturing, and refurbishment should be incorporated in the design concept to help open up discussions on the feasibility of product concepts and their related services in IT-supported co-design.
- Design for resource recovery: products need to be easily traceable and collected after their end of life or for maintenance.
- High significance of social, functional, and emotional utilities.
- Willingness to present product design concepts publicly.

- Products shall contain 100% Glebanite if possible.
- Continuous integration of feedbacks by end customers generated via digital tool Idea Manager by Holonix as part of the co-design approach.

Furthermore, the development of products and services follows open innovation principles. That means that product development does not necessarily follow close-to-market principles but focuses on finding optimal solutions for current or future problems in an open and interdisciplinary manner [6]. For the FiberEUse project it was commonly agreed to develop creative design products with high functionality, high value to public society, as well as of high aesthetic demands.

The second brief marked the beginning of the co-design process. Immediate actions after the design briefing comprised field surveys among potential customers differing in education, age, residence, and financial status, as well as desk research by designers.

2.3 The Co-Design Procedure

A multi-step approach employing analog and digital co-design encompasses ad hoc meetings (online and offline) for direct interaction with the designers. In FiberEUse, it was beneficiary to involve designers in all creation phases until the completion of prototype production.

The first offline co-design phase shall result in the collection of design drafts. Open innovation online tools such as the feedback software Idea Manager by Holonix benefit the collection of design concepts and feedback quantitatively and qualitatively.

The use of online feedback tools. In general, a web-based software tool for continuously gathering customer feedback and for supporting product co-design and improvement along the phases of the product design process assists in amplifying the coverage of a collaborative environment. It is recommended to make use of digital tools in a multi-step process including several pre-tests in order to provide time for further optimization, to build up acquaintance with the tool among potential participants, and to raise engagement in contributing.

In a first feedback collection, stakeholders can take part by registering and contributing comments. The detailed methodology of the Idea Manager will be described in more detail in Sect. 3.

The second phase of the online and offline co-design process can be rounded off by a presentation and feedback collection from a wider public, respectively customers.

Design concepts feasibility analysis. A feasibility analysis assists with the selection of the most promising product design concepts matching the aforementioned criteria and 'must-haves' of the design briefing.

Uploading the revised and adapted product design concepts by the design teams of the NDU to be presented publicly to the Idea Manager software kicked off the second IT-supported feedback collection. The link to access the Idea Manager was sent to the consortium to be passed forward to existing customers and relevant stakeholders and published in the various networks of FiberEUse. This procedure allows a broad definition of product-use-oriented customer groups to enhance feedback quality and reduce potential bias caused by the selection of narrow-defined target groups.

A public audience is invited to take part in the IT-supported co-design sessions evaluating selected product design concepts. The access to the feedback collection demands several promotional activities, such as social media activities. In general, frequent feedback collection activities must be part of an overall communication plan to raise the engagement of interest groups.

Recommended proceeding is the advance notice of release dates of preliminary and final results. A frequent and organized direct communication that informs about all aspects relevant for targeted stakeholder groups via convenient media channels is key to a sustainable success of dissemination achievements.

Analysis and operationalization of criteria and assessment scale. Pre-screening design concepts from an industrial point of view includes feasibility, potential business case, and their relevance to the focus of the project on circular economy principles. Accordingly, four criteria including a brief operationalization for each of them are defined for assessing the design concepts:

- Production feasibility
- Closeness to market introduction
- Potential volume of market
- Circularity.

In the following, a brief operationalization is given for each criterion to clarify the analysis (Table 1).

In order to quantify the assessment of the design concepts, each design concept is assigned a score on a Likert scale of 1–5 corresponding to each above-mentioned criterion (1 representing least promising to 5 as the most promising) (Table 2):

Subsequently, web-based tool Idea Manager for collecting and assessing creative product design concepts aiming to support the co-design process is introduced.

3   Digital Co-Design: The Idea Manager by Holonix

3.1   Defining the Scope and Objectives of the Idea Manager

The Idea Manager (IM) is a web-based software-tool designed to provide and support activities along the process of idea generation and co-design.

The IM can be accessed via: https://fibereuse.holonix.biz/ideamanager.

According to the recommendations on the user-centered software development that enhances usability and user experience, the following steps have been taken into account to implement and adapt the Idea Manager for the FiberEUse co-design and feedback collection:

Table 1 Analysis and operationalization of criteria for assessing the design concepts

Criterion	Operationalization of criterion
Production feasibility	This criterion is measured based on complexity as well as economic points of view in terms of production and packaging. In particular, the criterion is assessed depending on several factors such as the shape/properties of the product; the applicability of using semi-finished material for its production (which is cheaper and easier from the production point of view); the required technologies for its production; the need of manual input in the production process (related to production costs), etc.
Closeness to market introduction	This criterion is assessed based on the approximate level of the product's introduction to the market, which is evaluated mainly based on the benchmarking with products already on the market with the same/a similar functionality, as well as the usefulness/necessity of the product for final users
Potential volume of market	This criterion is assessed depending on the potential volume of the production, which is a factor depending jointly on the type of the market for the product and on the volume/amount of recycled material in each product piece. Basically, the larger the potential market of the product, the higher this criterion is assessed
Circularity	This criterion is assessed based on the potential of the product for improved circularity in the value chain. In this regard, diverse factors of circularity (e.g. life cycle of the product) as well as the efficiency of the take-back system for collecting the product after its life has ended are taken into consideration (ease refers to applicability but also the ability to offer services with the product for improved take-back system)

Table 2 Assessment scale interpretation

Criterion	Assessment scale interpretation	
	1	5
Production feasibility	Least feasible/expensive	Highly feasible and easy/cheap production
Closeness to market introduction	Far from market introduction	Close to market introduction
Potential volume of market	Low-potential market volume	High-potential market volume
Circularity	Most difficult take-back system/short product life cycle	Easiest take-back system/long product life cycle

- identifying and selecting appropriate dialogue techniques;
- defining the sequence and timing (dynamics) of interaction: order of actions as to register, submit contribution, search for content and edit it, verify if applicable to the role assigned;
- defining the information architecture of the user interface of an interactive system to allow efficient access to interaction objects: overall integration and implementation.

As the result of the design process and input gained during the software testing and refinement phase, the implemented features have been defined to face specific objectives of the software with respect to its purpose and FiberEUse co-design scope. Continuous testing and refinement of the software interfaces improves functionality and usability of the tool, e.g. registration system, submission of new content, and feedbacks.

3.2 The Functional Features of the Idea Manager

The IM provides a collaborative environment for sharing ideas on innovative products by means of publishing ideas and collecting reactions and comments on these submitted ideas. In the process of co-design, the aim of the tool is to provide qualitative and quantitative feedbacks from a target community.

The protection of IPRs (Intellectual Property Rights) must have priority. In order to support the protection of IPRs, the IM provides the option of restricted access to a company, an organization, or individual initiators. In that case, ideas can be shared only with a specific audience, for instance a certain group of designers. If restricted access is preferred, contributors—which can be end-users, designers, entrepreneurs—may see and contribute to the concepts of other designers, but results are not open to the wide public.

In any case, the IM always provides clear information about the type of publication (generally accessible or restricted) and asks its users during the registration process to agree to the publication mode when accepting the general terms and conditions.

In the FiberEUse project, the IM supported an open-access co-design following the open innovation principle, encouraging a wide public audience to take part in the development processes.

In general, collected feedbacks along the co-design process can be used to determine future refinements of the proposed design solutions and the selection of the most attractive and feasible design product concepts.

The IM enables the following stakeholders to participate in co-design.

- Organizations ('drivers') looking for innovative ideas, the most promising directions, informed decisions, and strategic solutions for product design and development.
- Idea creators ('contributors') proposing, sharing, and elaborating new ideas and solutions for design products.

- Moderators ('content managing users') who can decide on the appropriateness of ideas that can be shared within the community, whose members in turn generate ideas and provides feedback in the form of comments, votes, and further elaborations; moderators can also create surveys and polls, access statistics, and aggregate ideas into concepts.
- Evaluators ('viewers') participate in co-creation; they can directly express their appreciation and opinions by voting and commenting ideas or answering to surveys and polls.

The implemented functional features of the IM [7] are:

- Creation, search, and editing of ideas
- Management of ideas: approval of publication and removal from publication list
- Visualization of ideas statistics
- Creation of voting mechanisms.

Web-based software tools such as the IM need to be functionally and graphically designed to achieve two of the main objectives of a project:

- stimulating the (co-)creation and evaluation of ideas about new products from re-cycled materials, as well as
- enhancing people's awareness of how materials, e.g. fiberglass reinforced composites, can be used to give life to new objects.

The interfaces of the online tool need to be designed and implemented to provide information about the project and offer the necessary information to guide the stakeholders involved through the idea creation and evaluation phases. The graphic appearance of the site was adopted to FiberEUse's visual identity to provide visual brand recognition.

4   Creative Product Concepts Embedding Recycled Fibers

4.1   Production Technologies

The starting point of the selection of creative product concepts in the FiberEUse project is the identification of production methods for prototype production incorporating recycled fibers. Ten production methods are currently being used and/or tested:

1. open-mold casting
2. closed-mold casting
3. compression molding
4. vacuum casting
5. continuous lamination
6. continuous extrusion

Table 3 Technology in relation to recycled content of selected prototypes

Technology	rGFRP content wt%	Prototype
Open-mold casting	41%	Variously colored cylinder-shaped tests
Closed-mold casting	37% because fresh glass is added (to be confirmed during preliminary test)	Swing
Compression molding	37% because fresh glass is added (2×600 g/m^2 biaxial) for improved mechanical resistance 38.7% because fresh glass is added (1×375 g/m^2 CFM) for improved mechanical resistance	Yellow (stool) Lamp

7. 3D printing
8. rotomolding
9. centrifugal molding
10. filament winding.

Several prototypes and concepts of creative design products following the aforementioned design brief and demonstrating the manifold possibilities of applications of recycled fiber reinforced composites are described in detail in Chap. 15 'Use Case 1: Mechanical Recycling of Short Fibers' of this book.

4.2 Manufacturing Results

Table 3 sums up the production technologies for the prototypes and indicates the achieved percentage of recycled content per design product prototype. All of the rates have exceeded the initially targeted 30%.

5 Co-Creation | Co-Design: Creative Methodologies and Their Challenges

Co-creation and co-design as specific instance of co-creation in collaboration with designers [8], have always been tools used by design practitioners and have even become increasingly popular among wider groups over the past decades. The idea of co-creation differs in perception from a strongly rational to a spontaneous and almost playful approach [9]. Co-creation methodologies provide the possibility to engage entire stakeholder communities in developing activities to benefit from various user contributions and perspectives.

However, among the manifold advantages of co-creation and/or co-design, such as creating a better fit between the product or service offer and customers' or users' needs [10], the risk of negative impacts is evident: companies or institutions may abuse these approaches to exploit valuable ideas and insights supplied by customers and/or users spending time and energy in collaborating with these companies and institutions while companies may also violate the IPR of designers and co-creators. This danger can never be entirely prevented when concepts are (publicly) shared. Facilitators and users—if being aware of their IPRs—may be willing to share and exchange ideas in order to support the optimization of products and services when benefitting from solutions in their professional or private lives. It remains the responsibility of a co-creation and/or co-design initiator to point out those benefits and risks before participation in joint actions. Thus, it is of great importance to keep co-creation and co-design as open processes that allow the involvement of many groups and to ensure at the same time that those having the power to use this process are provided with enough incentives, such as potential market opportunities, to keep the process fair and sustainable.

It is necessary to question the strategy and method of co-design. Utilizing a co-design process is a preferable approach and a solution for a specific problem, question, or project (like FiberEUse) with respect to the necessity of:

- gaining information about some innovative recycled material (which can only come from the stakeholders involved);
- convincing designers and companies to use this material, provided its performance is satisfactory;
- changing mindsets of people/users to use (these/other) recycled materials: and
- building on awareness for circular economy.

The design profession is a problem-solving profession involving various disciplines (not only product and industrial design, but also communication design, social design, information design, interface design, etc.). Designers are also able to change the user behavior and to facilitate changes (e.g. social change) by means of their work. Questions and problems concerning our society and environment are increasingly posed to designers, who are able to not only improve and change a product or the usage of a product, but also a service and/or system. It is not only designers working on solving a problem, though. In order to minimize clients' risks and funds invested, and in order to influence customers'/users' attitudes and acceptance, more and more people (in organizations, companies, etc.) need to work from different sides and with stakeholder groups.

Hence, the profession of a designer has developed rapidly from problem-solver to consultant (who is expected to have more and more expertise in various disciplines like mechanics, engineering, computer science, packaging, logistics, etc.). Designers now may find themselves as leaders of processes, such as the design thinking approach, checking multiple possible solutions to a problem as to their feasibility and acceptance on the side of clients/users [11].

Going beyond feedback collection, one needs to ask: Has co-design really become increasingly relevant? Or is it just used more because clients/companies want to exploit various means and methods when creating a new product?

Two aspects of a company's behavior help understand the popularity of co-design methods even if sufficient copyright protection is at stake:

- Co-design contributes to finding out if a product/service really meets customers' and end-users' needs.
- Companies want to have guarantees before investing in/spending money on innovation.

Co-design methodologies are definitely a possibility to start influencing and changing mindsets, raise awareness, and move faster in the direction of a circular economy approach, as all stakeholder groups can be involved in the process and be made responsible for the outcomes. Still, their will and wish to bring about change and their positive attitude are required, as is, on the companies' side, a willingness to invest.

If…

- a designer as creator,
- a company as producer, and
- a client as user.

…act in a visionary way and feel responsible for our future, health, environment etc., they might already be very much aware of using recycled materials and/or products resulting from a circular economy approach. The methodology proposed is one of truly involving all stakeholder groups—in order to gain insight in all their aspirations and needs.

Still, it is necessary to keep in mind that IPR rights are involved and that they should be considered and protected by all means; the more people and stakeholders are involved, the more important this becomes.

6   Conclusions and Prospects

For FiberEUse the process of co-design and feedback collection of creative products embedding recycled composites utilizes analog and digital means aiming to involve stakeholders, in particular product designers and the community of end users. Their cooperation is essential for accomplishing a circular-oriented design of products including a dismantling and return systems for product recycling.

Analog (personal) knowledge transfer and IT-based feedback collection tools based on the example of FiberEUse's online tool Idea Manager as part of a strategically oriented communication is crucial for long-term success in co-design for circular economic transition. The advantages of co-creation and co-design for end users and production companies are pointed out and potential risks of copyright violations are discussed.

Future steps include the integration of data and knowledge of essential European stakeholders in cloud-based platform(s) for value chain integration to reach a wider number of potential users of recycled composites.

References

1. Ico-D—International Design Council, https://www.ico-d.org/connect/index/post/989.php. Last Accessed 6 June 2021
2. Greger, R.: Design im Marketing: Mittel zum Zweck. Echomedia Verlag, Vienna (2009)
3. Stahel, W., MacArthur, E.: The Circular Economy, 1st edn. Routledge, London (2019)
4. Kortmann, S., Piller, S.: Open business models and closed-loop value chains: redefining the firm-consumer relationship. Calif. Manage. Rev. **58**(3), 88–108 (2016)
5. Bonaiti, G.: How is made… Glebanite. Presentation (PDF) at FiberEuse Kick-off Meeting, Milan (2017). https://www.designaustria.at/wp-content/uploads/Riverasca-Glebanite-Presentation_DA.pdf. Last Accessed 16 July 2021
6. Chesbrough, H.W.: Open Innovation: The New Imperative for Creating and Profiting from Technology. Harvard Business School Press, Boston, Mass (2003). https://www.nmit.edu.my/. Last Accessed 17 July 2021
7. Arabsolgar, D., Musumeci, A.: FiberEUse: large-scale demonstration of new circular economy value chains based on the reuse of end-of-life fiber-reinforced composites—a circular it platform to manage innovative design and circular entities. In: Proceedings 65, 23 (2020). https://www.mdpi.com/journal/proceedings. Last Accessed 12 June 2021
8. Sanders, E.B.N., Stappers, P.J.: Co-creation and the new landscapes of design. CoDesign **4**(1), 5–18 (2008). https://www.researchgate.net/publication/235700862_Co-creation_and_the_New_Landscapes_of_Design. Last Accessed 23 June 2021
9. Coates, N., Ind, N.: The meanings of co-creation. Eur. Bus. Rev. **25**(1), 86–95 (2013)
10. Steen, M., De Koning, N.: Benefits of co-design in service design projects. Int. J. Des. **5**(2) (2011)
11. Schreckensberger, P. et al.: Design Management. Zwischen Marken- & Produktsystemen, 1st edn., BoD (2015)

Modular Car Design for Reuse

Justus von Freeden[ID], **Jesper de Wit**[ID], **Stefan Caba**[ID], **Carsten Lies**[ID], and **Oliver Huxdorf**[ID]

Abstract The design of reusable composite structures for cars needs high constructional effort. The car must be divided into separable modules meeting ecologic and economic requirements. Here, a battery containing platform and a seating structure were selected as large components with high potential for reuse. In a first step the desired car is described setting the basic scenario. A carsharing vehicle shows perfect conditions due to low logistics effort and the business model of the owner. This sets the boundary conditions for the design of the platform. Two different approaches were tested and merged into a concept ready for reuse. Simulations of the stiffness and the crash performance show good values. First large CFRP profiles were produced in a complex pultrusion process. An associated seating structure following similar design principles was constructed using profiles and nods. All load-cases that can occur during the utilization phase could be beared. Both modules together can form the basis of a reusable car. The design principles like detachable joints—in particular the utilization of detachable adhesive connections—can be adapted for any other technical composite product.

Keywords Fiber reinforced plastics · Reuse of composite components · Non-destructive testing methods · Composite repair · Detachable adhesive · Laser-based remanufacturing processes

J. von Freeden · C. Lies
Fraunhofer Institute for Machine Tools and Forming Technology, Chemnitz, Germany

J. de Wit · O. Huxdorf
INVENT Innovative Verbundwerkstoffe Realisation und Vermarktung neuer Technologien GmbH, Braunschweig, Germany

S. Caba (✉)
EDAG Engineering GmbH, Fulda, Germany
e-mail: stefan.caba@edag.com

M. Colledani and S. Turri (eds.), *Systemic Circular Economy Solutions for Fiber Reinforced Composites*, Digital Innovations in Architecture, Engineering and Construction, https://doi.org/10.1007/978-3-031-22352-5_12

1 Introduction: The Idea of a Reusable Car Structure

Todays cars are in use for only about 200,000 km [1]. This represents about 4000 h of utilization. A such short time is not common for machines with comparable costs. The use-phase of the car in majority consists of the time in parking spaces. Thus about 12–16 years for a privately owned car is common [2]. For business applications the value of the car sinks rapidly due to costs for inspection and repair. Thus about 160,000 km or 4 years usually lead to a sell-of of the car. The car is then owned privately or directly shipped to developing countries. When customized vehicles are produced this low time or distance will mean an even shorter utilization.

Carbon fiber reinforced plastics on the other hand show a very high durability [3]. This means they will always exceed the lifetime of the cars they are used in. The comparatively high emissions for their production raises the carbon footprint while the low weight reduces the footprint during the use-phase [4].

Taking these facts into account, it seems a good idea to enlarge the lifespan of composites to enable them to play out their benefits. Thus, a possibility should be found to enlarge the use-phases of composite parts. Here the idea of recovering the structural carbon parts comes into the game. If it is possible to regain the large CFRP components from used cars—in particular from customized cars—a lower carbon footprint and reduced costs over lifecycle can be expected. For that purpose, two large parts of a car were chosen as demonstrating objects. A platform structure that is invisible for the customer and a seating structure hidden behind the cushions. Both provide high mechanical performance and can reduce the weight of the car. To enable the idea of reuse, the platform and the seating structure must be redesigned.

1.1 Basic Requirements of the Car

The car is planned as a car for the coming years (Fig. 1). During the last years a steady trend is an increase in mobility as a service (MaaS) [5]. The most important type of MaaS is Carsharing which was selected as the main purpose. Cars will not be owned by the drivers any more. They will be owned by companies renting them for short times and distances. Passengers will rent the car via app and the car will have many different drivers during the day. The car will drive in the city for most of the time. It must cover multi purposes like people transport or shopping. At least 4 seats must be included. During the long use phase technologies will change. Autonomous driving is an example for the technologies that should be updateable into the concept and not be impeded.

The car must take into account that the drivers will not clean the car after renting it. So in particular the interior design must enable a simple cleaning, e.g. via vacuum cleaning. A flat floor design is a possible design element to support this. Also the cushions must be designed in accordance to this.

Fig. 1 Possible design of a car including the reusable parts

The future drives electric. It is clear that the car must be a battery electric car to take the change of the propulsion into account [5]. The energy storage is usually placed in the bottom of the car, because it is heavy. Due to better drive dynamics it should be placed between the axles. Thus it can only be placed in the platform. Electric drives do not need a gearbox. This also supports the idea of the flat floor design. The battery capacity is today part of numerous discussions. The range of the car is defined by the capacity of the battery. For the car sharing car, it is important that the car needs not to be useable for long distance travelling. On the other hand the users do not want to recharge the batteries after every rental. Thus a range of more than 450 km is seen as a good compromise between cost, weight and range today [6]. It is clear that updateability must be given since the research on new battery systems is still ongoing. The battery compartment must be exchangeable and repairable.

The powertrain of electric cars can be located next to the wheels. Thus the concept should include drive either in the axles or directly in the wheel. This is also a possible factor of improving the updateability by a modular design. The possibility to mount different drives with high or low power.

The weight of the platform is decisive. Only a lightweight vehicle can save energy in comparison to a steel or aluminium car. The reusability over a long time span will only be possible, if carbon fiber reinforced plastics are used. Only this material can provide the aimed high durability. On the other hand, these materials are expensive and their production emits high amounts of CO_2. This backpack must be equalized by low emissions during the use phase. It is also clear that the platform must be affordable. So a cost objective of $< 3500€$ is set. The production rates are estimated as 10,000/year.

Table 1 Basic dimensions of the car

Wheelbase	2670 mm
Length	4371 mm
Width	1830 mm
Height	1600 mm
Curb weight	1100–1600 kg

The basic requirements for reuse are a use phase of > 25 years. During this time more than 1,000,000 km will be driven. At least 5 reuse procedures must be covered. After the first use phases a remanufacturing procedure must be possible causing less than 7% weight increase. The overall cost reduction by using the reusable platform instead of 6 single platforms should be more than 50%. As well the energy consumption during production should be decreased also by at least 50%. This should be provided by a decrease in CF use of 70%.

Dimensions

The car must be ready for any type of use. However the platform must be designed for a particular car as a starting point to fulfil all requirements and to test the performance. Thus the dimensions will be fixed, but they must be chosen in a manner that other vehicle dimensions will be possible. The dimensions are derived from numerous multi-purpose cars (Table 1).

The dimensions fulfil the aspect of multi-purpose use. Also the new possibilities with autonomous driving are included. The dimensions of the platform and the seat are defined by the external dimension. The selected medium class vehicle dimensions with a height slightly above average enables many different variants of a car (Fig. 2).

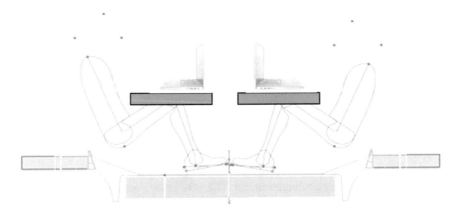

Fig. 2 Possible use-cases of the interior

1.2   Basic Requirements of the Platform

The selection of the part to be designed for reuse was influenced by the knowledge that logistics and rework processes will require cost intensive manual work. Hence a large part of the car body had to be chosen. The mechanical properties of CFRP enable high performance parts. Thus a part with high stiffness requirements could be provided. The platform fulfils these basic requirements. The CFRP section of the platform does not provides the basic functions of the platform like the safety function of the battery compartment. It is the basis for a whole car. The name of this element was chosen to be CFRP foundation.

Boundary Conditions

The platform provides all basic mechanical properties of the car. The whole drivability but also the crash safety and the ergonomics for the passenger are mandatory. Car design usually begins with the ergonomics. Based on the EDAG Light Car, a former show car, the ergonomics and the construction was set up. The car was stretched into the defined dimensions (Fig. 3).

Also the legal requirements must be fulfilled. This covers the visibility of other cars in traffic. The mirrors must be placed and the windows must be dimensioned and positioned. The wheels must be spring-mounted to achieve a high level of comfort and the steering must be possible. Thus the wheels need space. The auxiliary units also need a suitable design space. Again the spaces must be large enough to position any future auxiliary unit (Fig. 4).

The reusability is the main innovation of the platform. The long utilization phase and a changeable design means that the platform will be totally invisible to the customer. The crash safety is the main driver for the construction of the body in

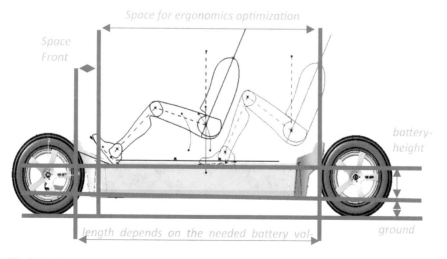

Fig. 3 Basic dimensions for the ergonomics optimization

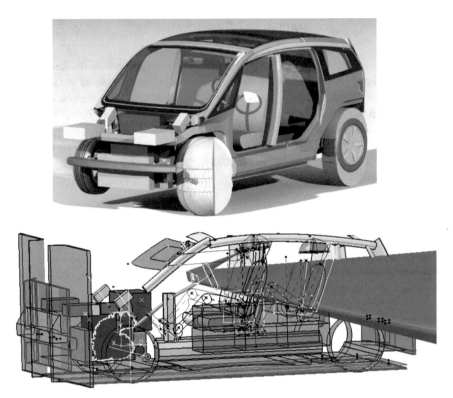

Fig. 4 Ergonomics, legal and space investigation of the basic car model

white. The platform must be able to fulfil the EURO NCAP requirements. This means a front crash and a side pole crash are the most important scenarios.

1.3   Basic Requirements of the Seating Structure

For demonstrating a further component within automotive engineering, an interior component was selected in addition to the platform. In order to be able to demonstrate the variety, the interior should be represented by a seating structure. A seating structure is available for all models in a wide variety of variations, from a simple seat to a bus seat to a bench seat. This is how the potential of a modular design can be exploited.

The clear main objective of the seating concept is the reusability of the individual components. In order to receive as high a degree of utilization of the resources as possible, as many exchangeable components as possible are to be represented. Only so it is possible to accomplish a re-use of individual components. In order to reach this, it is necessary to ensure a fast assembly as well as disassembly. Attachment

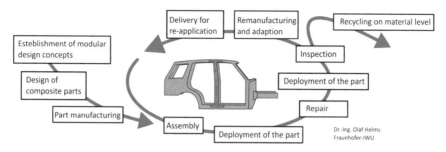

Fig. 5 Concept for reusing an automotive component

elements must be separable and/or removable beyond that. If these requirements are possible, it can be assumed that the seat structure has a life expectancy of at least 30 years. This period includes at least three life cycles or reuse scenarios. Inspection and repair scenarios are also included here.

Requirements beyond life expectancy concern functionality. For example, a corresponding structure should be movable on the platform to be fastened and the seat itself should be adjustable. Since repairs may be required in the event of reuse, the weight of a reused seat structure may be 110% of that of a new seat. The cost of a new seat may be 150% compared to a reference seat structure. Since this seat structure is largely reused in a further life cycle, the cost of such a reused seat must be only 50% compared to a new seat.

With these requirements and boundary conditions, the scenario of use-case 3, an application for a car sharing fleet, is possible (Fig. 5).

2   Design of the Platform

The design of the platform using CFRP in the foundation is possible using different construction methods and manufacturing processes. The processes must be mass productive and create a very high strength laminate. Thus resin transfer molding and pultrusion processes are seen as the most promising. Molded parts can be very complex but require an expensive mold while pultruded profiles can provide high stiffness but the connection of sill requires node elements. The conception stage was divided by these two basic processing ideas between IWU and EDAG. Both worked on the same basis of dimensions and requirements to compare these technologies.

2.1 New Concepts for Reusable Platform Structures

(a) Profile concept

The profile concept is based on the idea of creating a framework with profiles and connecting elements as the basic structure of the platform. The pultrusion process is used to produce the profiles. Pultrusion is very well suited for the cost-effective and continuous production of fiber-reinforced profiles with high lightweight potential. The use of profiles and connecting elements makes the vehicle platform scalable. Thus, different vehicle sizes can be easily developed by using shorter or longer profiles.

For the design of the connecting elements two different variants were developed: shell structures and node elements. The first design concepts are shown in Fig. 6.

The left design in Fig. 6 shows the front and rear side members, which are joined to two big shell structures (front and rear wall). The shell structures are connected through the left and right outer sills. In comparison to the shell structure concept the right design shows another solution to connect the profiles by nodes. The nodes connect the front and rear side members with the left and right outer sills by using some crossbeams. The advantages of the shell structure compared to the node elements are the lower number of interfaces and the resulting improved stability. However, the large and complex shaped front and rear wall have a high production costs and tooling. On the contrary, the nodes have significantly lower production

Fig. 6 Design concept for platform with profiles and connecting elements (left: shell structures, right: node elements)

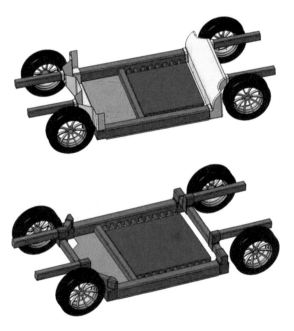

Fig. 7 Advanced concept of reusable platform using profiles

costs and tooling costs. Nevertheless, the load transfer over the many interfaces is a challenge, especially in case of a crash.

For these reasons a new design for the platform using profiles was developed. This design is shown in Fig. 7.

The new design concept is a combination from the previous ones. To connect the different profiles smaller shells are used, which means lower production and tool costs. The interfaces between the profiles and connecting elements are designed with large surfaces to ensure optimum load transfer. In addition, two large crossbeams in the form of tubes to accommodate crash loads were inserted at the front and rear. On these crossbeams, the front and rear side members are supported. Two shells enclose the cross member and the longitudinal members and thus ensure the connection and fixation. An adhesive bond is planned to use as the joining technique at this point. The sills of the vehicle platform are divided into an inner profile, which together with the crossbeams forms the battery box, and an outer profile for the absorption of crash energies. Moreover, the platform was designed symmetrically to save costs by using the same parts in the front and rear area.

The second platform design also has weak points and challenges that need to be solved. Complex geometries can only be realized with great effort and peak loads in critical areas can only be achieved using massive elements. However, this leads to a strong increase in weight of the vehicle.

(b) Molded concept

The molded concept is based on former cars already on the road just like BMW i3 or monocoque cars like the McLaren models. The basic idea is that all fiber reinforcement sheets are combined in a mold and after hardening the injected resin combines these sheets into a single part with all functions. The design is seamless and a high flexibility in fiber layup is possible.

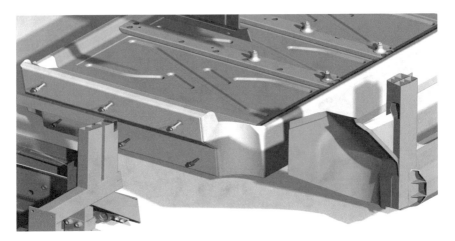

Fig. 8 Design of the platform with a molded foundation

In the molded concept the side beams are formed by a sandwich element consisting of a foam core and CRFP sheets. The front and the rear part of the car are constructed in a similar manner. The combination forms a frame similar to the profile design containing the batteries (Fig. 8).

The preforms needed to form the part play an important role to lower costs. Cut-off is one of the major factors to optimize the product. Material costs for carbon fibers are high. On the other hand the recycling needs high effort and the mechanical properties of the waste are usually worse than with the virgin material. Thus a concept reducing the cut-off during preforming was developed. It takes into account the foam cores in the side bars. These can be used to wrap the fabric. The profile parts are connected via simple geometries with only few single preform geometries (Fig. 9).

The top is closed via aluminium formed sheet. The reason to use aluminium is not the stiffness, but the electric shielding and flame retardancy requirements. The bottom is also formed using a sandwich part. So the stone chipping as well as the bollard test can be passed without any damage to the battery (Fig. 10).

(c) Merged Design

The two presented concepts show individual strengths and weaknesses. Even though both can provide a technical feasibility, the economic and ecologic assessment would differ. While the profile concept shows lower cost in particular for the profiles, the molded concept can provide all functions in one part. These functions are mainly connected to the joining technologies necessary for reuse.

Thus both concepts were evaluated and combined to the final design of the platform. The advantages of the profile should be used. In addition Fraunhofer IWU researches bended profiles that could be useful in the front and the back of the car to lead the forces from a crash into the sill. The profiles used for the sill were too simple. To provide more functions and a higher crash performance, they will become more complex. The possibility to produce hollow profiles will be used to replace the

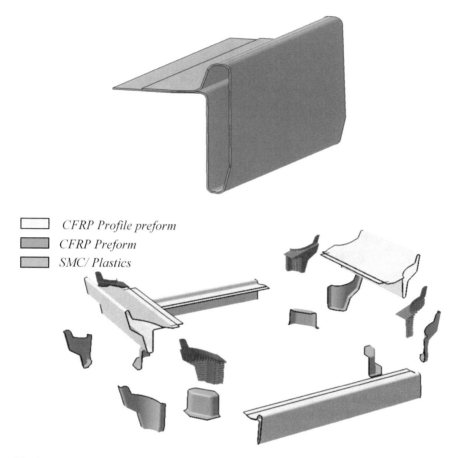

CFRP Profile preform
CFRP Preform
SMC/ Plastics

Fig. 9 Preform concept of the molded design of the foundation part

foam cores with bars. The connection of the profiles is critical, because the forces must be lead into the profile. The profiles can only be constructed using a single geometry over the whole length. So the loads must be transferred softly. On the other hand the thickness of the hollow profile is hard to translate to a molded part. Thus a 2 component connection was designed. Large areas with an adhesive will be used to provide a large cross section. The parts are designed demoldable (Fig. 11). The main component will be joined from outside of the battery compartment. It will be produced with SMC/BMC. Recycled fibers could be used. To connect the bars the profiles will be cut in this area to achieve more surface for the adhesive. From inside of the battery compartment a formed sheet will be used that can be molded easily.

In an earlier version it was planned to close the corners by putting the profiles into a molding process. But it was shown that it will be too complex to close the profile to prevent the resin from flowing inside the profiles.

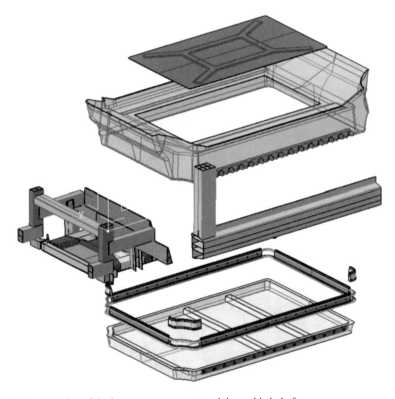

Fig. 10 Basic design of the battery compartment and the molded platform

Fig. 11 Demoldable design of the corner elements

The whole design of the foundation is now aligned to the profiles and the weaknesses of profile constructions are addressed. This is the basis for further developments concerning reuse and safety of the concept (Fig. 12).

The front and the back of the car will be made of metals due to a shorter life-cycle and the higher requirements for crash. Here, a design proposal was made taking into

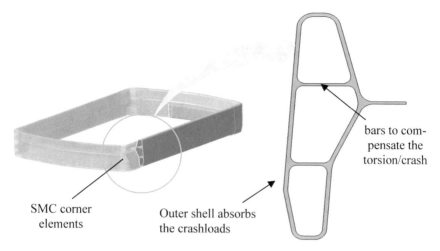

Fig. 12 Final basic design of the foundation

account reuse, safety and functionality. The basic idea is that the CRFP foundation is covered by metallic crash elements to form the platform.

2.2 Crash Simulation

The basic design can provide stiffness and strength for the platform. The crash loads require a high energy absorption during the deformation of the parts. Here, composites show lower values in comparison to metals. Hence, crash elements are designed for this special purpose. A material providing crash performance as well as a low density is aluminium. Thus all parts of the platform that are not intended to be reused can be produced using this metal.

The design is based on experiences made with previous vehicles. The crash load paths are defined to lead the loads around the passengers and the battery. No intrusion of the battery compartment or the passenger cell is allowed.

The most important crash cases are the front crash and the side pole crash. Both must be passed and the intrusion must be prevented. For the front crash after the Euro NCAP the car must be able to withstand a crash against a steady wall with 50 km/h. Thus the kinetic energy of the car must be transformed into the deformation of the aluminium crash elements.

The simulation showed that no intrusion could be detected, but in a first step the accelerations were too high. Usually a human can stand 30 g, but in the first version for a very short time more than 50 g could be measured. So the crash elements behind the bumper were reinforced and the beams in the subframe were shortened. The effect was a higher acceleration at the beginning of the crash and the deletion of the peak. The bended profiles around the battery were able to lead the loads into

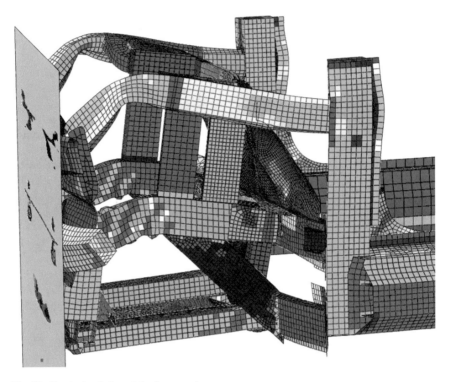

Fig. 13 Crash simulation of the front crash

the sills. The CFRP profiles were connected directly to an aluminium profile. The combination showed good performance (Fig. 13).

The side pole crash after Euro NCAP requires a crash into a steady pole of 254 mm. The speed is 32 km/h. Again no intrusion of the battery is allowed and the accelerations of the driver must be reduced under 30 g. The side pole crash is more critical, because the car must be stopped in a very short time and a short deceleration length. The maximum intrusion is a very good measure for a comparison. It is not possible to measure the acceleration of the passenger directly because the platform is only one part of the car. Thus the criterion was the intrusion into the sill.

In a first version an aluminium profile with bars was defined. It will be attached to the CFRP profile via an adhesive. Both profiles were dimensioned during the crash simulation. While the CFRP profile should not be destroyed during the crash, the aluminium profile should show as much deformation as possible to take up the crash energy. The battery compartment was also constructed using a wet pressed sandwich floor element. It can contain up to 28 battery modules. During the side pole crash the deformation will be high. Thus two aluminium profiles were placed right under the seat carriers. They transfer the loads to the other side of the car and improve the stiffness of the platform. The seat carriers also provide the connection between the seat system from INVENT and the platform. The connection between the aluminium profiles and the pultruded profile was achieved using a very simple molded element. It provides stiffness and carries the loads from the CFRP into the profile (Fig. 14).

In 5 iteration steps the dimensions of the profiles were defined. The aluminium material was selected as a high ductility material with high energy absorption. This enabled the reduction of the bar thicknesses. On the other hand the directions of the bars were adapted to strengthen the profile for the first contact with the pole. With this procedure the weight of the sill could be lowered by 30 kg. The pultruded profile could be designed with feasible fiber directions and wall thicknesses. The idea behind

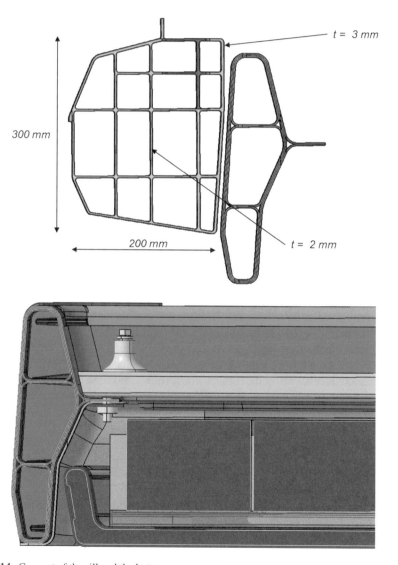

Fig. 14 Concept of the sill and the battery compartment

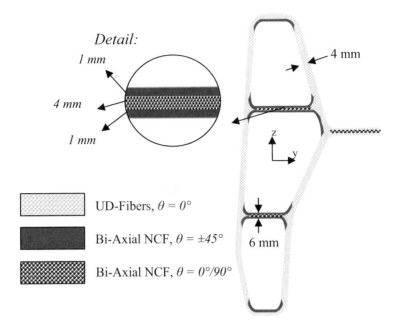

Fig. 15 Dimensions of the pultruded profile in the sill

the fiber lay-up was to achieve two objectives. On the one hand, the platform must provide much stiffness for a high driving comfort on the other hand the crash loads must be withstand. The main fiber direction is 0° in x-direction. This maximises the bending stiffness of the platform. The fibers in the bars provide stiffness in the crash load case. A simple 90° lay-up is not possible due to manufacturing restrictions. The fibers must be pulled into the mold. This is only possible for fabrics in combination with fibers in 0° direction or symmetric ± 45° (Fig. 15).

The crash simulation of the final set-up shows the expected result. The maximum intrusion into the sill is 173 mm. This is 77 mm less than the comparison vehicle. The intrusion does not affect the battery. The acceleration of the battery after a strong slope stays on a steady level of 30 g. This means that the deceleration of the vehicle works as desired (Figs. 16 and 17).

2.3 Manufacturing Technologies

The platform design consists decisively of four profiles (aluminium and CFRP), four connecting elements on the platforms corners and the floor of the battery compartment. For these components, low-cost and mass-production methods are necessary to fulfil the economic requirements of vehicle production.

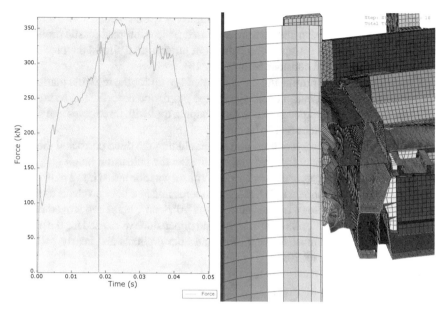

Fig. 16 Crash simulation after 17.5 ms

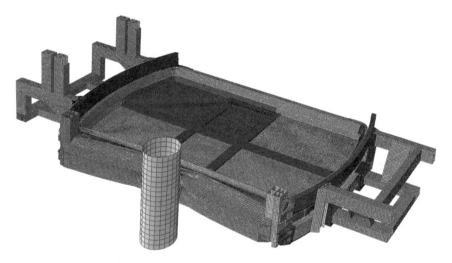

Fig. 17 Crash simulation at maximum intrusion

Pultrusion is such an economic and continuous process for the production of straight and low weight profiles made of fiber-reinforced plastic. Therefore, fibers and additional fabrics are pulled from bobbins and impregnated with the thermosetting resin. The wet reinforced fibers are pulled through a heated tool where the curing process of the plastics starts. By increasing the temperature in different heating zones of the tool, the thermosetting plastic cures completely within seconds. Two alternate working puller which grasp and forward the cured fiber-reinforced plastic profile are after the tool. A special set-up of the pultrusion process also allows the production of curved profiles with a constant radius (Fig. 18).

The geometry of the profile and the calculated lay-up for the reusable platform is very complex and represents a challenge for the pultrusion process. To produce this profile in the pultrusion process, a multipart shaping tool with three cores and panels for a reliable fiber guidance had to be developed (Fig. 19).

Based on the CAD data the three-part tool including the three cores and the fiber guidance panels were manufactured and installed in the pultrusion line.

The calculated fiber set-up of the platform profiles was converted into a pultrusion-compatible lay-up. As shown in XXX specific semi-finished fiber products as unidirectional fabrics (0°, green), triaxle fabrics (+45°/0°/−45°, blue) and woven fabrics (0°/90°, orange) with different fiber weights and dimensions were used for the pultrusion process. In total almost 100 semi-finished fiber products and nearly 140 fiber roving were required (Fig. 20).

Fig. 18 Pultrusion process

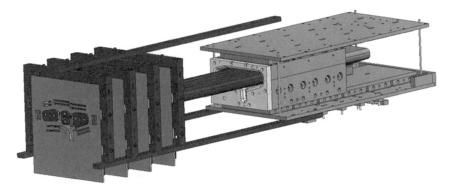

Fig. 19 Pultusion tool for the platform profiles

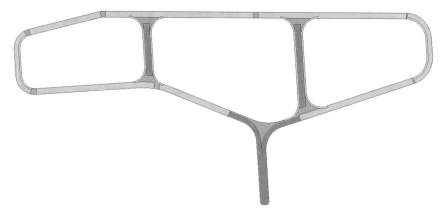

Fig. 20 Fiber lay-up suitable for pultrusion process

The pultrusion process was prepared using the semi-finished products. The fibers and fabrics were positioned in the bobbins stock and passed separately through all fiber guidance panels. The fibers were pulled through the tool and led to the pullers. These pullers were equipped with contour-specific hardwood fittings for gripping the profile. In the next step the resin system (unsatured polyester) were mixed and filled into the resin bath.

The semi-finished products were impregnated with the resin and pulled through the specific heated tool (including the cores). The resin systems reacts and the profile is formed because of the heat treatment in the tool. With this installment a few meters of the profile could be pulled. Further optimization loops regarding the fiber guidance and the resin formulation were done. As result requirement-specific profiles were produced for the platform and part of the further demonstrator production (Fig. 21).

The aluminium profiles will be produced using extrusion. The heated melt will be pressed through the dies and cooled down. This technology is predestined to produce

Fig. 21 Demonstrator—pultrusion profile

profiles. It offers low costs for medium part numbers and gives good mechanical properties. The materials are usually wrought alloys perfectly fitting to the purpose of high crash performance. The floor of the passenger cell is produced using aluminium sheets. They are necessary to fulfil the requirements for the reaction to fire.

The corner elements of the platform are made of a bulk molding compound (BMC). BMC is a composite material, which consists of reinforcing fibers, (mostly) polyester resin, fillers and additives. The mineral content is often very high, including glass fibers, the mineral content can rise to over 80%. So it is a comparatively small amount of resin necessary. The recipes are easy to adapt to customer requirements. The material is one of the simplest materials to process, is suitable for mass production and allows short cycle times. Since heat must be supplied during processing, the cycle times are in the minutes range and not in the seconds range. The compound is put into the open mold, pressed and heated. It contains fibers with a length of up to 30 mm to achieve high stiffness and strength. High impact strength combined with inherent corrosion resistance of this material ensure good appearance even at high mileages, which is important for the longlife platform. The technology is capable to produce more than 10,000 parts per year. The same manufacturing technology can be used to produce the connection element between the pultruded profile and the aluminium profile in the battery compartment. In Task 4.2 the use of thermal recyceled carbon fibers is investigated. If the mechanical properties are sufficient, a BMC material with these rCF can be used.

The floor of the battery compartment will be produced as a wet pressing part. The geometry is quite simple and the fibers will not be formed in small radii. The preforms are wet out with the resin and transferred to the mold. Thus the impregnation is very simple through the thickness of the fabrics. A higher performance is reached by

Fig. 22 Demonstrator of the reusable vehicle platform

implementing a sandwich core into the part. This maximises the stiffness by adding just few weight. A very simple foam sheet can be implemented directly into the press. It must be taken care of the pressure inside the mold to assure that the core is not crushed during the process.

For showcasing the use-case of the reusable platform a demonstrator was manufactured using prototyping technologies. It is used to represent the idea of reusing structures. It is shown in Fig. 22 and includes a touch screen giving information about reuse.

3   Design of the Seating Structure

The starting point of the development of a reuseable seating system was a concept survey, asking about the aspects of the relative target segments of the automotive sector inter alia: pricing, seat arrangements, and user profiles. The survey result assessment helped to reveal that cheap and minimalistic city-cars have the biggest potential for circular reusable seating applications in (shared) car fleets in urban environments. These cars should be used by individuals as well as families. The result of the survey and the following concept-freeze is visible in Fig. 23.

The above described concept was the starting point to design a manufacturable seating structure assessing possible technologies and materials. The first step in a detailed design of the seating structure is to split the described frames of backrest and seat shell into different parts. In this way an easy manufacturing and demanufacturing of these parts is possible. Figure 24 shows this early design stage.

To get a modular seat structure, the backrest has an identical design as the seat shell. If the individual segments are designed as profiles, only two different types are needed, which must be connected to each other: Longer lateral profiles as well as short horizontal profiles. As the backrest has the same geometry as the seat shell, a middle element must be included to reach the correct height of the headrest. As

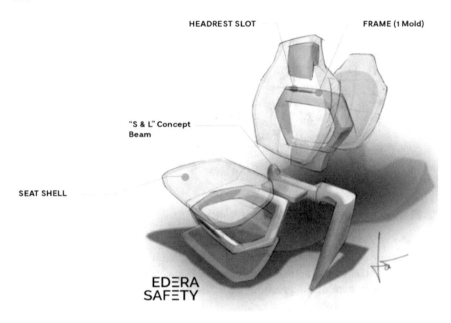

Fig. 23 Final concept for seating structure

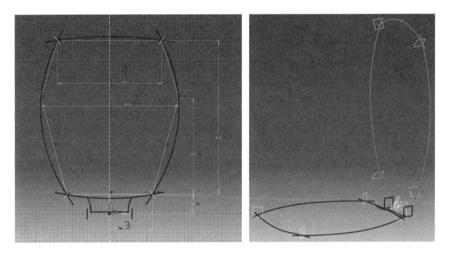

Fig. 24 Skeleton design of backrest and seat shell

shown in the final concept (Fig. 25), the seat is mounted only on one side to the platform.

For this early model a first FEM analysis was performed to verify that the design of the seating structure meets the deformation requirements for all load cases. The applied load cases are mainly loads caused by misuse or static deflections (Table 2).

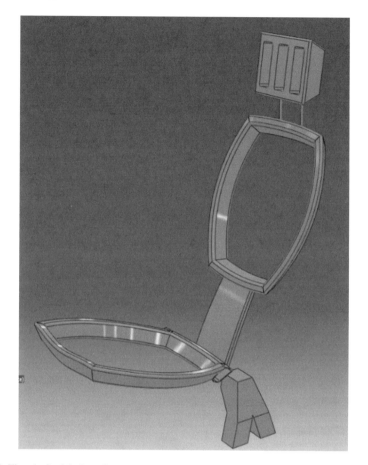

Fig. 25 First draft of design of seat structure

Load cases caused by a crash are not considered due to the fact, that the platform will absorb these loads mainly. In Fig. 26 it is visible that the seat area between the seat carriers is not affected during crash. The seat only has to carry the loads caused by the passenger. Figure 27 shows the positions of the applied loads.

Unfortunately, this first FEM-analysis showed too high deformations which means that the initial design of the seat structure had not the required stiffness. The design had to be optimized for fulfilling the given requirements and tolerances in deflections. In order to increase the stiffness, several changes were implemented:

- The wall thickness of the CFRP-profiles was increased.
- A foam core was implemented for all CFRP-profiles.
- The intermediate element between backrest and seat shell was removed.
- This causes a necessary extension of the profiles towards the middle cylinder.

Table 2 Applied load cases [7]

ID	Load case	Force [N]
1	Static deflection (load which the entire seating structure shall withstand)	1.600
2	Submarining	4.000
3	Unrestrained cargo impact	6.250
4	Passenger pushing against seat back	3.600
5	Driver throwing weight against seat rest while sitting down	7.000
6	Rearend car-crash	8.320

Fig. 26 Top view of the seat area during side pole crash

Fig. 27 Positions of applied
load cases

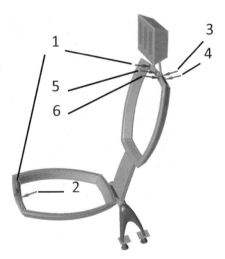

Fig. 28 Optimized design

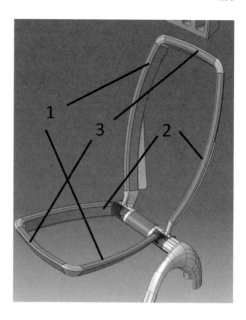

- Shear panels were implemented to increase the stiffness additionally, visible in Fig. 28 (brown-coloured parts).

Due to this design, the geometries of backrest and seat shell are not identical anymore. The initial idea of only two different kinds of profiles is no longer feasible. But a small amount of identical profiles is possible: Regarding the curvature and cross-section, only three pairs of profiles are sufficient for the manufacturing of this seating structure. The numbers in Fig. 28 are marking the three kinds of profiles with same curvature and cross-section. A FEM-analysis of this design showed that the deflections for the described load cases (Table 2) are small enough to proceed with this design.

In addition to the profiles, the following subchapters describe in detail other necessary components.

Corner elements

The CFRP-profiles will be connected by these elements. The corner elements are inserted into the hollow ends of the profiles. An adhesive is injected which will bond the parts together. As material a sheet moulding compound (SMC) or aluminum is chosen (Fig. 29).

Middle cylinder

The concept for a middle connecting element has been developed. This part is a metallic component, to which the profiles of backrest and seat shell will be mounted. Aluminum was chosen due to the complex geometry. The space is large enough to implement possible adjustment mechanisms. The middle element is subdivided into

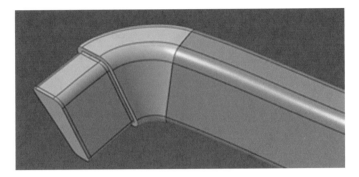

Fig. 29 Corner element (blue)

Fig. 30 Middle element (grey)

five parts for a simplification of the design for a cheaper and easier manufacturing. Additionally, changing the position of the sidearms (connection area of the profiles) to each other makes a smooth adjustment of the seat possible (Fig. 30).

Mounting element

The mounting element is the part which connects the seating structure to the platform. The element is designed by using a topology optimization software (TO). This part is a CFRP element with included foam-cores. The TO showed, that a single element in the middle area of the platform is sufficient, as shown in the early concept-phase. The mounting area for the element on the platform includes appropriate slides for a slidable seating structure (Fig. 31).

After describing all the single parts of the seating structure, the final design is visible in Fig. 32. It represents a novel styling supporting the requirement of a car sharing vehicle as well as the option for reuse. The seating structure was manufactured in prototyping processes. The mounting element was realized using bearing

Fig. 31 Mounting element
(black)

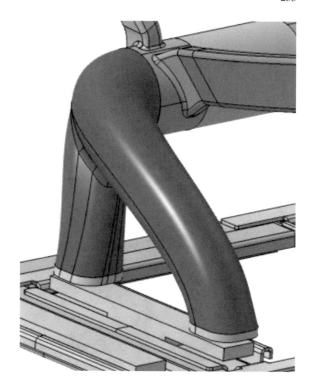

for machine constructions. It is movable and bears the loads for passengers in use
and crash.

4   Conclusions

The change of the boundary conditions for the design of cars to reuse of mechanically
highly loaded lightweight design structures requires new concepts. The structures
on the one hand must be light and cost effective, but on the other hand durable and
reusable. Taking these ideas into account, new designs of a platform and a seating
structure were found.

4.1   *Platform*

The platform must provide stiffness and crash safety. It bears all major loads during
driving and saves the battery compartment from intrusions that would lead to a catas-
trophic developments. Two basic concepts were designed with different construction
philosophies. A profile based solution with as many pultruded profiles as possible

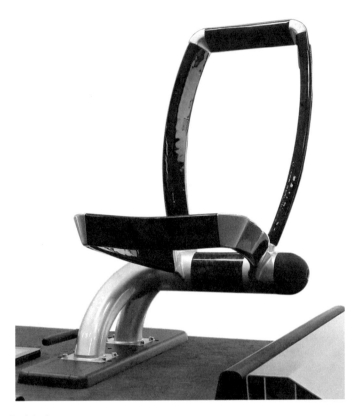

Fig. 32 Final design

showed low costs for molds and good lightweight performance, but a the connection to other components and the reduced design space for the batteries were negative aspects. A molded concept providing all connection possibilities and complex structures was the second concept. Here, the stiffness was found to be very high, but the costs for molds and the resulting high investment costs reduced the possibilities in particular for medium series production rates that are expected to be the starting point.

With the experiences made with both concepts a merged concept was derived providing the lightweight and low investment cost design of the pultruded profiles in combination with the simplified connection and the high design freedom of the molded concept. The merged concept uses profiles for the rocker panels and the major beams while the corner elements are molded with a SMC/BMC. Here, also recycled fibers could be applied to lower the carbon footprint. Crash simulations showed good results, thus the concept is seen as ready for detailed developments and first prototypes.

4.2   Seating Structure

Besides the platform the design of a seating structure was presented. The seating structure is designed with the main targets reusability and remanufacturability. The design was optimized regarding highest possible strength with sufficient small deflections. It was shown that an easy separation process is feasible. By these measures the scenario of reusing the parts at least three times becomes realistic. The functionalities described above are implemented in the presented design, so that these requirements are fulfilled as well.

In addition to the design shown, a detailed FEM analysis was carried out to optimize the profiles regarding wall thickness and manufacturability. It could be shown which materials for which components are necessary to bear the loads.

References

1. Heycar Magazin, Was ist die typische Lebensdauer eines Autos (2020). [online], Available: www.hey.car/magazine/was-ist-die-durchschnittliche-lebensdauer-eines-autos
2. Statista.com, Typische Lebensdauer von Autos in Deutschland nach Automarken (2014), [Online], Available: http://de.statista.com/statistik/daten/studie/316498/umfrage/lebensdauer-von-autos-deutschland/
3. Szabo, R., Szabo, L.: The cars need lightweight composites to reduce the CO_2 emissions ToT improve the fuel efficiency. Express (2018)
4. Al-Lami, A., Hilmer, P., Sinapius, M.: Eco-efficiency assessment of manufacturing carbon fiber reinforced polymers (CFRP) in aerospace industry. Aerosp. Sci. Technol. **79** (2018)
5. Bundesverband Carsharing, Fact sheet car sharing in Germany (2022)
6. Horvath & Partners, Faktencheck E-Mobilität: Status quo der E-Mobilität in Deutschland—Update 2020 (2020)
7. Steinwall, J., Viippola, P.: Concept Development of a Lightweight Driver's Seat Structure and Adjustment System, Combining Optimization and Modern Product Development—Methods to Achieve a Lightweight Design. Master's thesis in product development, Chalmers University of Technology, Gothenburg, Sweden (2014)

Product Re-Design Guidelines

Severin Filek, Harald Gründl, Viktoria Lea Heinrich, Marco Kellhammer, and Tamara König

Abstract 'Product re-design guidelines' outline the rethinking of consumer and capital goods design in a context of a transition from a current linear economy towards a circular-oriented economy. These guidelines deliver generalized quality standards addressing various target groups who are to be engaged in order to achieve sustainable results. The definition of crucial counterparts of a system or beneficiaries, respectively, changes the guidelines' discuss factors that are fundamental for the integration of sustainability aspects into existing and/or novel business models. With regard to the circular progression in future, awareness building for improvements in disassembling and remanufacturing will remain essential among designers and producers.

Keywords Re-design · Design · Guidelines · Quality standards · Awareness building

1 Introduction

As part of 'Closing the loop', an EU-action plan for the circular economy, the European commission stated in 2015:

> The circular economy will boost the EU's competitiveness by protecting businesses against scarcity of resources and volatile prices, helping to create new business opportunities and innovative, more efficient ways of producing and consuming. It will create local jobs at all skills levels and opportunities for social integration and cohesion. At the same time, it will save energy and help avoid the irreversible damages caused by using up resources at a rate that exceeds the Earth's capacity to renew them in terms of climate and biodiversity, air, soil and water pollution.

S. Filek · T. König (✉)
Designaustria—Knowledge Centre and Interest Organization, Designforum Wien, Vienna, Austria
e-mail: tamara.koenig@designaustria.at

H. Gründl · V. L. Heinrich · M. Kellhammer
IDRV—Institute of Design Research Vienna, Vienna, Austria

© The Author(s) 2022
M. Colledani and S. Turri (eds.), *Systemic Circular Economy Solutions for Fiber Reinforced Composites*, Digital Innovations in Architecture, Engineering and Construction, https://doi.org/10.1007/978-3-031-22352-5_13

The transition to a circular economy is a system change, which calls for a rethinking of linear processes in favor of recyclable products and product service systems. In order to ensure a sustainable development on economic, social, and ecological levels, the United Nations have agreed to reach a total of 17 Sustainable Development Goals (SDGs) by 2030. The responsibility to implement these goals—including well-being, quality education, gender equality and others—resides equally with both designers and manufacturers. Designers mediate between all actors involved in the development process and can thereby create transparency. For this (information) exchange to be successful generalized quality standards for circular design aim to provide stakeholders, in particular designers and manufacturers, with a circular re-design manual including comprehensible standards and tips helping to realize the sustainability goals.

The Ellen MacArthur Foundation, launched in 2010 as one of the leading organizations to accelerate changes in our current linear system states in order to 'make a transition to a more circular economy, and in a broader perspective society, possible, circular economy was defined as transformation from products turning into waste to products being reused to their maximum value before being returned to the biosphere in a productive way' [1]. This concept offers an alternative to the linear 'take-make dispose/waste' model of consumption and production: products and materials are designed circular and are being kept in a closed loop by recycling, recovering and regeneration at the end of each service life.

Within the circular design eco-design strategies are combined with the aim of keeping as many resources as possible in closed loops and transferring them quickly and efficiently from a state of uselessness to a state of use.

As Hora/Tischner [2] state, 'Even though eco-design aims above all at the best possible design of ecological and economic benefits, the social dimension should also be taken into account as far as possible in the product development process.' Knowledge about sustainability standards becomes a decisive innovation tool. Knowledge about material use and efficiency, energy output and input and product durability is combined with design strategies, such as reparability, modularity and other strategies of ecological responsible design. Circular design not only includes transformation in ways of ecological change but social development as well. Social Design Lab [3] solidifies the term 'circular societies' by incorporating societal changes into the concept of circularity. Thus, transition in terms of circularity comprises both ecological and societal changes.

2 Generalized Product Re-Design Guidelines

Societal changes are realized in a cooperative manner, which includes various counterparts, such as designers and creatives, manufacturers and producers but also consumers, users and policymakers.

The following addresses the benefits of the use of these circular product re-design guidelines, describes their role, responsibilities and potential for a transition in reference to circular design.

2.1 Counterparts of Circular Design

Designers and Creatives. Designers and those working in creative fields acquire orientation with regard to the realization of recyclable products and product service systems, receive support in the search for more detailed information, and take first steps toward implementing the standards. All design sectors are addressed in the circular design manual: Functioning material loops require the involvement of graphic and interactional design to organize information and interactions along the complete value chain. Sharing concepts in product (and textile) design offer new usage possibilities and forms of material use and procurement.

Design will no longer be defined and used as marketing strategy. In Circular Design the principle of beta versions will define the understanding of design. Based on the concept of beta versions in software development, partly finished or uncompleted versions of products will be released. Developed in its basic principles these versions are open to changes and adaptions by its users. Hence, the user turns into a 'prosumer' (producers + consumers) by contributing to the development of a product or service. The advantages of beta versions are wide-ranging feedbacks with the possibility of further adaptions or developments, thus resulting in a more sustainable service or product. Furthermore, products will become more resilient; have longer lifespans and the possibility to carry out updates. Modularity as well as decentralized production and design development will be part of the design work in a circular economy. Instead of disregarding the extension of the product life cycle, reuse is planned and all resources ideally remain in a closed loop.

Manufacturers and Producers. Producing industry specialists receive sets of criteria and inspiration related to potential circular economic models, which are to be addressed and fulfilled in collaboration with designers. The manual brings both parties (designers and manufacturers) to the same level of negotiation and aims to help to understand each point of view.

Products and the possession of products are no longer the source of added value. Instead, its functionality and use define the source of added value. The focus shifts from products without service to product service systems that consider users' needs and include the principles of a circular model: repair, reuse, and remanufacturing. Globalized, cost effective production is no longer desirable. Alternatively, the 'global/local" principle will be complemented into manufacturing processes: knowledge generation takes place on a globalized level while production and distribution is realized locally.

Waste Managers. As a general supplier of waste removal and recycling services, as well as dismantlers and logistical operators, waste managers provide evaluation of optimal logistic scenarios for specific supply/value chains by taking into account different phases of waste management. Since waste management as a whole is developing towards resource management, it is becoming more important to recover, reuse and recycle as much raw material as possible from waste products.

Consumers and Users. Consumers and users are no longer passive members of consumption processes. As 'prosumers' they actively use and test products and turn into experts while taking part in development stages. Instead of owning products, prosumers use products. Products are no longer globally produced and cheap but locally manufactured and available. Local resources and skills are being used. They are characterized by durability, reparability and are service friendly.

Policymakers. Manufacturing processes will no longer be determined or dominated by cost efficiency. Henceforth, the focus lies on quality in material as well as in human resources. Glocal structures in terms of insourcing replace global structures where production and qualification are being outsourced. Furthermore, resources are considered as common goods. The responsibility for closed material and production loops is a rule and not the exception. The focus shifts to promoting individuals and not solely the economy.

2.2 Guidance to Circular Design Processes

What questions need to be asked before starting circular design projects?

The following notions and principles are to be considered for one's own project work in the spirit of ecological sustainability. They serve as guidelines for the circular design process and as first steps into the direction of a circular economy.

Purpose: Is the design of a service or system possible? Can the design of a product be avoided? How useful is the purchase of a product?

Service Units: How can one successfully design service and utilization units? Which services does the product provide?

Attitude (personal, corporate): How can attitude be refined and awareness be raised toward ecologically sustainable concepts and systems, and how can this be shared with other actors?

Information and Research: How and where do I find information and figures related to the environmental aspects of my project? Are there existing environmental declarations on comparable products and services? How do I inform users and educate them in the resource-efficient use of a product or service?

Optimizing Use: Which requirements apply regarding service life, intensity of use, life cycle, and disposal scenarios? Do the rules and standards correspond with the actual use? Where do I find information about the existing rules and standards?

Resilience: How do products and services become resilient and independent from external factors like short-lived fashions and trends? Which tools can help to achieve circular design products/projects?

2.3 Mapping the Circular Design Process

Sketching out the product life cycle

Where does it come from? Where does it go? Which resources does your product/service consume? Which requirements apply for this product/service category?

Disassemble products to understand functions and processes.

Working from the future to the present

Sketch out the ideal usage scenario and break it down into real possibilities.

Prioritizing possibilities

Create new contexts instead of products. Select appropriate design strategies.

Identifying the impact

Illustrate the impacts of actions. How do the materials affect the environment (soil, air, water)? Which influences does your design have on society?

Recognizing hurdles and barriers

Clarify the limitations of your actions: What effects can I provoke? Where do I need help?

Planning next steps

Who do I want to collaborate with in the future?

2.4 The Role of Life Cycle Assessment (LCA)

Whether and to what extent the abovementioned strategies have an effect on the environmental balance of a product or service can only be determined with the help of effect categories. The figures are derived from the results of an LCA. Different tools can be used for a general estimate and the comparison of environmental impacts. Checklists functions as reminders of which steps need to be completed and which requirements have to be made. Material categories may give an overview of materials and other resources. They provide a simple first analysis for designers becoming acquainted with the concept of circular design. Material databases (analogue or digital) provide raw numeric values.

2.5   Circular Design Criteria

Knowledge about sustainability standards and circular design criteria become a decisive innovation tool in future design projects. As described in this report various ecological aspects and requirements have to be considered when designing circular services and products. The following seven circular design criteria have been selected in order to give an overview of requirements that have to be met by designers and manufactures. They do not claim to completeness but are defined as first steps into a circular design process and further a circular economy. For further criteria and possibilities of establishing the standards in the work of designers, manufacturers and producers, we recommend the different label types ISO 14020 provided by the International Organization for Standardization.

Type I (ISO 14024) Environmental Labelling. These certifications are made in the framework of an independent review by a third party. Products with one of these labels have better environmental performance than comparable products (e.g.: EU Ecolabel, Austrian Ecolabel, Blue Angel).

Type II (ISO 14021) Self-Declared Environmental Claims. This certification is made by the manufacturers themselves. They stipulate criteria for a comparison. Products that bear this label exhibit an improvement in terms of environmental aspects, for instance, in comparison to previous models (e.g.: Recycling code).

Type III (ISO/TR 14,025) Environmental Declarations. This certification is based on the results of an eco-balance and comprises environmental data and information about the complete life cycle of the product. The review of various criteria enables a differentiated product comparison (e.g.: EU Energy Label, EPD—Environmental Product Declaration).

2.6   Circularity Potential—Thinking in Loops

Linear economic models—from the cradle to the dump—are being replaced by planned closed loops as demonstrated in Circular Economy System Diagram (Fig. 1) by the Ellen MacArthur Foundation. Following this concept, there is no longer waste produced. Natural loops are created for natural materials, for example, compostable packaging. Technical products and components are routed into technical loops. External impacts, such as emissions, are taken into consideration throughout the complete life cycle as stated by inter alia Ellen MacArthur Foundation and IDEO.

Effective: Effectiveness is defined as an increased resource productivity. More benefits with less resources and materials needed.

Consistent: Instead of using new resources, materials and services of existing ecosystems (technical as well as natural) should be used and sustained.

Sufficient: Sufficiency describes the absolute reduction of resource usage through changes in behavior and decreased demand. Planned obsolescence is not an option having a sufficient and working technical or natural loop in place. Generally speaking,

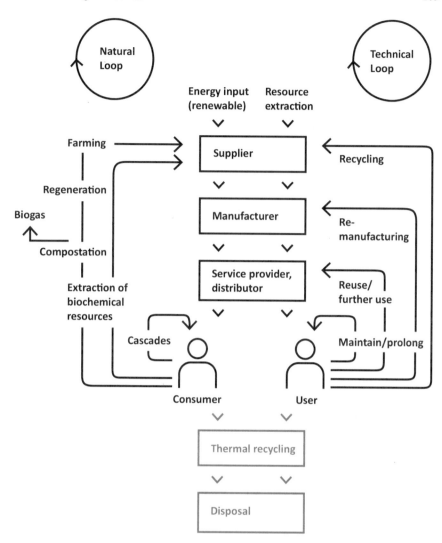

Fig. 1 Ellen MacArthur Foundation, 2015: circular economy system diagram

circulation in a loop without additional external energy or processing aid is impossible. It is important to identify these flows of materials. A distinction is made between emissions in the air, water, and soil and effects on biodiversity.

As first steps or starting points we recommend deciding for a technical or natural loop when designing products and services. In which loop will your product or service fit in? A distinction has to be made between materials in terms of natural resources and those of technical characteristics. When designing a product, responsibility for the afterlife and return of the product has to be taken.

2.7 Determination of the Optimal Lifespan

Designers define the service life of products and services. The technical lifespan of all components accounts for the real service life and intensity of use. A long lifespan in terms of durability guarantees the longest possible usage that corresponds with the usage context. Emotional product bonding can be named as an important factor for durability. Besides the technical service life, a high appreciation and bond with the product on the user's side should be attained in order to extend the real service life—regardless of trends and fashions. A shorter lifespan of a product is only aspired if it is ecologically justifiable and particularly circular. It is advisable to create incentives for longer use of products and materials. Generally speaking, the assumption holds that the cumulative expenditure of energy and resources is amortized with a longer lifespan. In rare cases do other factors achieve a better environmental impact. For example, one forgoes reparability when using materials made of carbon fiber but attains a more efficient product. As first step we recommend researching the average service life of the product that is to be designed. Second, an appropriate product lifespan has to be designed. When possible, designing a low-maintenance product is a recommended option.

Reparability and Modularity—Universal comprehensibility. Preferably use standardized connections and components. The product should be comprehensible for as many people as possible. At least experts should be able to perform repairs, extension, and disassembly.

The following criteria in reference to the Directive 2009/25/EC of the European Commission should be kept in mind:

Reparability: If necessary, the product can be opened without destroying it. The individual components are accessible and can be replaced.

Modularity: Components can be replaced, extended, or updated.

Disassembly: The use of different materials and connections is reduced; nondetachable connections are a taboo.

Open interfaces allow adaptation to technical innovations. Replacement parts should be available not only for the manufacturer but also for the user of the product. The manufacturer assumes a broader manufacturing responsibility, guarantees and warranties should be adjusted to the actual lifespan. The possibilities to repair and maintain are an incentive for a longer product use cycle. Attention should be paid to usability and product safety standards and regulations.

As first steps it is recommended to provide a usage and repair manual for the users. When the possibility to include reparability as criteria in the design process is given, all components and wear parts should be design exchangeable.

Energy Use—Increase efficiency per energy unit. Energy efficiency is the relationship between the energy output and input. Effective use of energy sources increases economy and reduces its environmental consumption. Energy from renewable energy sources is preferred. A distinction is made between thermal and electric energy. The EU energy label provides information about the energy consumed during usage; the cumulative energy expenditure (CEE) describes it over the entire life cycle.

A longer lifespan has a positive effect on the CEE, as does the use of recycled materials. Intelligent and customizable control systems facilitate a more efficient use of energy. As first steps we recommend using renewable energy resources exclusively. It is advisable to reduce the energy consumption per service unit and to create incentives to reduce the energy consumption in general. One option is to enable feedback about the real energy consumption in order to include the user into the usage circle.

Material Use—Increase efficiency per material unit. Material efficiency refers to the relationship between the resulting product and the employed raw materials. The material use is optimized over the complete life cycle in terms of its impacts on health and the environment. Waste materials are avoided and fed back into the loop. The use of new, non-renewable resources is reduced. When selecting materials, attention should be paid to the origin, environmental impacts, and production conditions. Renewable resources are preferred. Materials should be able to be separated according to type and be marked as such. Reparability should be guaranteed [4]. Avoidance of problematic substances is preferred: Substances and materials should not have an adverse effect on health or the environment. Hazardous substances and materials should be reduced and marked as such in all cases. The first step is the replacement of resources with recyclates. In any case, it is reasonable to mark the materials used. This provides full transparency for manufactures as well as users and designers.

Design of Services—Offer product-service systems. Service units and business models should be conceived and organized, so that services and products can be made available to as many people as possible. In this way, on the one hand, resource efficiency increases (see: 4. Energy Use and 5. Material Use), and on the other, it facilitates products of a higher quality with improved serviceability. In this context the advantages of the digitalization (e.g. online networks, artificial intelligence, block chain technologies) can be applied expediently and strategically. Manufacturers and service providers work together closely in order to provide a closed natural or technical loop. Another option is to intensity the use of products. Products for sharing concepts are used considerably more intensively over the same time frame. When designing sharing concepts incentives for usage than property have to be thought of. Even more consistent would be the replacement of the product with a service, such as car sharing. First step here is comparing the actual use of the product in relation to the non-use.

Society and Circular Design—Make progress in society. Design decisions have a significant influence on who uses the product and how. In circular design the principles of reuse and social interaction are intrinsic to all product life cycles. Not only ecological but also socially sustainable and participative design strategies are important. Production processes are to be developed that take social values into account, create possibilities for further education, and facilitate social exchange.

3 Conclusions and Future Perspectives

Key target groups of generalized product design guidelines are practitioners experienced with circular economical systems and inexperienced people, as only a symbiosis of material providers, manufacturers, designers and consumers, respectively 'recyclers' can guarantee a successful transition to a circular economy.

All challenges and solutions experienced during production processes were integrated in these design guidelines, thus providing focal target groups of designers and manufacturers/investors with a practical guidance on how to potentially improve the product applications.

In the course of further steps business strategies will be determined, explored and developed with the aim to increase interest from European companies for an accelerated transition of their value chains towards a circular economical orientation. The inclusion of design guidelines for product re-use may provide (first) comprehensive information to stakeholders on the essential changes from linear to circular design in production.

In the long term, incessant development of product re-design guidelines by detailing information for industrial value chains needs to be established. Public dissemination and circulation of guidelines among pivotal stakeholders will support the crucial action of raising attention towards circular economically oriented product/industrial design.

References

1. Ellen MacArthur Foundation: Towards the circular economy. Opportunities for the consumer goods sector (2013)
2. Hora, M., Tischner, U.: Mit Ecodesign zu einer ressourcenschonenden Wirtschaft, p. 11. Hessen Trade & Invest GmbH, Wiesbaden (2015)
3. Social Design Lab, Hans Sauer Stiftung (eds.): Wege zu einer circular society. Potenziale des Social Design für gesellschaftliche Transformation. Munich (2020). https://www.hanssauer stiftung.de/inhalt/uploads/200420_HSS_Paper_CircularSociety_online.pdf. Last Accessed 29 June 2022
4. Club of Rome. https://clubofrome.org. Last Accessed 8 May 2020; Institute of Sustainable Economy. https://www.ioew.de/en/. Last Accessed 8 May 2020

Cloud-Based Platform for the Circular Value-Chain

Dena Arabsolgar and **Andrea Musumeci**

Abstract The FiberEUse IT platform is a tool that enables the exchange of information among stakeholders working into and crossing the glass and carbon fibers value chains. The list of stakeholders had been identified and is listed into the document. The list of users of the platform is also listed into the document, although it is not fixed. During the collection and analysis of the requirements, the circularity of the approach had been in the core focus. This drove to the identification of solutions able to cover all the different and innovative aspects of a Circular Economy Systemic Innovation. New products, new processes, new materials, new connections, quick and unexpected introductions of innovations, methodological approach, end of waste concept, availability of objects for dismantle, not-standardized parameters, are only some of the aspects arisen analysing deeply the requirements. They have all been approached in the platform which is by definition circular, dynamic, expandable and polymorphic. In the FiberEUse project, the validation activities of the IT platform had adopted an agile approach involving demonstrators during the preliminary phases of analysis, and mock-up creation. An internal technical validation had been completed. The methodology had been adopted and main results are reported in Sect. 5.

Keywords Circular IT platform · Circular entity · Information management · Matchmake · Enable value chains · Search

1 Information Exchange Needs in Circular Economy

The IT platform will support the companies in scouting new potential market applications which, in the new circular economy system is an aspect which needs to be analyzed and managed many times from scratch.

Thanks to FiberEUse partners there was the chance to analyze the stakeholders growing amount of new needs, to group them and to address them, starting though

D. Arabsolgar (✉) · A. Musumeci
HOLONIX Srl, Corso Italia 8, 20821 Meda, MB, Italy
e-mail: dena.arabsolgar@holonix.it

© The Author(s) 2022
M. Colledani and S. Turri (eds.), *Systemic Circular Economy Solutions for Fiber Reinforced Composites*, Digital Innovations in Architecture, Engineering and Construction, https://doi.org/10.1007/978-3-031-22352-5_14

from supporting the creation of connections among companies of different sectors and different geographical areas.

The IT solution developed in the project supports different value chain stakeholders during the search and acquisition of product lifecycle information about companies and objects. Those objects had been defined "circular entities", and can be products, materials, processes and treatments, wastes, or more.

The information available for each circular entity is the one which the company itself wants to share and can propose: description, informative details, technical details, processes adopted to create it, consultancy services used, etc.

1.1 Linear Economy Information Exchange Limits

To support companies in the creation of the new value chain in a circular economy environment, this solution overtakes the limits of the management of not circular value chains and provides the users with the possibility to work on each and every circular entity impacting its value chain and visualize it at a glance.

In the management of a linear system there are some aspects which needs to be considered. Among them, the fact that linear economy manages objects in a single life cycle, from raw material still end of life, which is generally expected to be a waste. Both aspects of "single life cycle" and "waste" are barriers for circular economy, where there should be no waste at all, and were lifecycles should be many and interconnected among each other.

Another relevant limit of the linear economy, is the fact that the lifecycle management today is generally thought into one single sector. This means that cross-sectorial interconnections can happen, but are not considered as a native and needed input to the common business of the company.

1.2 Circular Economy Information Exchange Objective

Companies need a virtual place to share information among those who are approaching the new circular economy system. It has to consider the involved companies, to describe them, to share the details of the products, to catch the history of their cycles, displaying the key steps in the transformation journey and taking into account the involved ecosystem parties.

The solution proposed is based on a list of functions that allow the users to manage circular entities along the steps of their circular transformation and to handle them in all the interactions with different processes and transactions with other companies.

Beside the functionalities provided in the standard menu, it is worth highlighting the introduction of three pivotal features, namely:

- Circular Entities

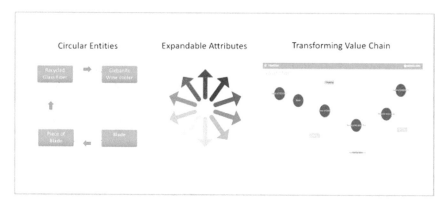

Fig. 1 FiberEUse IT solution pivotal features

- Expandable Attributes
- Value Chain Ecosystem chart.

Those are values that provide innovative characteristics to an object, making the platform substantially different from a common product lifecycle management platform.

Those features required an "ad hoc" architectural design which is able not only to represent the key differentiating functionalities but also to the IT complete support (Fig. 1).

1.3 Positioning of the Solution in the FiberEUse Project

The FiberEUse IT platform will support the project to answer to the project objective:

> The development of an innovation strategy for mobilization and networking of stakeholders from all the sectors related to composites.

The objective had been reached through the development of a user-friendly, cloud-based IT platform for engagement and collection of a network of stakeholders across the whole value chain, from composite manufacturers to re-demanufacturers, users associations, policy makers, investors, and so on.

The solution will allow companies to get in touch, creating new connections, enabling new value chains, and creating new market opportunities. The deals which are closed to effective exchanges are tracked and shown into a chart representing the entire eco-system generated.

The Value Chain Ecosystem chart represents graphically the connections which are being generated through the usage of the platform. The growing network will be shown, becoming useful for all the stakeholders, including companies, policy makers and authorities.

The solution delivered at the end of the project targets FiberEUse objective, and it had been developed in order to be scalable, extensible and expandable. It could therefore reach a wide number of new potential users working with glass and carbon fibers but also working with different circular entities in different domains and sectors.

The FiberEUse IT platform supports the identification and share of the physical availability of circular entities (products, waste, materials,…) to be re-used, recycled or remanufactured, and also to explicit their demand.

External services are searchable and available through the IT platform, as for example consultancy services. They include specific support on specific subjects or choices: the co-design module described in Chap. 11, the wind blade disassembly decision support system described in Chap. 3, a Life Cycle Assessment and Life Cycle Costing tools developed during the project, and eventual additional services as proposed by partners during the project activities.

1.4 FiberEUse Requirements Collection Methodology

The requirements collection and the software development could have used standardized methodologies and proceed across standardized steps designed for consistency [1].

But, being FiberEUse an innovative project, and being innovative also its context of application, it had been a must to choose a requirements collection process equipped with built-in flexibility and oriented to the new group of stakeholders and stakeholders' needs.

According to Mihai Liviu Despa, "no methodology will fit perfectly the profile of a specific project. Then the best matching methodology should be used or in the case of experienced project teams and project managers a combination of methodologies could be introduced. In the case of innovative software development projects a new methodology is required" [2].

In FiberEUse, an additional point of complexity had been given by the nature of the innovative domain of Circular Economy, which is mainly unknown and provides inputs which are not common in IT developments.

All those details needed to be brought out and checked carefully.

FiberEUse therefore adopted existing methodologies, but went further proposing a flexible, agile and user centred application example of requirements collection—development and testing of innovative solutions.

The main objective of the activity of Requirements Collection was the identification of all functional and not-functional requirements. Being an innovative context, it was a must to reach the objective through a cyclic validation of the requirements collected through different stages of validation and approval by the end users and by the stakeholders.

The approach adopted in all the validation and approval steps had been user-centred, creating at each stage workshops, focus groups, and active participation to the activities.

Stage by stage, at each physical project workshop, had been collected or re-checked all the functional and not-functional requirements.

Functional requirements collected had included the analysis of:

- Needed platform capabilities
- Minimum expected usability
- List of expected features
- Selection of the Minimum Features Proposition
- List of expected Operations.

Not-functional requirements had been proposed at the beginning of the project and then adjusted according to the final functional requirements list. They had been then defined by Holonix, according to the expected result and TRL at the end of the project, and checked with project partners:

- Performance
- Stability
- Security
- Technical specifications, including technological choices.

1.5 FiberEUse Stakeholders Needs List

Being an innovative context, it had been a must to reach the definition of the stake-holders needs through a cyclic validation of the requirements collected through different stages of validation and approval by the end users and by the stakeholders. The approach adopted in all the validation and approval steps had been user-centred, creating workshops, focus groups, and active participation to the activities, with the intention to integrate User Experience Methods into an Agile Software Development [3, 4]. During FiberEUse project many workshops had been conducted. At each of them the requirements collected and the structure proposed had been tested and veri-fied with stakeholders, until a final validation and the development of the platform. The stakeholders needs had been grouped as follows:

1. Waste provider and ByProduct provider:

   - Need to dispose the end of life of the products:
     - Need to reduce costs
     - Un-allowed landfilling of materials
     - Desire to end product life in a sustainable way

   - No knowledge about:
     - Who to offer or send products
     - If objects should be preprocessed
     - How objects should be treated

2. Producer, manufacturer:

- Interest in producing new products with different materials having specific characteristics:

  – Less expensive materials
  – Different material with same property values
  – Not traditional materials
  – Right availability of materials according to needs

- No knowledge about:

  – Possibility to compare traditional and new materials
  – How to purchase
  – Who can provide the material

3. Process provider who have the possibility to process end of life products:

   - Capacity to create new materials

     – Innovative processes
     – Innovative materials
     – Same processes or materials used in an innovative way

   - No knowledge about:

     – Who can provide the waste, or material to be processed
     – Who can be interested in the new material obtained with the new process
     – Who can be interested in the new technology to be used as third parties

4. Service and logistic operators:

   - Capacity to pre-process end of life products
   - Capacity to organize on EU wide level
   - Knowledge about the materials, processes and sector
   - No knowledge about:

     – Who is interested in the new services.

2   A Circular Economy Virtual Platform

The two main elements which emerged from the workshops with stakeholders, were the need to manage circularity of products keeping up their history, and the interest in collaborating and sharing information among different stakeholders.

2.1 Circular, Collaborative and Open Information Management

Through the Fig. 2 is shown how the management of the information becomes circular: objects are no more managed from the concept of a new product to the end of life, but they are managed through cycles from object to object. From the functional point of view, this means that the solution have to create knowledge and match needs of stakeholders to let them know if, how, what, when, they can use, re-use, re-cycle or re-manufacture again. This can be obtained through functions which allow people know, and be aware, search, enter in touch with others, comment, match needs, suggest themselves or others, visualize, and keep history tracked. From the IT point of view, this is reflected in the use of the recursion [5] which adopts the Matrioska approach [6].

The FiberEUse IT solution had been developed into two different period of time, according to the project needs:

- a first solution, called Idea Manager, to support stakeholders in the co-creation of new products in collaboration with designers,
- and the FibeEUse IT platform, to give users a virtual place to share information among those who are approaching the new circular economy system.

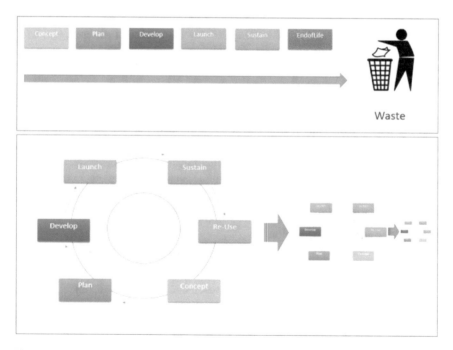

Fig. 2 Linear versus circular information management circle

2.2 Idea Manager Solution

The Idea Manager is a web-based software-tool designed to provide and support activities along the idea generation process, which is part of the open innovation approach. It is not company-centric, but it is thought to be multi-actor, being able to involve employees, designers, customers, providers, and the interested community in general.

Figure 3 shows the part of the tool referred to the functionality of searching and processing different ideas coming from designers (up) (interesting, public, private, under evaluation, etc.) and collaborating on three designer's ideas (Modul X^3, Fiberbench, Bicycle Rack), giving feedbacks, votes, and answering to surveys (bottom). For more details on Idea Manager please refer to Chap. 11.

2.3 FiberEUse IT Platform Solution

The FiberEUse IT platform is a Multi-Sided platform [7]. This means it is a solution which connects two or more interdependent user groups, by playing an intermediary or a matchmaking role. This kind of solution plays an important role throughout the economy, as they minimize transactions costs between market sides. In FiberEUse one of the innovative aspects is the fact that the platform is intended to be also cross-market and cross-sector, involving directly companies of different and wide extraction. In the case of FiberEUse, the solution is based on the type of material, and this is the element able to connect different users from different sectors in the same place.

It has to consider the involved companies, to describe them, to share the details of the products, to catch the history of their cycles, displaying the key steps in the transformation journey and taking into account the involved ecosystem parties. The solution proposed is based on a list of functions that allow the users to manage circular entities along the steps of their circular transformation and to handle them in all the interactions with different processes and transactions with other companies.

Beside the functionalities provided in the standard menu, it is important to highlight the introduction of three pivotal features, namely:

- Circular Entities and recursion
- Expandable Attributes and Polymorphism
- Value Chain Eco-System chart.

Those are values that provide innovative characteristics to an object, making the platform substantially different from a common product lifecycle management platform. Those features required an "ad hoc" architectural design which is able not only to represent the key differentiating functionalities but also to the IT complete support.

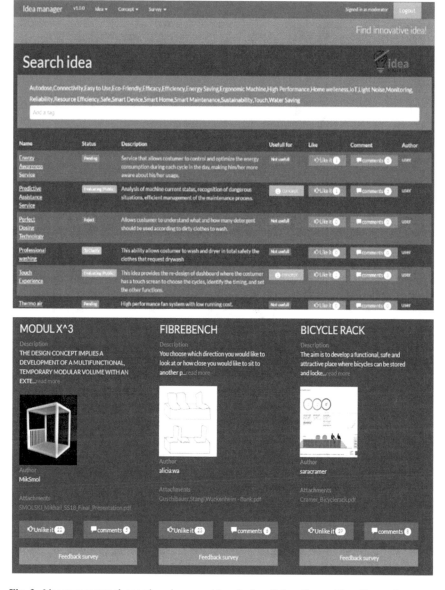

Fig. 3 Idea manager tool: search and process ideas (up), collaborating on ideas (bottom)

2.3.1 Circular Entities

In modelling an application or an information system, "Entities" are defined as representation of categories of objects, or set of objects with similar characteristics.

Fig. 4 FiberEUse circular entity: theory (left) and implementation (right)

In traditional applications for product lifecycle management, i.e. applications that follow the lifecycle of a product from the prototype to the end of life, the entity is the product itself, that doesn't change its characteristics during its lifecycle.

In our case, in Circular Economy, we are also managing objects which are generally unknown, because usually they are new. Those objects can be waste (if thought in linear economy), materials, pre-processed material, new processes and machinery, new final products, prototypes, etc. According to the circular economy definition, all those kind of objects are part of a cycle and cannot be classified as waste, or material, or final product anymore [8]. They are all Circular Entities.

In our IT platform the concept of circular entity replaces and goes beyond the concept of a static product entity. It is designed as circular entity, i.e. an entity that can be whatever, even though it has not been defined before from anyone else, and which can be described through its history and through its parameters.

The history of the circular entity can be reported through its internal resources, coming from the same company, or external resources, coming from other companies. In case of external resources they require a previous negotiation and a contract with the parties providing the resources, enabling an advancement in the value chain and a more sustainable path. From an IT standpoint such kind of circular entities are designed to satisfy specific architectural requirements in order to be dealt with the needs of changing parameters. Information can be public or private, according to the company.

In Fig. 4 on the left is represented the theory behind the recursion of the circular entity, while in Fig. 4 on the right is shown the correspondent functionality in the FiberEUse IT platform.

2.3.2 Polymorphism of Data and Expandable Attributes

Circular Entities and free attributes which describes the circular entity in the way the company prefers, are together a fruitful and complete combination to describe every kind of object. This reflects the concept of Polymorphic data [9].

Polymorphism of data defines data as extensible, able to serve different product histories without the ambition to create a common and classic entity relationship

Fig. 5 Polymorphic data example

scheme but thinking by high level object oriented design. Polymorphism usually is the way by which an object can be described through its different interfaces of use. But also, taking in consideration attributes sets (subsets of all attributes of the same object class), we can define polymorphism of data saying that an entity has many point of view considering the actor which observes it. For example, for a producer a shirt has different attributes than the same shirt has for a logistic or waste collector (Fig. 5). In FiberEUse IT Platform this is given by the possibility to insert free additional attributes to the Circular Entities, described as the company prefers through an attribute name and one or more values.

While managing products in a traditional lifecycle with static entities, database are designed for static attributes, i.e. attributes that contains, store and track a set of historic data coming from events related to a product that doesn't change during its life, are, roughly speaking, the fields of a table.

The FiberEUse IT platform is designed to give the users full freedom to provide attributes and specific characteristics of a product, process or service according to how they better want to use the platform itself and according to their business model. The underlying design objective, in fact, is to facilitate the Matchmaking.

Mandatory Attributes are recommended by the FiberEUse platform (Name, category, description). Expandable Attributes can be added freely and in free format, and are characteristics deemed relevant to the producer in the light of a potential matchmake. Such attributes can be, for instance, length, weight, colour, thickness, or results of LCC-LCA study, or any other feature that can be of interest for the next transformation.

2.3.3 Value Chain Eco-System Chart

To achieve and sustain a competitive advantage, and to support that advantage with information technologies, companies should understand every component of their value chains and value systems. A value chain is commonly described as a set of activities that a company operating in a specific industry performs in order to deliver a product (or a service) for the market. A company's value chain forms a part of a larger stream of activities, named "Value system" after Porter [10].

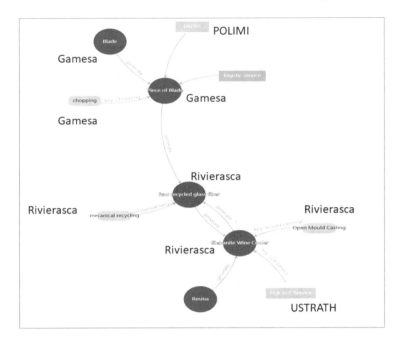

Fig. 6 Value chain eco-system chart extract (circular entities: red, processes: grey, services: light blue)

Traditional non circular value systems, take into account the suppliers and the ecosystems of third parties that provide the inputs needed to the company to understand their value chains (i.e. logistics, distribution,…). All parts of these chains add value to the value system.

To support companies in the understanding of the value chain in a circular economy environment, FiberEUse IT platform overtakes the limits of not circular value chains and provide the users with the possibility to visualize the ecosystem of circular entities, processes and services, and to visualize them at a glance. A small extract of the Value Chain Eco-system chart is in Fig. 6.

3   FiberEUse It Platform

The software solution developed in FiberEUse is a cloud based multi-sided platform where multiple potential users can be enabled to manage and access relevant information.

A Multi-Sided platform is a solution which connects two or more interdependent user groups, by playing an intermediation or a matchmaking role [7]. This kind of solution plays an important role throughout the economy, as they minimize transactions costs between market sides. In FiberEUse one of the innovative aspects is the

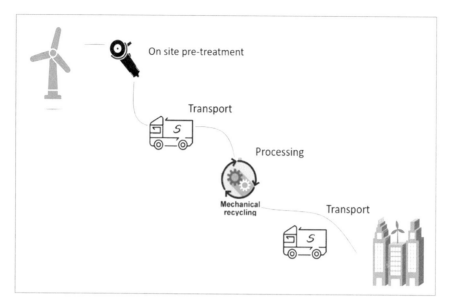

On site pre-treatment

Transport

Processing

Mechanical
recycling

Transport

Fig. 7 Example of connections across a multi-sided platform

fact that the platform is intended to be also cross-market and cross-sector, involving directly companies of different and wide extraction. To work well, the platform must achieve a network effect, involving many users and stakeholders in general (Fig. 7).

Although they have the same architecture, platforms can be different with respect to their objectives and context of use. In the case of FiberEUse, the solution is based on the type of material, and this is the element able to connect in the same place different users from different sectors. The platform is described in following paragraphs through three UML charts: Use Case diagrams, Class Diagram, Deployment Diagram [11].

3.1   Use Case Diagram

The Use Case diagram of the FiberEUse platform released at the M40 is the following (Fig. 8).

The Use Case diagram had been divided into two different sections, one related to the front end and the other one related to the back office. The Backoffice considers stakeholders which can be users of the platform but also receivers of information collected through the platform:

- The Administrator of the platform is the user with major rights.

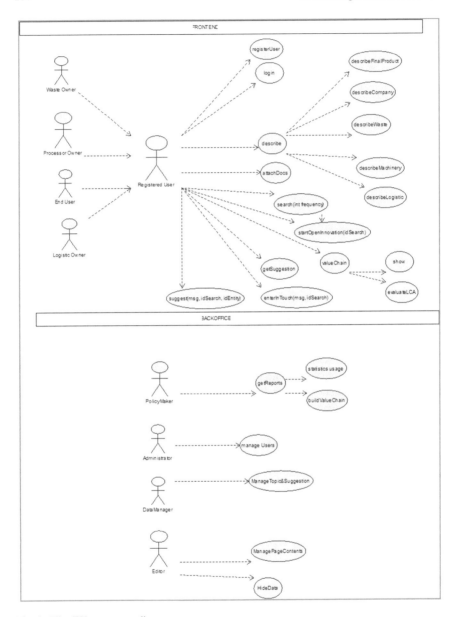

Fig. 8 FiberEUse use case diagram

- The DataManager is a person which will regularly take a look to the content of the information exchanged through the platform and will manage suggestions and topics.
- The PolicyMaker is an external user which is interested into aggregated information useful to take political decisions, as the European Commission and Administrators.
- The Editor is a user, which in the actual release is overlapping with the Administrator, which has the right to manage content and hide not coherent data.

The front end is accessed by all the users able to provide content and interested in using continuously the platform. For example waste providers, process owners, end users, consultants, and so on. All those users are interested in the same group of Use Cases, which includes:

- Registration
- Login
- Description of: products, companies, waste, machineries/processes, logistics
- Attachments of documents
- Search
- Receives of suggestions
- Value Chains visualization
- Value Chain LCA calculation
- Enter in Touch
- Suggestions
- Open Innovation.

As those users are interested in the same group of use cases, they had been gathered into one single user type which is the comprehensive "registered user". This solution improves flexibility and makes the solution more versatile. Addition of new kind of users will be easy.

3.2 Class Diagram and Polymorphic Data

In the complete Class Diagram of Fig. 9 it is possible to see how to describe:

- A Registered User (not completed as commonly known, not innovative)
- A Company (not completed as commonly known, not innovative)
- A Circular Entity (which can be an Object, a Process, a Waste, a Final Product, etc.)
- An Entity (Services, Logistic…)
- Attributes, which can be available for detailed search, to describe parameters specific of the entity
- Values of the attributes, which include the capacity to understand if two values should be considered as a range or punctual values

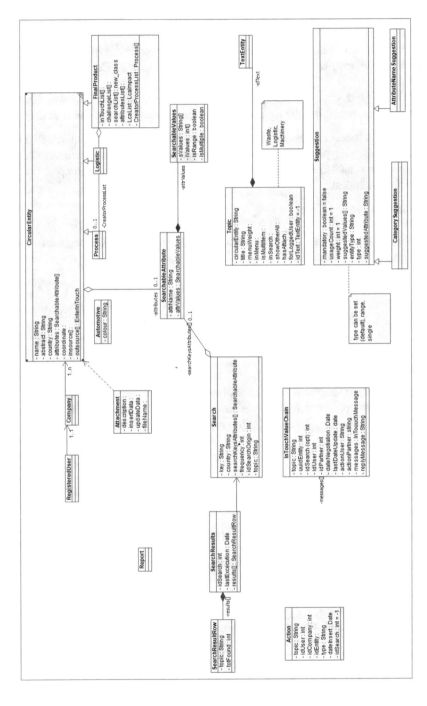

Fig. 9 FiberEUse class diagram

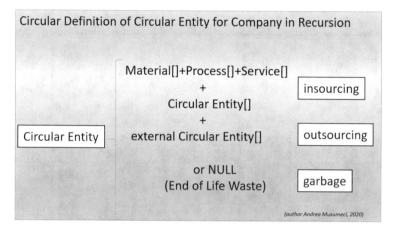

Fig. 10 Circular entity definition through recursion

- A Get in Touch message
- Action, to collect all the actions occurred during the platform use, in order to give a correct order to the search results
- Topics
- A Suggestion, to suggest attributes values
- With same structure of a suggestion, also a category suggestion or attribute name suggestion.

The new concept of Circular Entity had been introduced in the definition of the Classes, according to the IT definition of recursive object. This reflect the fact that a Circular Entity is made of another circular entity plus additional entities/processes/services which allows the existence of modifications, or null (Fig. 10).

As shown in the class diagram, attributes can be as many as required. This means that the Entity Data Model is not predefined but it can be created according to the needs of the user. The attributes describe the parameters which are useful to define the entity. As it is possible to add many attributes, they have been defined Expandable Attributes. The class diagram reflects the concept of Polymorphic data [9, 12].

3.3 Deployment Diagram

In Fig. 11 is reported the Deployment Diagram of the FiberEUse platform, reporting all the components of the platform. It is structured through the three levels view-model-control pattern, which separates the tasks of the components into three roles:

- Model: gives the methods to access data useful for the application

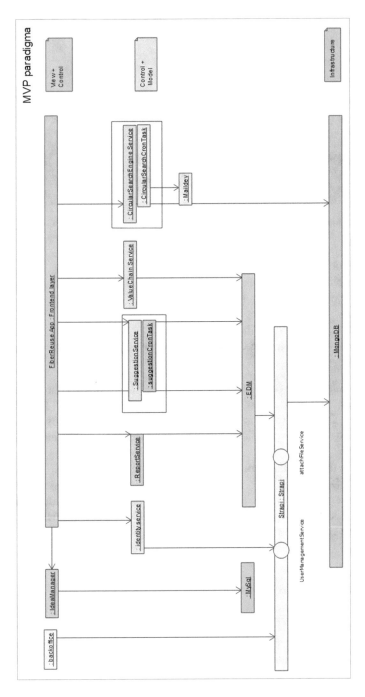

Fig. 11 FiberEUse deployment diagram

- View: visualizes data contained in the model level, and manages the interactions among users and agents
- Control: receives users inputs, through the view level, and actuates them modifying the status of the view and model level.

In the FiberEUse platform, the control level is split among the view and model levels. View level includes:

- The Idea Manager tool, as described in Chap. 11
- The New FiberEUse front end
- The visualization, for administrators, of the Strapi backoffice.

Model level, which functionalities are described in Sect. 4, includes:

- Identity Service
- Report Service
- Suggestion Service
- Value Chain Service
- Circular Search Engine Service
- The Mail Delivery Service.

The Infrastructure is made of:

- MySQL database used by the Idea Manager tool
- The Eco-System Data Manager
- Strapi
- MongoDB used by the FiberEUse platform.

The color code meaning is the following:

- Light blue for new components of the view level
- Orange for new components of the model level
- Yellow for open source components adopted to control the platform
- Pink for Holonix proprietary components adopted in the platform, including:

  - Idea Manager solution
  - EDM, which is the Ecosystem Data Manager able to manage data and coming from ManuSquare project [13].

- Violet for the database, which for FiberEUse platform is a MongoDB and for the Idea Manager was a MySQL database.

3.4   FiberEUse Front End Layer

The FiberEUse frontend is a web application that use VueJs and the Quasar framework for composing the interface. The web app has been designed with flexibility in mind. All the pages and menu are loaded at runtime shaping the GUI. The next image will show the frontend structure (Fig. 12):

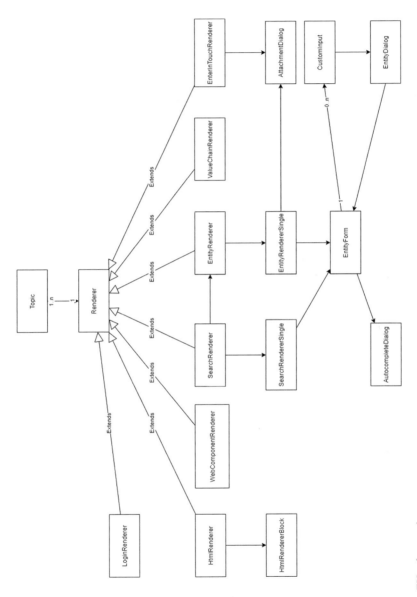

Fig. 12 FiberEUse front end structure

The principal entity is the topic that has a renderer. A renderer is a component that draws graphic elements. Every renderer has a specific implementation and can produce a set of pages and menu voices. The most important renderer are:

1. EntityRenderer

   (a) Responsible of managing entities composed by dynamically rendered form
   (b) Every attribute of the form can be mandatory, shown behind condition or as a free input from user
   (c) Attribute can be of different type:
       (i)   text
       (ii) textarea
       (iii) set of string
       (iv) set of other entities (which can be filtered from configuration ex. only my entities that are plastic)
       (v)   a combination
   (d) Attribute value can be also conditionally available, i.e. it appears a nation if you select a specific material from another attribute
   (e) It has attachments of various kind: image, pdf, csv, xlsx, etc.

2. SearchRenderer

   (a) The search form is able to search on every topic that has a "inSearch" attribute and give the possibility to search custom attribute decided by the user

3. EnterInTouchRenderer

   (a) Create an interconnection through entities

4. ValueChainRenderer

   (a) Is calculated from Entities and EnterInTouch and gives a graph view of the value chain

5. WebComponentRenderer

   (a) Able to display a webcomponent and make the web app a microfrontend integrator able to show different vertical application.

The EntityForm component is used by different topic and it is able to draw a form dynamically.

3.5   FiberEUse Infrastructure

FiberEUse platform infrastructure is made of the components listed in Sect. 3.3: Strapi, Mongo DB, Ecosystem Data Manager.

3.5.1 Strapi

Strapi is an Open Source Content Management Solution 100% Javascript, fully customizable and completely flexible. All details are available at: www.strapi.io.

Strapi had been adopted to allow the configurability, at platform level, of mandatory/optional attributes and topics at circular entity level.

3.5.2 Ecosystem Data Manager

Eco-System Data Manager (EDM) is a middleware which had been chosen to allow the dialogue among FiberEUse client and the repository on MongoDB. It had been chosen a microservice to enable to future versions of FiberEUse to add new levels of persistency of the data. For example, in case of need to add the saving of components on blockchain or of need to transfer content to another platform, through this solution it will be possible only changing configuration parameters, and not changing services at GUI level.

In this way, the GUI does not communicate directly with the database, and in future it will be possible to add new persistency levels. The EDM is a Holonix solution, which had been released in its first version in ManuSquare project [13].

3.5.3 MongoDB

It's the de facto standard in the Open Source, to manage object oriented and polymorphic data persistency. In FiberEUse it had been used in free community edition.

All details are available at: www.mongodb.com.

4   FiberEUse IT Platform: Innovation Developed

The FiberEUse IT platform was designed since the beginning with the exploitation in mind.

Easy for the users and conceived to be managed by administrators with zero programming skills, embeds a great deal of innovative elements.

Hot-Shaping Data Model and GUI, allowing the handling of entities that change during their re-use circles:

- Designed for dynamic entities—not static products
- Circularity of objects—no more waste but new entities in different forms and states, no more end of life concept
- Capability to dynamically describe entities.

Advanced Search algorithms suitable for objects that change and transform during their life:

- Advanced search on parameters that can structure in unexpected or unforeseen ways.

Innovative representation of the Value Chain:

- Build the Value Chain
- Follow the Value Chain as it changes.

Display and track key interactions in the Ecosystem:

- Show the ecosystem created after the deals
- Track and provide data to other platform for insight advanced analysis.

Self-definition of Standards:

- Self-definition of Standards, by majority of usage (the more a certain unit is used the more it will be suggested to become a standard)
- Suggestions support the continuous targeting of the more frequently used terms and definitions.

Ease of use, handling and personalization:

- Almost no training required for users
- Easy to customize
- No developer skills required for Administrators.

5   FiberEUse IT Platform: An Application Example

Among many different potential software tests methodologies, Holonix adopted the following internal testing processes, with the main intent to find software bugs (including errors and defects) and to verify that the solution fits for its use. The correspondence with the functional and not functional requirements had been tested and verified during the agile processes of creation and mock-up validation with partners.

5.1   FiberEUse System Test

At high level, the internal tests were all system testing, intended to verify:

- That the solution meets requirements
- Responds to the given inputs
- Is sufficiently usable for the adoption, even though the TRL expected is low, and usability could be improved
- Achieves the main results that the stakeholders desire.

5.2 FiberEUse Technical Test

Holonix proceeded with a high level phase of test about technical characteristics:

- Check of correct installation process
- Scalability of the solution had been verified also creating content for different potential sectors
- Security had been verified, updated, and is under continuous monitoring
- Performance had not been analysed deeply according to the TRL of the solution and the number of values expected which is not very high. Experience from previous projects and commercial solutions in Holonix are enough relevant to affirm that the team has enough experience to improve performances according to growing needs.

The system test had been performed according to the sample testing cycle process. First of all, a plan had been proposed to generate a complete usage cycle of the solution:

- Registration with different users: Gamesa, Rivierasca, Polimi
- Log In
- Creation of a circular Entity "Blade" with the Gamesa user, describing the exact attributes proposed by the company during the Bregenz meeting
- Creation of a second Circular Entity "Piece of Blade" with the Gamesa user
- Creation of a "chopping" process with the Gamesa user
- Creation of a "Disassembly Service Decision Support System" consultancy service with the Polimi user
- Logging in as Gamesa user, searching a "Disassembly consultancy service" on the platform and receiving the Polimi service as result
- Entering in touch with Polimi
- Logging in as Polimi and accepting the connection
- Closing the deal to confirm the connection
- Connect the created object closing the recursion, by adding the "chopping", the "Piece of Blade" and the "DSDSS service" to the "Blade" circular entity
- Logging in as Polimi and create the consultancy service "LCA LCC calculation"
- Logging in as Rivierasca, and creating a Circular Entity "construction waste" with the details received by the partner at the Bregenz meeting
- Creating a process "mechanical recycling"
- Entering in touch with Gamesa, which could be a provider of material to be recycled
- Connecting the circular entities
- Create a "Wine Cooler" Circular Entity by Rivierasca, with the described parameters
- Search the LCA/LCC service and find Polimi
- Test the LCA LCC integration and report results into the Wine Cooler circular entity

- Enter in touch, completing again the connection and the deal, but this time sending an attachment to the consultant through the Enter in touch message
- Complete the Wine Cooler description by verifying the connections with the closed deals
- Check the Value Chain chart created from the Wine Cooler
- Check the Value Chain chart created from the chopping process
- Check the Value Chain chart created from the piece of Blade
- Log in as Rivierasca and close the recycling circle of the Wine Cooler by reporting that a Wine Cooler can be done starting from another Wine Cooler.

The first complete cycle test had been reported through a Word document listing the process of the test, screenshots, notifications, feedback and comments. It had been shared with the developers in Holonix.

Main errors identified were related to the dimensions of the attachments, loop recursion of the value chain chart in the last case of a wine cooler as part of a wine cooler, visualization of different LCA/LCC results which was not allowed, useless "deactivate" button on circular entities.

A second round of the same complete cycle test had been completed with positive results for each of the listed points above.

6   FiberEUse IT Platform: Functionalities

Besides the above described innovative entities, the user is provided with a set of functionalities easily accessible through an appropriate menu bar. The list of functionalities follows, described at high level with the intention to highlight where innovation is.

The design of the platform doesn't include developments for capabilities already covered by state of the art application as it focuses on aspects that are new, innovative and not covered by other applications at the state of the art.

Capabilities already covered by applications available on the market can be easily integrated in a further exploitation phase.

Therefore the following aspects were not developed, but can be exploited further with few development effort:

- Localization data on geo maps
- Data input of company basic data
- Registration of multiple users for the same company.

6.1   Log in, Home Page, Project Description Page

The Log In function is set to protect access from unauthorized users. Entering the appropriate User name and Password, the Application takes the user in the FiberEUse web application home page.

The home page is designed for immediate use with no training required, designed to provide an overview "at a glance" of the functions made available to the user, plus call to actions icons.

The Project description page is designed to describe goals, values and key features of the project.

6.2   Company Description

The "Company" menu is designed to allow companies to provide any detail to describe firm's activities, products, process, services, and all characteristics deemed useful to participate to the project and be recognized in the ecosystem. The user has to fill mandatory attributes and can optionally fill in custom attributes and add attachments to improve public details about its company.

6.3   Circular Entity Menu

The "Circular entity" menu entice a full set of core functions in order to handle appropriately the circular entities in their transformations. This menu aims at providing the user with full autonomy in adding, listing, describing and visualizing its own and third parties' connected entities (Figs. 13, 14):

- Load attributes of the circular entity (name, location, category, description)
- Expand to free attributes as needed (name of the attribute, value of the attribute, type of numeric value as list or range)
- Load External resources related to the circular entity (Another Entity, Process, Logistic, Service)
- Load Internal resources related to the circular entity (Another Entity, Process, Logistic, Service)
- Load LCC-LCA results
- Load Attachments
- Provide Suggestions
- Show the Value Chain Eco-System chart
- See other companies' circular entities
- Enter in Touch with other companies providing other circular entities.

Fig. 13 FiberEUse circular entity menus

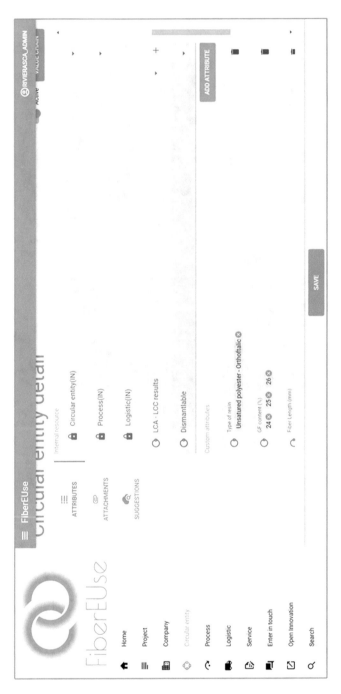

Fig. 14 FiberEUse circular entity custom attributes

6.4   Process, Logistic, Service Menu

The "Process" menu, "Logistic" menu, and "Service" menu are functionalities designed with the same logic of the Circular Entity menu. They provide users with the opportunity to see other companies' processes, logistics, services, and to input and give a detailed description of new once. Also for them is possible to:

- Add new process/logistics/service, with details, new attributes and attachments, if any
- Visualize and list process/logistics/service of other providers belonging to the ecosystem
- Visualize the value chain
- Enter in Touch.

6.5   Enter in Touch

The "Enter in Touch" function allows the user to get in touch with other parties in the ecosystem, send and receive requests in order to offer entities to be put in the re-use virtuous cycle or to spot opportunities coming from other parties. The user can:

- Ask for a new connection sending a message
- Receive a message for a new connection
- Visualize details about those companies which are already connected
- Register a deal, to let the platform know that a match had been made.

By design, in order to protect the privacy of negotiations between parties and also to be open for future different levels of exploitation the platform does not handle the intermediate phases of the negotiation and the name of the company is hidden and will be shown only after the reply to the "enter in touch" request. When the agreement is set, the application require the parties to record that there is a deal.

6.6   Open Innovation

The Open Innovation Menu is designed to integrate the Idea Manager tool, to give opportunity to co-create new ideas, in an environment of Open Innovation, taking advantage of an IT platform.

6.7 Search Engine

The Search engine function is designed to allow users to searching for entities, processes, logistics, services, or specific key words. This function provides the capability to retrieve quickly within the application what the company is looking for. This is beneficial in order to save time spent in searches and leave time for other added-value activities. From this menu is possible to proceed with a generic search or with a specific search on specific kind of entities and specific attributes:

- Start a new search (by keyword, entity, process, logistic)
- Add filters to refine and speed up specific searches
- Save searches for future queries
- Visualize former searches saved
- Show other searches.

6.8 Suggestions

The "Suggestion" functionalities are powerful. With this function, any viewer of the entity can suggest the company' service, product, and process to that specific entity. "Suggestions" offers the capabilities to:

- Receive suggestions on similar or interesting circular entity, processes, and services.
- Add/suggest attributes names and parameters guidance.
- Suggest yourself to other companies.

This function opens potentially a full world of options in terms of future developments and exploitation.

6.9 Value Chain Eco-System Chart

The Ecosystem chart function is a powerful, dynamic, and effective representation of the Value Chain and of the entire Value System. The Value chain graph is dynamic, as it is able to vary as new players, new products or new services enter in the chain. It is well described and shown in Sect. 2.3.3. It:

- Gives visual evidence of the Ecosystem that generates during the product transformation phases
- Allows to represent at a glance the value created by the ecosystem
- Allows to represent connections among partners and activities
- Provides a simple ready to use demonstration of how product generate a re-use ecosystem

- Opens to new possibility to give evidence of the sustainability of the products and of the much lower impact on the environment.

7   Conclusions and Future Research Perspectives

This chapter has reported the development and the results achieved by the FiberEUse platform and the connections with the other project activities. The platform had been presented through a technical point of view and mentioning at high level all the functionalities available. Functionalities responds to the stakeholders needs and requirements collected through the workshops.

The internal test using Polimi, Rivierasca and Gamesa users had reached the expected results, all the circular entities, processes and services proposed had been easily described and search, and the match-make had worked.

After the adoption at pilots, the IT platform will be ready to reach European level, and also to be open to additional sectors which could be crossed. The analysis on how the management of information in Circular Economy needs to be done is a result which will go further beyond FiberEUse project and will be reused by partner in future activities and exploitation.

One of the main results is the fact that End of Waste and Circular Entities are two concepts which the FiberEUse IT platform is natively able to manage.

Other functionalities are relevant and innovative, as the advanced search based on attributes, and the Eco-System value chain representation. Both can be further improved. The adoption phase with pilots will be useful to improve some awareness about the developed functionalities and on future developments which are expected and planned from now on.

References

1. Saeed, S., Jhanjhi, N.Z., Naqvi, M., Humayun, M.: Analysis of software development methodologies. Int. J. Comput. Digit. Syst. (2019)
2. Despa, M.L.: Comparative study on software development methodologies. Database Syst. J. (2014)
3. Hinderks, A.: A Methodology for Integrating User Experience Methods and Techniques into Agile Software Development. http://ceur-ws.org/Vol-2370/paper-04.pdf
4. da Silva, T.S., Martin, A., Maurer, F., Silveira, M.: User-Centered design and agile methods: a systematic review. In: 2011 AGILE conference. IEEE, pp. 77–86 (2011). https://doi.org/10.1109/AGILE.2011.24
5. Rubio-Sanchez, M.: Introduction to Recursive Programming. CRC Press (2017)
6. Corballis, M.C.: The Recursive Mind: The Origins of Human Language, Thought, and Civilization—Updated Edition. Princeton University Press (2014)
7. Sánchez-Cartas, J.M., Gonzalo, L.: Multisided platforms and markets: a literature review (2018)
8. https://ec.europa.eu/jrc/en/research-topic/waste-and-recycling
9. Stroustrup, B.: Polymorphism—providing a single interface to entities of different types (2007)
10. Porter, M.E.: Competitive Advantage, Ch. 1, pp. 11–15. The Free Press (1985)

11. Rumbaugh, J., Jacobson, I., Booch, G.: The unified modeling language reference manual. Comput. Sci. (1999)
12. https://en.wikipedia.org/wiki/Polymorphism_(computer_science)
13. EU H2020 ManuSquare project (GA 761145)

Use Case 1: Mechanical Recycling of Short Fibers

Herfried Lammer, Tamara König, Giacomo Bonaiti⬥, and Roberto Onori

Abstract The main objective of Use Case 1 is the development of industrial demonstrators of new products incorporating mechanically recycled glass fiber composites. These demonstrators will determinate the technical feasibility and cost effectiveness for glass and carbon fibers recycling solutions. The demonstrators include structural parts like a ski by HEAD Sport and sanitary products like shower trays by Novellini where the recycling fibers are used for existing products. A series of design concepts have been developed supported by a design briefing and a co-design methodology for street furniture and similar products, where the recycled materials are already considered from the start of the design of the product.

Keywords Ski · Shower tray · Creative products · Mechanical recycling process · Recycled glass fiber · Recycled carbon fibers

1 Introduction, Motivation and Objectives

The main objective of recycling fibers is the development of industrial new products incorporating mechanically and thermally recycled Glass Fiber Reinforced Plastics (rGFRP) and Carbon Fiber Reinforced Plastics (rCFRP). Following the product co-design, demanufacturing, material re-formulation, remanufacturing and re-use solutions developed by the research institutes, different potential applications for rGFRP and rCFRP are demonstrated at a large industrial scale on real-case scenarios.

H. Lammer (✉)
HEAD Sport GmbH, Wuhrkopfweg, 1, Kennelbach, Austria
e-mail: h.lammer@head.com

T. König
designaustria, Museumsplatz 1, 1070 Wien, Austria

G. Bonaiti
Riveriasca SPA, Via Strasburgo, 7, 24040 Bottanuco, Italy

R. Onori
Novellini SPA, Via Mantova, 1023, 46034 Romanore, Italy

303

M. Colledani and S. Turri (eds.), *Systemic Circular Economy Solutions for Fiber Reinforced Composites*, Digital Innovations in Architecture, Engineering and Construction, https://doi.org/10.1007/978-3-031-22352-5_15

The overall functional validation of the single demonstrators is performed by the individual standards and criteria by the different products.

Three different uses cases are demonstrated by the company partners Novellini Group, HEAD Sport GmbH and designaustria.

Novellini Group is the 1st European producer of shower cabins and bath tubs with around 1 million pieces per year. The amount of virgin GFRP consumed annually by Novellini Group is around 1.400 ton. HEAD Sport GmbH is on a mission to bring high-performance gear to every athlete, professional and amateur, to allow them to be the best they can be in racquet sports, winter sports, or swimming. Its product like the skis are structural composite parts, and produced in high quantities. The amount of virgin GFRP consumed annually by HEAD Sport is around 400 ton.

designaustria represents the interests of around 1300 members from various design backgrounds. The knowledge bundled in the organization is complex and widely networked. designaustria represents the Austrian design scene on both national and international levels, seeking to highlight its achievements. designaustria raises awareness for Austrian products and services and thus contributes to strengthening Austria as an industrial location. As a platform, designaustria encourages discussion and opens up new possibilities for designers. In this sense designaustria invited its members to design new products from recycled fiber composites. The demonstrators where then realised in cooperation with Rivierasca. Since 1963 Rivierasca has been producing laminates in fiberglass reinforced polyester. From the most varied application possibilities of the traditional roll, they have moved on to special planes, to the whole range of corrugated and corrugated translucent sheets for covering and infill, up to the opaque sheets for various applications. The amount of virgin GFRP consumed annually by Rivierasca is around 25.000 ton.

The use cases, which will be shown in this chapter, are based on fibers recycled within the FiberEUse project. The recycling process used from the waste to the raw material, that will be used industrially, goes through several intermediate stages of transformation. The key technologies exploited for the realization of demonstrators is the mechanical recycling: it consists of obtaining a defined particle size of shredded particles, with a minimum of energy and resources. This value chain is essential to ensure the quality of the materials that will reach the industrial process, both in automotive and construction sectors. This mechanical recycling process is followed by the adaption of production processes used in the relevant industries.

In the following tables the demo-cases and the materials used for their realization are shown. Table 1 shows the materials used for the demo cases and that have been previously shredded to obtain a suitable pellet for the relevant industrial processes.

2 Democase 1: Sanitary Furniture

Showertrays provide the foundation of a shower encloser, providing a safe and sturdy base to stand on. Technical requirements for the products are mechanical behaviour to

Table 1 Materials used for automotive demo cases

Partner	Demo case	Matrix	Fiber
Novellini	Shower tray	UP	rGF
	Bath tube	UP	rGF
HEAD	Ski	Epoxid	rGF
Design Austria	Creative products	UP	rGF

UP Unsaturated polyester

such loads, chemical resistance, and cleanability tested according to UNI EN 14527 standard for shower trays and UNI EN 14516 for bathtubs [1–4] (Fig. 1).

Fig. 1 Bath tube and shower tray as demonstrator for the use of recycled material

The top layer of a showertray or a bathtube is an acrylic sheet, with a high gloss and scratch resistant. These sheets are thermoformed. To increase the stiffness of the tray, glassfibers and resins are sprayed on the backside, the sheet acts as an open mould. Local reinforcements like metal frames can be used and finally holes for the connection are milled. The project target in this first demo-case is the replacement of at least 40% of virgin GFRP composite with mechanically recycled and reformulated material for sanitary manufacturing.

The final composite must fulfil all the technological specifications reported above, saving costs and bringing environmental benefits. The mechanical recycling process reduces the fiber length and also the mechanical properties of the material.

During the mechanical recycling process, the fiber glass is not separated from the master resin, and a mix of both, fiber glass plus original resin is obtained. The dimension of the fibers inside such recycled granules is less than 4 mm. Yet for the shower tray the stiffness is not only defined by the Youngs modulus of the material, but also from the moment of inertia and the thickness. Recycling granules from own production scrap, from windblades and from laminate roofing have been tested. After shredding the average granules length was around 2 mm (Fig. 2).

Figure 3 shows the production process of a bathtube. The reinforcement thickness is made by polyester resin with glass fiber. This thickness is obtained by spray lay-up technologies. In some part of the product, inside of the reinforcement thickness, there are wood or metal insert to increase the rigidity.

A solution for the use of recycled material was developed where onto the acrylic sheet a first thin layer of virgin long fibers was placed, then a core layer of recycled granules, and on the outside again a layer of virgin long fibers. With this process the adhesion between thermoformed sheet and reinforcement can be guaranteed and the rigidity could optimized by adaption of the thickness of the reinforcement layer introducing the granules in the middle.

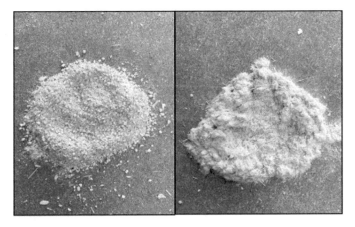

Fig. 2 Different types and sizes of mechanical recycled granules from shower trays windblades and GFRP roofs (< 2 mm particle size above, > 2 mm below)

•*Process As-is:*

- *Thermoforming*
- *Spray lay-up*
- *Rolling*
- *Curing*

Fig. 3 Production process of a bathtube

By this process a sort of sandwich was obtained, where on the top and the bottom there are the "virgin" material and in the middle the recycled granules (Figs. 4 and 5).

The issue is how to transport and distribute the recycled granules in order to make the bathtub or shower tray. For this the production process of spray lay up needs to be adapted (Fig. 6).

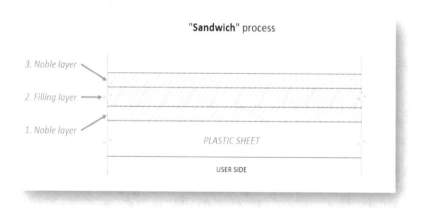

Fig. 4 Sandwich structure of a shower tray or bathtube with the use of recycled material as a core layer

Fig. 5 Cross section of shower tray with rGFRP

Fig. 6 Modified device for
spraying granules

A special nozzle was positioned over the spray-layup gun. In this way we could
mix resin, fiberglass and granules. The viscosity of this mixtures is especially
important for the production of the vertical bathtub walls (Fig. 7).

The overall thickness was about 7 mm, which was an increase of about 50% versus
a reference product. The slight increase in weight is not relevant for the performance.
The resin for spraying the recycled granules was adapted regarding rheologic properties and shrinking behaviour during curing. Overall, by this solution 40% of glass
fibers could have been replaced by recycled granules. Relevant to specific standards
[1, 2] all key performance indicators like geometrical deformation, resistance against
impact and temperature and permitted deflections have been achieved.

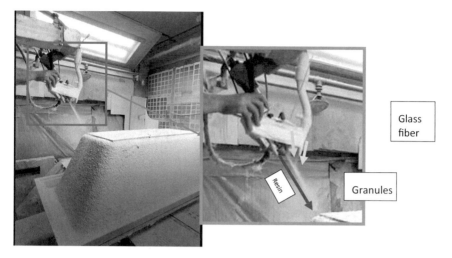

Glass
fiber

Resin

Granules

Fig. 7 Production of bath tub demonstrator

3   Democase 2: Skis

A ski is a very complex composite part which combines many different materials and which is highly stressed typically at temperatures below 0 °C. Any material delamination could lead to a severe safety issue.

Figure 8 shows a cross cut of a ski, consisting of a base made of polyethylene, a special aluminum alloy metal reinforcement, steel edges, glass fiber reinforcements and a top layer. The core material can be either wood or polyurethan foam with a density of appr. 700 kg/m^3. The different core materials define different production processes. For the demonstrator a solution was developed where a fraction of GFRP mechanical recyclate was be used as filler to strengthen the polyurethane (PU) core.

In addition to the mechanical recycling types mentioned above and shown in Fig. 2, uncured prepreg of production waste of HEAD was also used for the demonstrator (Fig. 9).

Conventional molding technologies were adopted for their realization Uncured prepreg and also windblades had been shredded to a fiber length of appr. 6 mm.

In case of prepreg this fraction was pressed into a rectangular sheet of the size of the center part of a ski, appr. 50 by 8 cm, with a weight of 50 g each. The remaining reactivity resulted in a porous and fragile sheet. In case of recyclates of windblades some resin was added. To improve the handlings trials have been made with a bottom layer of glass fiber fleece (Fig. 10).

This part was then put in the ski, in the area, where the binding is mounted. During the production process, it had to be fixed, as the polyurethan is injected from the top of a ski. And due to the low viscosity during the injection, the polyurethane fills out all pores of the part so that it works as a reinforcement (Figs. 11 and 12).

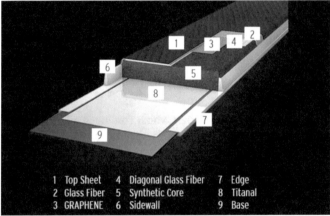

1	Top Sheet	4	Diagonal Glass Fiber	7	Edge
2	Glass Fiber	5	Synthetic Core	8	Titanal
3	GRAPHENE	6	Sidewall	9	Base

Fig. 8 Skis as demonstrators for the use of recycled materials and a schematic cross section of such a structural composite part

Fig. 9 Shredded HEAD prepreg to a size of appr. 6 mm

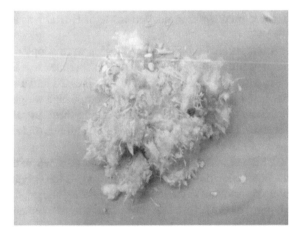

Fig. 10 Reinforcement part made of rGFRP to be used in the center area of a ski

Fig. 11 Manufacturing of a demonstrator ski with the reinforcement parte made from recycled material is located in the center area

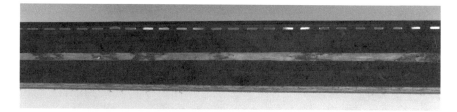

Fig. 12 Crosscut of a ski with the reinforcement in the center area made from rGFRP

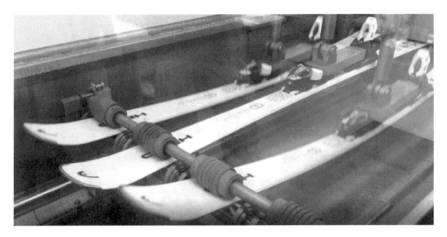

Fig. 13 Durability test of demonstrators as part of the KPI

Table 2 Suitability of different recycled GFRP types for Democase ski

Variant	KPIs	Comments
Fluffy material	Yes	Difficult to work at ski press, time consuming inlay work
EoL wind blade	Yes	Easy to handle at ski press
Construction scraps	Yes	Too thick, bubbles at the end in PU layer
OSB type of plate	Not always	Holes in OSB type of plate do not fulfill KPIs, but happen for 2 mm plate
Pressed fluffy material	Yes	Best approach taken for further investigation

This reinforcement is acting on the screws, to increase the screw pull out force, relevant for a crash, and the maximum torque moment, relevant for the mounting of the screw. And the amount of polyurethane could be reduced down to a density of around 500 kg/m^3, so that the overall weight remains unchanged and the standards, a ski has to fulfil in these categories, are still fulfilled [5] (Fig. 13).

Several types of recycled GFRP have been evaluated for the demonstrators. Table 2 shows the summary of these trials.

4 Democase 3: Design Products

Designaustria invited its members to take up the challenge to design products from recycled material. An iterative co-design process of the remanufactured demonstrators involving end-users was implemented, so as to improve the perceived value and the market acceptance of such rGFRP-based products. Both conventional (i.e. molding or extrusion) and innovative remanufacturing technologies (3D printing)

Table 3 Recycled content per production method of Rivierasca (status September 2021)

Technology	Max. recycled GFRP (wt%)
Open mould casting	50
Closed mould casting	80
Vacuum casting	44
Continuous lamination	39
Continuous extrusion	72
Filament winding	35

were adopted, depending on the product requirements, and appropriate process optimization were carried out. A series of design concepts have been developed supported by a design briefing and a co-design methodology. The challenges and solutions encountered in the production are a substantial part in the design guidelines in order to give crucial target groups of designers and manufacturers/investors a practical insights in the current possibilities of product applications for rGFRP. The concepts have been developed according to the guidelines developed in Chap. 13 In addition to the real physical demonstrators, a virtual concept was developed and published in the online FiberEUse Library to be found here: www.fibereuselibrary.com.

The initial target goal of FiberEUse was the use of at least 30% w/w of GFRP recyclates in the manufacturing process. This amount exceeded during the research and testing within the project as shown in the following Table 3. Mechanical recycled (grinded material) is not only glass fiber but also cured resin, thus the achievement of 80 wt%.

The production methods for the five demonstrators were open mould casting, closed mould casting with and without pressing and plastering technique. The following Table 4 indicates the technologies being tested, the recycled content in percentage by weight and the tests performed.

Design briefing for products incorporating rGFRP

A design briefing is the essential communication tool for clarifying the assignment between client/company/producer and designers to meet user needs. It contents all technical, aesthetical and functional details for a sustainable production. Initial position of the design briefing was the prerequisite of the use of grinded scraps and fresh unsaturated polyester (UP) resin (Fig. 14) called "Glebanite©".

The grade of "Glebanite©" is the percentage of powder (grinded scraps) compared to the percentage of UP resin in a mixture. With a 50% of fiber content by weight in the virgin material, grade 50% means that in the recycled product fiber weight is 33% on total weight.

To the basic mixture, further additives may be used to improve special properties:

- Pigments: to give color to the product (see Fig. 15);
- Fresh glass fibers: to improve mechanical resistance (CSM, CFM, fabrics, milled fibers);

Table 4 Design demonstrators summary

Technology	Recycled content wt%	Protype	Test performance
Open mould casting	41.2% grinded scraps of glass fiber composites, 58.8% unsaturated polyester resyn (UP)	"Wine Cooler" by Martina Hatzenbichler	Overturning: no breakage or chipping
Closed mould casting without pressing	37% grinded scraps of glass fiber composites including fresh glass added for mechanical resistance, 63% unsaturated polyester resyn (UP)	"Swing" by Barbara Gollackner	Static load of 200 kg applied: no breakage
Open mould casting without pressing	44% grinded scraps of glass fiber composites, 56% unsaturated polyester resyn (UP)	"Social Furniture 04 Stool" by EOOS	Static load of 200 kg applied: no breakage Dynamic load of 95 kg applied: no overturn
Closed mould casting with pressing	38.7% grinded scraps of glass fiber composited including fresh glass added for mechanical resistance, 61.3% unsaturated polyester resyn (UP)	"LAMP" by Louis Funke	The lamp was assembled by the designer itself; no test was performed by Rivierasca Note: the lamp is sufficiently stiff for handling without particular attention
Plastering technique on 3D-printed model	35% grinded scraps of glass fiber composites, 65% unsaturated polyester resyn (UP)	"Trash Pot" by Valentinitsch Design	Static load of 200 kg applied: no breakage Dynamic load of 95 kg applied: no overturn

Fig. 14 Grinded scraps and fresh UP resins

Fig. 15 Colouring of Glebanite © by using pigments

- Additives: for fire protection, antimicrobical, dispersing agents, flexible behavior …
- Other filler: for example, glass bubbles (or PU foam powder) to have a very light material.

The initial target goal of FiberEUse was the use of at least 30% w/w of GFRP recyclates in the manufacturing process. This amount has been already exceeded during the research and testing within the project for various production technologies. e.g. for open mould casting it was possible to use up to 50%, for closed mould pressing even up to 80%, for vacuum casting up to 44%, continuous lamination 40%, continuous extrusion 70%, 3D printing 50%, rotomolding and centrifugal 50% or for filament winding up to 60%.

Design examples of different production methods

The following examples of material selection and production choice describe the current status of prototype production incorporating recycled fibers. The prototypes were design by designers form the designaustria network who worked closely with partners in the production following the design brief.

"Swing" by Barbara Gollackner design

For "Swing" Rivierasca decided to create a prototype in polystyrene milled with a 5-axis milling cutter which was then used for the construction of a rigid fiberglass mould divided into two parts. This swing needs a high mechanical resistance, it must be able to withstand a load given by the weight of a person located between the two anchoring points of the rope used to suspend the seat. It is therefore not possible to make a casting piece which would be too weak respectively too heavy. Rivierasca used the plastering technique, since it was necessary to reinforce the seat with glass fiber material combining woven roving and chopped strand mat (Fig. 16).

Fig. 16 "Swing" by © Barbara Gollackner (concept view, exhibition view)

"Yellow (stool)" by © EOOS

For the realization of the first stool Rivierasca prepared a 1 cm thick reinforced slab. The slab was made with a 44 wt% mixture spread by hand on a glass biaxial non woven 0/90° and then closed with another biaxial non-woven 0/90° in order to leave them outside, then inserted in a press and heated up to 90 °C for curing. Using a two external biaxial (0–90°) 600 g/m^2 each (same grammage is due to have a balanced laminate so to avoid bending) to have a very high stiffness and strength, containing the thickness and weight. Having a greater thickness would result in too heavy weight. The current piece (see Fig. 17) weighs 9.5 kg.

The challenge encountered is given by the thickness lower than the original design, which entails the use of smaller screws for the assembly of the pieces, and a greater difficulty in making the pre-holes. The external surface gives a glimpse of the mesh of the biaxial giving an industrial look to the panel.

"Lamp" by © Louis Funke design

For "Lamp", Rivierasca prepared two 6/8 mm sheets with same technology of "Yellow (stool)" by EOOS (closed mold pressing). One object in white and one in black colour were produced with a very rough aesthetic effect. The sheets were

Fig. 17 "Social Furniture 05 Stool" by © EOOS (exhibition view)

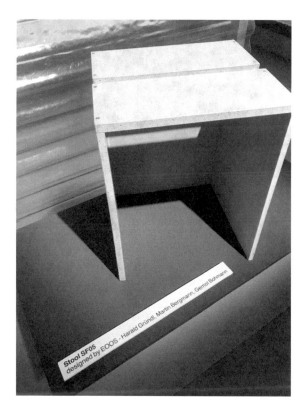

made by hand with a single layer of continuous fiber glass mat inside, leaving rGFRP (Glebanite©) on the outside in order to "read" the movement of the waste used during the pressing phase. For this object, single plates are glued together. Glebanite© is suitable for being glued with most common glues (Figs. 18 and 19).

"Wine Cooler" by Martina Hatzenbichler

According to the concept statement by designer Martina Hatzenbichler the starting point of the design of this Wine Cooler (Fig. 20) was the fusion of two contrary initial forms and materials. The volume of a raw and unfinished monolithic body is broken up here through the circular segments of the organic form of a glass bottle in order to create openings. These, however, are not arranged arbitrarily, revealing

Fig. 18 "Lamp" by © Louis Funke design (product view, exhibition view)

Fig. 19 "Lamp" by © Louis Funke design (detail material sample)

Fig. 20 Wine Cooler
Martina Hatzenbichler
(exhibition view, 2021).
Photo: T. Koenig

their design concept primarily through the addition of several wine cooler elements. The outlines and edges are extended as the respond to one another.

Turning and stacking the individual elements offers a multitude of possible combinations (Fig. 21). This results in an entirely new overall structure, with the Wine Cooler turning into a wine shelf or even into wine architecture (Fig. 22).

Wine cooler consists of 41.2% grinded scraps of glass fiber composites and 58.80% unsaturated polyester resin (UP) and was manufactured by open mould casting.

"Trash Pot" by Valentinitsch design

Trash Pot by Valentinitsch Design consists of 35% grinded scraps of glass fiber composites and 65% unsaturated polyester resin (UP).Valentinitsch Design created an elaborated concept that turned the object made of waste again into an object for collection of waste. "TRASH POT is multifunctional sitting device that invites you to sort your trash in correct manner and allows you to recycle your trash the way it deserves." (Fig. 23).

Fig. 21 Wine Cooler Martina Hatzenbichler (exhibition view, 2021). Photo: Martina Hatzenbichler

Fig. 22 Martina Hatzenbichler, Wine architecture

Fig. 23 Trash pot
(exhibition view, 2021).
Valentinitsch design Photo:
T. Koenig

5 Conclusions and Future Research Perspectives

The democases Showertray and ski in the project FiberEUse proved the feasibility of using mechanically recycled glass fibers, including resin in existing products. The key performance indicators for customers and for standards in each category could have been fulfilled with some modifications and adaptions of the specific production process. Putting this in future and with some further development into serial production will allow a reduction in the use of virgin material and also a reduction in waste of the companies.

These prototypes of the democase designaustria made of mechanical recycled fibers exemplify the wide range of options for designers and production companies. More prototypes need to be further explored and optimization for industrial serial production. The determination, exploration and development of business strategies to raise interest of European companies in an accelerated transition of their value chains towards circular economical orientation need to be further investigated. Design guidelines will provide valuable information to stakeholders on the necessary changes from linear to circular design in production.

References

1. EN 14527—2016: Shower tray for domestic purpose
2. EN 14516—2015: Baths for domestic purpose
3. EN 15719—2015: Sanitary appliances—baths made from impact modified coextruded ABS/acrylic sheets—requirements and test methods
4. EN 15720—2010: Sanitary appliances—shower trays made from impact modified coextruded ABS/acrylic sheets—requirements and test methods
5. ISO 8364:2017 Alpine skis and bindings — Binding mounting area — Requirements and test methods

Use Case 2: Thermal Recycling of Long Fibers

Sonia García-Arrieta◯, **Iratxe López Benito**◯, **Marta García**◯,
Giacomo Bonaiti◯, **Olatz Ollo Escudero**◯, **and Cristina Elizetxea**◯

Abstract This chapter describes the industrial demonstration of the reuse of recycled fibers obtained by a thermal process. Four demonstrators are described in which both recycled carbon fibers and recycled glass fibers have been incorporated into different matrices. The automotive sector proposes 3 demo cases (Pedal Bracket, Front-end carrier and Cowl top support) with demanding mechanical and thermal requirements. These components were manufactured by injection molding with thermoplastic matrices. The construction sector proposes 1 demo case (Light transmitting single skin profiled sheet.) with mechanical and light transmittance requirements that was manufactured by continuous lamination. It is demonstrated that the incorporation of recycled fiber for these applications is technically possible, fulfilling the requirements demanded by each sector.

Keywords Thermal recycling process · Pyrolysis · Recycled glass fiber · Recycled carbon fibers

1 Introduction

Glass fiber (GF) and carbon fiber (CF) reinforced polymer composites are widely used in sectors like transport (automotive, aircraft, railway, boats) and construction

S. García-Arrieta (✉) · O. O. Escudero · C. Elizetxea
TECNALIA, Basque Research and Technology Alliance (BRTA), Mikeletegi Pasealekua, 2, 20009 Donostia-San Sebastián, Spain
e-mail: sonia.garcia@tecnalia.com

I. L. Benito
BATZ S.Coop., Torrea 2, 48140 Igorrem, Spain

M. García
MAIER S.Coop., Pol Arabieta S/N, 48320 Ajangiz, Spain

G. Bonaiti
RIVIERASCA S.P.A., Via Strasburgo 7, 24040 Bottanuco, Bergamo, Italy

© The Author(s) 2022
M. Colledani and S. Turri (eds.), *Systemic Circular Economy Solutions for Fiber Reinforced Composites*, Digital Innovations in Architecture, Engineering and Construction, https://doi.org/10.1007/978-3-031-22352-5_16

(building and infrastructures, plants, wind turbines), because of their attractive properties such as light weight, good corrosion resistance, high strength, high fatigue resistance and good dimensional stability [1]. Nevertheless, there are some disadvantages in the use of composite materials which include the high raw material and fabrication cost, susceptibility to impact damage and a greater difficulty in repairing a recycling.

Between the different available reinforcements, GF is the most used in several industrial applications, mainly due to its low cost but also because it's good thermal and electrical insulation. On the other hand, CF is characterized by better mechanical properties combined with a lower density, which makes possible to manufacture lightweight and very robust parts. The biggest disadvantage of CF is the raw material and manufacturing cost, which is much higher than GF.

In the case of the automotive sector, apart from using steel or aluminum to produce different parts, the main reinforcement used combined with polymers is GF. It is well known that in automotive sector, cost and weight reduction are imperative. Although GF is cheaper than CF, there is a growing interest for replacing virgin GF with recycled carbon fiber (rCF) in some automotive applications, due to its interesting properties mentioned above.

In the construction sector, some lightweight applications exist as high dimension roofs, where the low weight is an important requirement together with appropriate mechanical properties. In some applications, translucency is also important. Therefore, composite materials in this sector are widely used. The use of these laminates is especially high for industrial applications like pavilions, sports center, shelters, car park canopies, etc. These products at their end on life produce a high amount of waste, which is interesting to reuse; if possible, in the same application thus, closing the loop in a circular economy perspective [2].

The use cases which will be shown in this chapter, will be based on fibers recycled within the FiberEUse project. The recycling process used from the waste to the raw material that will be used industrially, goes through several intermediate stages of transformation. This value chain is essential to ensure the quality of the materials that will reach the industrial process, both in automotive and construction sectors. The key technologies exploited for the realization of demonstrators are:

- Pyrolysis or thermal recycling (Chap. 5): it consists of obtaining liberated clean glass or carbon fibers (rGF or rCF) by means of a thermal treatment which eliminates completely the matrix and other organic components of the composite residue.
- Re-sizing (Chap. 7): it consists of applying a chemical treatment on the surface of the recycled fibers, to increase the fiber-matrix interfacial adhesion for future applications.
- Compounding (Chap. 7): it consists of preparing a semi-finished product made of fibers and thermoplastic matrix in pellet form through an extrusion process.

In the following tables the demo-cases and the materials used for their realization are shown. Table 1 shows the materials used for automotive demo cases and that have

Table 1 Materials used for automotive demo cases

Partner	Demo case	Matrix	Fiber
BATZ	Pedal bracket	PA6	rCF short
	Front end	PP	rCF strand
MAIER	Cowl top support	PP	rGF short

PA6 Polyamide 6, *PP* Polypropylene

Table 2 Scheme of the materials used for construction demo case

Partner	Demo case	Matrix	Fiber
RIVIERASCA	Light transmitting single skin profiled GFRP sheet	UP	rGF mat

GFRP Glass fiber reinforced plastic, *UP* Unsaturated polyester

been previously transformed by a compounding process to obtain a suitable pellet for industrial injection molding process.

Table 2 shows the materials used for the construction demo case and which fibers have been previously processed to obtain a suitable mat for the industrial laminating process.

2 Definition of Demo Cases

The demo cases based on thermally liberated recycled fibers are characterized by demanding requirements, both mechanical and aesthetical, affected by the quality of the fiber. The recycled fiber must be clean and resized in order to be compatible with the new matrix, providing good properties [3, 4]. Automotive OEMs (Original Equipment Manufacturers) are looking for sustainable plastic materials (so called "green plastics") to decrease the carbon footprint during the production of vehicles.

From the automotive sector parts manufactured by classic thermoplastic injection process or Injection Moulding Compound (IMC) process have been selected, based on the use of thermoplastics and fiber pellets. In the case of constructions, the laminating process will require a thermoset matrix which is deposited on the GF mat.

In this project, the challenge is to introduce rGF and rCF in structural components. The target output demonstrators are going to be a Pedal bracket, Front-end carrier, Cowl top support and industrial roof.

2.1 Demo Cases 4.1 & 4.2: Pedal Bracket (A) and Front-End Carrier (B)

The demo-cases selected by BATZ to work with rCF reinforcements are:

Fig. 1 Pedal bracket

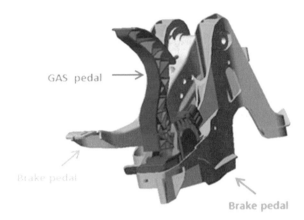

A. The **pedal bracket** is a component in which different pedals are assembled (mainly gas and brake pedal). The pedal bracket is a security part with high structural requirements, both stiffness and strength due to driving conditions and emergency issues. The lightweight solutions are made of reinforced thermoplastic but there are also made of metal (Fig. 1).

B. The **front-end carrier** is the frontal structure of the vehicle where the front-end module system relied on. Different components such highlights, radiator, latches, cross members, etc. are assembled on it. The front-end carrier has to support different loads due to their weight as well as different usage scenarios. Adding to that, the front-end carrier contributes to the torsion performance of the car. Due to the high requirements that this kind of components have to withstand, the full plastic variants are made of high percentages of GF reinforcement (up to 40–50%) (Fig. 2).

2.2 Demo Case 5: Cowl Top Support

The demo-case selected by MAIER to implement rGF is a cowl top support. The cowl over is the part which is positioned between the windshield and the front hood, while the Cowl over support is the part that gives structure to the part. In Fig. 3 both components can be seen. The functionality of this part is to stand stiff and integer at static conditions, and break after absorbing a certain amount of energy at dynamic conditions. This part is produced by injection moulding process.

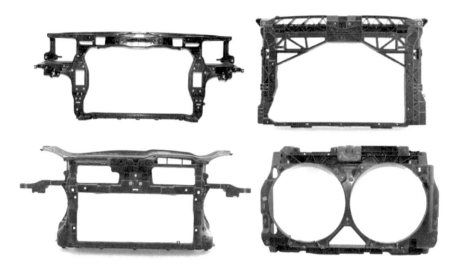

Fig. 2 Front-end carrier examples [5]

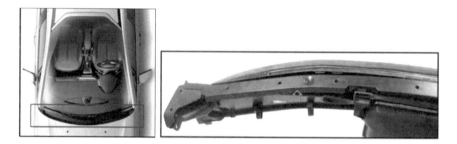

Fig. 3 Cowl over (left) and cowl over support (right)

2.3 Demo Case 6: Light Transmitting Single Skin Profiled GFRP Sheet

The Rivierasca's demo-case is a roof. It has to cover up any type of building but also to fit some requirements like, for example, a good acoustic or thermal insulation, good fire resistance or light transmission to indoor environments [6]. So, in this case, a light transmitting single skin profiled GFRP sheet has been developed (Fig. 4).

Fig. 4 Light transmitting
single skin profiled GFRP

3 Pedal Bracket and Front-End Carrier Demo Cases

3.1 Pedal Bracket

Materials

The objective of this demo case is to replace a pedal bracket currently manufactured with PA6 + 40% of short GF by a new design based on PA6 + 20% of rCF (PArCF20). The pedal bracket must support the mechanical requirements defined for its use with the new material.

Manufacturing process

The pedal bracket demonstrator is processed with a conventional injection machine using pellets of PArCF20 developed in the FiberEUse project. The machine used is the same than used for usual GF reinforced plastics. The injection process is similar for both materials, but some parameters, such as melting temperature and injection velocity had to be adapted to the rheological characteristics of the new material (Figs. 5 and 6).

After its geometrical verification, the brake and gas pedal are assembled in the bracket for validation test (Fig. 7).

Validation test

In the validation test, after an aging pre-treatment, the part is subjected to different loads applied in different areas of the brake and gas pedal. The applied load

Fig. 5 Raw material (left) and injection machine (right)

Fig. 6 Pedal mould (left) and prototype manufactured with PArCF20 (right)

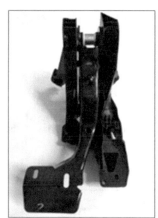

Fig. 7 Pedal bracket prototype ready for validation test

must produce a component deformation less than a vehicle manufacturer defined requirement. In this case the component is validate for its use.

One of the more restrictive requirements is to withstand loads up to 3000 N, applied in the brake without any break exposed at extreme temperatures (80 °C and −35 °C). The Fig. 8 shows the test set-up. The new bracket manufactured with recycled fibers fulfils the requirements without suffering any damage. Table 3 shows the results of the validation test where the load values supported by the part are shown, demonstrating the technical viability of this prototype manufactured with rCF.

(a)   (b)

(c)

Fig. 8 **a** Validation test setup in climatic chamber **b** pedal bracket set-up **c** load applying on the frontal brake

Table 3 Testing results

Load case name	Load (N)	Requirement		Result	
		Elastic deformation	Plastic deformation	Elastic deformation	Plastic deformation
Frontal brake	2.300	< 17 mm	< 5 mm	OK	OK
	3.000	No breakage		No breakage	
Frontal accelerator	200	< 6 mm	< 2 mm	OK	OK
	1.000	No breakage		No breakage	

3.2   *Front-End Carrier*

Materials

The front-end carrier used to evaluate the performance of the new material is currently manufactured using PP + 40% of long GF (PPGF40) by IMC process, and it is replaced by PP + 20% of rCF (PPrCF20) in strand format. The prototype manufactured with the new material must fulfill the defined mechanical specifications. In addition, the change from GF to rCF will reduce the total component weight because the material has a lower density.

Manufacturing process

The process technology used for this demonstrator manufacturing was IMC. IMC is a direct inline-compounding that can be used to process highly filled and/or long fiber-reinforced thermoplastics in the injection-moulding process. IMC is characterized by the inclusion of a twin-screw extruder with a high plasticizing performance into the injection concept. The injection-moulding compounder allows feeding of endless fiber roving into the melt near the end of the twin-screw extruder. This configuration allows having longer fibers in the final part, achieving better mechanical properties compared to the commercial long fiber compounds available in the market in pellets.

The main elements of the IMC are gravimetric metering, the twin-screw extruder and the clamp unit. A co-rotating twin screw is used for plasticizing and a piston for injection, instead of a single screw for plasticizing and injection used in a conventional injection machine.

In the context of using rCF, where they have strand format, the endless fiber roving feeding system is replaced by a specific dispenser for feeding this kind of fibers (Fig. 9).

The demonstrator was injected in 2 materials: the current PPGF40, and PPrCF20 for performance evaluation. The process parameters were about the same for both materials, only the temperature of the barrel should be controlled better with the rCF compounding due to conductivity effect of the CF (Figs. 10 and 11).

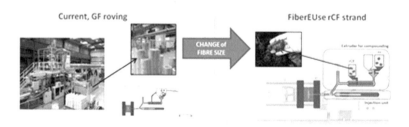

Fig. 9 Configuration to dispense rCF strand

Fig. 10 **a** Specific dosing unit for fiber strand **b** rCF in the hopper **c** rCF being dispensed by the twin screws of the feeder **d** Detail of the dosing unit and the funnel of the IMC machine

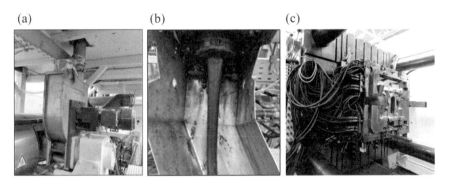

Fig. 11 IMC process during front end carrier manufacturing: **a** Lateral feeder where the recycled fiber is taken to the main extrusion screws of the IMC machine **b** melted PPrCF20 material flowing through the nozzle **c** front end carrier injection mould

Validation test

The validation test is focused on static load cases based on the use case of the bonnet latch overload. A compression and a traction load are applied in the latch area and the component must withstand a force of 2.300 N without any brake and achieving a specific stiffness.

With the aim to compare the component performance with different materials, a specific test was designed. This test is able to evaluate not only the component structural performance but also each material itself. Figure 12 shows the test set-up and Table 4 shows the loads on the component, the test conditions and the results obtained for the material developed PPrCF20.

The material developed in the FiberEUse project (PPrCF20) has been validated using the minimum value required for the component (Table 4), but it has also been compared with the results obtained in the same test for the original material that is proposed to be substituted (PPGF40). Table 5 shows the comparison between the two materials.

It can be observed that the experimental material has better physical properties than the current one in 3 of the 4 measured properties. In any of the cases, the values obtained always fit the specifications reported in Table 4.

Fig. 12 Front-end carrier validation test

Table 4 Testing parameters and results

		Stiffness	Strength	Results
Static	Bonnet latch traction	420 N/mm	2.300 N	OK
	Bonnet latch compression	350 N/mm	1.500 N	OK

Table 5 Comparison between current material and new material for Front-end carrier application

		Stiffness (%)	Strength (%)
PPrCF20 increment over PPGF40	Bonnet latch traction	5.3	−34.6
	Bonnet latch compression	18.2	7

Table 6 Material properties for prototype manufacturing

Physical properties	Test method	Unit	Optimum	Minimum
Melt flow rate (230 °C, 2.16 kg)	ISO 1133	g/10 min	14	10
Melt volume rate (230 °C, 2.16 kg)	ISO 1133	cm³/10 min	15	10
Density (23 °C)	ISO 1183-1/A	g/cm³	1.14	

4   Cowl Top Support Demo Case

In this case, the innovation is to introduce recycled material as a reinforcement. Currently, to produce this type of parts, PP + 30% of virgin short GF (PPGF30) is used. The objective of this development is to substitute the reinforcement, at least partially, with thermally recycled GF.

4.1   Materials

The cowl support manufactured by Maier is made of PPGF30 with conventional injection moulding and in the FiberEUse project, it has been replaced with PP + 30% of rGF (PPrGF30). In table 6 it can be seen the performances that the material must fulfil for the prototype manufacturing.

4.2   Manufacturing Process

The manufacturing process of this part is plastic injection moulding (conventional injection of plastics).

Plastic injection moulding is the process of melting plastic pellets, in this case pellets of PP + rGF, that once melted enough, are injected into a mould cavity (under pressure and thermal conditions), which fills and solidifies to produce the final product. Injection moulding machine is composed by parts: the injection unit, the mould and the clamp unit.

The Fig. 13 scheme explains the injection moulding process. PPrGF30 pellets are fed into a heated barrel (4) and mixed using a helicoidal screw (6). Then it is injected

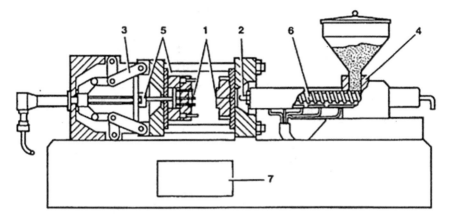

Fig. 13 Injection moulding machine scheme

into a mould cavity (1) though the nozzle (2). The other component of the injection machine is the closing group (3 and 5).

Parts to be injection-moulded must be very carefully designed to facilitate the moulding process. In this context, there are some aspects that have to be taken into account: material, desired shape and features of the part, mould material and the properties of the injection machine.

After the compounding of the PP material with the recycled glass fiber, there has been a first industrial injection trials to manufacture testing plates in order to analyze the differences between the injection process parameters with PPGF30 and PPrGF30. The parameters that have been studied are the screw temperature, the different injection pressure and the time of dosing and cooling.

After different injection trials the conclusion was that the injection process parameters needed are quite similar in both cases.

The demonstrator was injected in 2 materials: the current PPGF30 and PPrGF30 for performance evaluation. The Fig. 14 shows the material, the Fig. 15 the injection process and the mould of the cowl top support. The prototype obtained is shown in Fig. 16.

4.3 Validation Test

As mentioned before, the cowl top support has to give the needed stiffness to the part but also it has to break at determinate conditions. This characteristic must be maintained in different environmental and operating conditions. For that reason, the demo case is tested under different thermal and cycle conditions. The part also is tested by impact test method. The Table 7 shows the summary of the validation test methods and the requirements after the test to assure the suitability of the cowl over support of rGF.

Fig. 14 Recycled material in pellets

Fig. 15 Cowl top support injection process

Fig. 16 Prototype cowl top support with rGF

Table 7 Validation test conditions

Functionality	Test method	Requirement
Impact resistance	(1) 200 g/50 cm/−20 °C	Without break and without cracks
	(2) 1000 g/50 cm/23 °C	Without break and without cracks
Weathering	(1) 4 cycles. One cycle being: 16 h at 40 °C and 95%HR + 3 h at −20 °C + 6 h at 100 °C	Without deformation. Without peeling off
	(2) 3 cycles. One cycle being: 22 h at −40 °C + 2 h at 25 °C + 22 h at 95 °C + 2 h at 25 °C + 22 h humidity + 2 h at 25 °C	Without deformation. Without peeling off
Thermal stability	(1) 30 min at 110 °C	Without deformation. Without peeling off
	(2) 22 h at 90 °C	Dimensions between tolerances
	(3) 7 days at 70 °C	Dimensions between tolerances
	(4) 24 h at −40 °C	Dimensions between tolerances

HR Humidity relative

Table 8 Summary of the results of the tests performed

	PPGF30	PPrGF30
Weathering	OK	OK
Thermal stability	OK	OK
Impact resistance	OK	OK

The summary of the results is shown in Table 8.

5 Light Transmitting Single Skin Profiled GFRP Sheet Demo Case

The main objective to be reached with this demonstrator is to substitute virgin glass fibers (chopped roving) with a CSM (Chopped Strand Mat) made with rGF (so called TrGF Mat—Thermal recycled Glass Fibers Mat) to produce a light transmitting single skin profiled GFRP sheet. A Chopped Strand Mat is basically a non-woven fabric where a "binder" keeps the chopped fibers together.

Fig. 17 Cormatex industrial production of rGF mat

5.1 Materials

To manufacture this demo case, rGF from pyrolysis/depolymerization process have been used, combined with UP. The objective was to maintain the properties when the 25% of virgin GF coming from roving would be substituted totally by rGF coming in mat format. The rGF mat required in this demo case has been produced by an external company (Cormatex S.r.l.) in its facilities. The company developed an air-lay technology based on an aerodynamic fiber batt formation system, capable of replacing traditional mechanical processes using carding machine and cross lapper, in the nonwoven sector (Fig. 17).

5.2 Manufacturing Process

A light transmitting profiled GFRP sheet is usually produced with continuous lamination plant as shown in Fig. 18.

On a polyethylene terephthalate (PET) film (from 20 to 125 μm usually) catalyzed resin is spread with different systems as reciprocator, hopper, squeegee or rotating shaft. Over this first resin layer, a second glass fiber layer is added. Typical glass fiber layers are made directly by an in line cutting roving machine (so called chopped roving) or adding glass fiber mat as a chopped strand mat (CSM), continuous filament mat, woven roving or fabrics. These two layers goes under a calendar where a second PET film is coupled. As an option a thin layer of resin can be spread on the upper film, so called gelcoat finishing (Fig. 19).

Test specimens were manufactured by a continuous lamination plant at standard conditions (lamination speed) validate the use case.

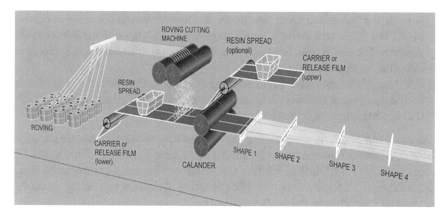

Fig. 18 Basic scheme of a continuous lamination plant

Fig. 19 rGF Mat processing

5.3 Validation Test

The objective of these tests has been to evaluate the processability in a continuous production line. The description of the test carried out are the following:

- A correct resin/GF ratio allows the surfaces to be regular and smooth. rGF mat has a resin/GF ratio 3.1:1 very similar to used virgin GF for Rivierasca application, that is 3.0:1. This means that a rGF mat can easily substitute a virgin CSM in standard application without varying raw material consumption.
- rGF mat wet out time: Wet out time is the time the resin needs to wet the GF. The GF is considered completely wet when it changes its colour from white to translucent. This test is carried out to verify the processability of the rGF mat in an industrial lamination plant without bubbles. The GF must be completely wet before entering the calander, so the wet-out time is closely related to plant speed. At Rivierasca plant the speed is 7.5 m/min and the area where the resin wets the GF is 8 m long. It has been verified that the wetting time of rGF is less than 63 s.
- rGF mat stability in continuous lamination (unbending and linear flow): The objective of this test is to verify that the rGF mat remains flat during lamination thanks to

smooth unwinding and linear flow under the calander without bending or moving aside. The folds will cause a non-homogeneous thickness of the laminate and aesthetical issues. Again, this test verifies the processability of the rGF mat in an industrial lamination plant.

- rGF mat visibility result (translucency) (Fig. 20): translucency is measured according EN ISO 13468-2 (light transmittance). The defined target for the "Light transmitting single skin profiled" was ≥ 60%.

Sample manufacturing and testing to avoid resizing of the rGF mat are shown in Table 9. Resizing was done through adding an organic functional silane to the UP resin system before laminating with the aim to improve the fibers-resin compatibility. The result from testing campaign was that re-sizing of fibers is mandatory to recover mechanical properties to standards values.

Besides a thermal demanufacturing rGF mat, a mat from directly shredded GF composite (Sh-rGF) has been also produced and tested. Such Sh-rGF mat shows an interesting solution for the composite industry especially for non-transparent applications (Fig. 21).

Samples of the results of the industrial lamination tests are shown in the Figs. 22 and 23. Sheet "A" is a test reference. It is a basic corrugated sheet made with virgin

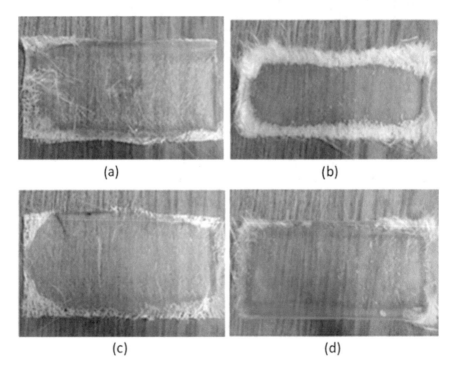

Fig. 20 Preliminary samples without sizing for testing translucency properties. **a** Test reference sample **b** test 1 sample **c** test 2 sample **d** test 3 sample

Table 9 Mechanical properties of specimens

Sample manufacturing

	Test reference	Test 1	Test 2	Test 3
Kind of fiber	Virgin GF	rGF	rGF	rGF
Sample thickness (mm)	0.66–0.74	1.2–1.23	0.73–0.79	0.62–0.78
GF content wt%	25	25	25	12.50%
Additive	Styrene 3%	Styrene 3%	Styrene 3%	Silane 3%
Tensile test				
Modulus (MPa)	8310	6449	8378	6429
Stress yield (MPa)	112	34	45	29
Strain (%)	1.7	0.62	0.59	0.47

Fig. 21 rGF Mat (on the left) and Sh-rGF Mat (on the right)

roving and UP resin. Sheet "B" was produced by substituting the virgin glass fibers (roving) with the rGF mat (the rGF mat substituted the 100% of virgin roving). Sheet "C" was produced by substituting the virgin glass fibers (roving) with the Sh-rGF mat (the Sh-rGF mat substituted the 100% of virgin roving).

Sheet "D" (Fig. 24) is another test reference of a standard product from Rivierasca. It is a basic corrugated sheet made with virgin roving enclosed by two layer of virgin woven roving (on top and bottom side) and UP resin. This kind of multilayer laminate aims to provide higher mechanical performance of the sheets. Sheet "E" (Fig. 24) was produced by substituting only the virgin roving in the middle of the laminate with the rGF mat (the rGF mat substituted the 100% virgin roving). Sheet "F" (Fig. 25) was produced with the same layering of sheet "E" but using Sh-rGF mat instead of rGF mat with the same 100% virgin roving substitution.

The summary of the results is shown in Table 10.

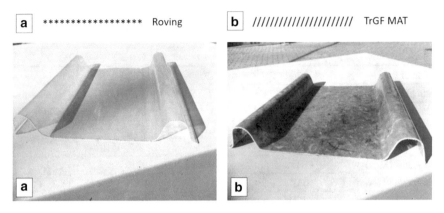

Fig. 22 Standard sheet and its equivalent made with rGF mat

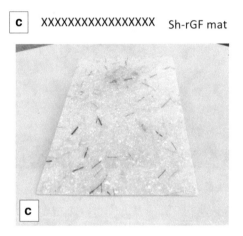

Fig. 23 Sample made with Sh-rGF mat

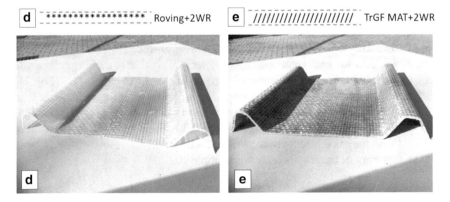

Fig. 24 Standard sheet (3 GF layers) and its equivalent made with rGF mat

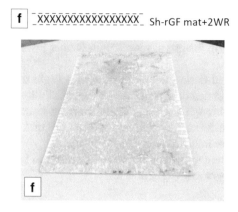

Fig. 25 Sample made with Sh-rGF mat and two woven roving

Table 10 Mechanical resistance of sheets

	Test ref. A	Test B	Test C	Test ref. D	Test E	Test F
Kind of fiber	Virgin GF	rGF	Sh-rGF	Virgin GF	rGF	Sh-rGF
Sample thickness (mm)	0.79–0.87	1.69–1.79	2.58–2.86	1.27–1.32	2.33–2.45	2.76–2.79
GF content wt%	25	25	17	14	13	13
Tensile test						
Modulus (MPa)	7765	5196	3302	16,018	15,317	7361
Stress yield (MPa)	81	41	20	216	73	97
Flexural test						
Modulus (MPa)	9994	6343	3378	14,651	14,802	14,257
Stress yield (MPa)	277	102	52	443	259	353

6 Conclusion

As a final technical conclusion after analysing the validation test of the different demo cases, it is summarized that:

- Pedal Bracket manufactured by BATZ passed the validation test, which consisted on an aging pre-treatment of the part, to subject later to different loads applied in different areas of the brake and gas pedal. Thus, it gives rise to the use of new materials in the automotive sector.
- Front-end carrier also manufactured by BATZ, has also been technically validated, fulfilling above the mechanical requirements defined for bonnet latch traction and compression tests. In some of the cases, the results obtained for the new material with recycled fiber overcome the values of the original material.
- Cowl top support also passes the validation test. After the results of the validation carried out on the component, it is proven that the weathering and thermal stability

tests have been passed. In addition, the impact resistance test, being the most critical due to the use of recycled glass fiber, has also been successfully overpassed. These results confirm the use of recycled glass fiber instead of virgin glass fiber for this application.

- The results obtained during the production of a Light transmitting single skin profiled GFRP plates demonstrated the technical feasibility of glass fiber recycling and reprocessing through a process chain from waste or EoL products to new semi-finished products such as rGF mat and Sh-rGF mat. These results are very promising for future industrial uptake and business.
- The reduction of the mechanical properties of rGF can be improved by using a mix of recycled fibers and virgin fibers.

References

1. Åkesson, D., et al.: Glass fibres recovered by microwave pyrolysis as a reinforcement for polypropylene. Polym. Polym. Compos. 21 (2013)
2. A critical review on recycling of end-of-life carbon fibre/glass fibre reinforces composites waste using pyrolysis towards a circular economy. Resour. Conserv. Recycl. **136**, 118–129 (2018)
3. Matrenichev, V., Lessa Belone, M.C., Palola, S., Laurikainen, P., Sarlin, E.: Resizing approach to increase the viability of recycled fibre-reinforced composites. Materials **13**(24), 5773 (2020)
4. Acres, R.G., Ellis, A.V., Alvino, J., Lenahan, C.E., Khodakov, D.A., Metha, G.F., Andersson, G.G.: Molecular structure of 3-aminopropyltriethoxysilane layers formed on silanol-terminated silicon surfaces. J. Phys. Chem. C **116**(10), 6289–6297 (2012)
5. https://portal.a2mac1.com/
6. Olivares Santiago, M., Galán Marín, C., Roa Fernández, J.: Los Composites: Características y Aplicaciones en la Edificación; Informes de la Construcción, Vol. 54 (2003)

Use Case 3: Modular Car Parts Disassembly and Remanufacturing

Stefan Caba⊙, Justus von Freeden⊙, Jesper de Wit⊙, and Oliver Huxdorf⊙

Abstract Cars designed for reuse are a novelty. The disassembly and remanufacturing of major vehicle structures is not an established process yet. This means new tasks and within that, new business models must be found to close the loop. New facilities and logistics are part of the process chain. In this chapter the processes will be outlined on basis of a car-sharing vehicle as an example of a fleet ownership. The reusable platform and the seat structure are the basis of this model-based consideration. The technical issues and obstacles will be discussed as well as economic and ecologic questions.

Keywords Remanufacturing · Inspection · Repair · Logistics · Detachable joints · Car-sharing · Fleet owned vehicles · Life-cycle assessment

1 Introduction: New Opportunities for Car-Sharing Vehicles

Cars are one of the best examples of progress over the last century. Today about 1.2 Billion cars are in use [1]. The numbers are growing fast due to a growing middle class in the developing countries. Buying a car is still a major step in life for most people, showing independency and comfort. Nevertheless, this trend has also led to increased traffic jams and high cost for parking spaces, because the infrastructure cannot grow accordingly. Today none of the cities with the highest traffic jam level is to be found in the western world [2]. This means the problem has already become

S. Caba (✉)
EDAG Engineering GmbH, Fulda, Germany
e-mail: stefan.caba@edag.com

J. von Freeden
Fraunhofer Institute for Machine Tools and Forming Technology, Chemnitz, Germany

J. de Wit · O. Huxdorf
INVENT Innovative Verbundwerkstoffe Realisation und Vermarktung Neuer Technologien GmbH, Braunschweig, Germany

© The Author(s) 2022
M. Colledani and S. Turri (eds.), *Systemic Circular Economy Solutions for Fiber Reinforced Composites*, Digital Innovations in Architecture, Engineering and Construction, https://doi.org/10.1007/978-3-031-22352-5_17

worldwide and it cannot be solved due to high demand for space in the city. Today, parking slots take up about 10% of traffic areas in cities [3]. Since vehicles are not in use for the major time of the day, the whole system is not very inefficient. The vehicles in parking spaces are blocking spaces for buildings, green areas and streets. The low utilization rates of cars in combination with the fast loss of value show that the private owned car in many cases is not the predestined solution for cities.

Classic solutions to improve mobility are to expand public transport like busses and metro lines. However, these solutions are not addressing the independency and comfort. Car-sharing as an alternative to owning a car could help to improve this challenge. This means the ownership of the car goes to the car sharing provider together with many changes of valuation and utilization.

For the FiberEUse project, sharing vehicles also show another benefit of high importance. These vehicles are not privately owned which means, the owner is always interested to earn money with these vehicles. The owner will act as a homo economicus and always decide the option for himself. This makes the decisions calculable and comprehensible. Another major aspect is logistics and knowledge. While a privately owned car can be sold to a new owner at any time, the car may be used anywhere in the EU or in other countries. To close the loop, the vehicles must be brought to a collection point with a very high logistic effort. For the car sharing cars this factor is different. Their position and sometimes even their condition are known and they can be collected comparatively easy. This lowers the cost from the start and it is seen to be a precondition for a first implementation of reusable car structures.

The idea of reusing structures led to the design of a vehicle platform and a seating structure visible Fig. 1 with special purposes that will be discussed in the following. Both together are demonstrating how reuse of vehicle parts made of CFRP will become possible to achieve improved life cycles.

1.1 Use-Phase of a Sharing Vehicle

The car sharing provider follows a simple basic business model. The user must rent the cars he possesses for a long time and distance. This means the cars intendedly designed for private utilization will be used for a much longer time per day resulting in a much younger age of the car when it reaches the maximum distance it was designed for. This distance varies from producer to producer. An experience value is about 160,000 km.

The owner of the car—this could also be a leasing company—will only keep it in the portfolio while it makes sense in an economic way. The cost for inspection repair is growing with more years and kilometres. The design of the car on the other hand must be kept young. The users want to drive a modern car with new technologies. These technologies are not only the powertrain, but also the electric components and connectable devices in the interior. The users are not the owners of the car. They have low experience in driving it. Not owning the car also leads to lower caution and

Fig. 1 Demonstrator of use-case 3—reusable structures for vehicles

sometimes vandalism. These effects lead to a comparatively short phase of utilization in car-sharing of about 4 years [4].

The overall distance of a carsharing car is not the only driver of value loss. The fleet management will always try to keep the fleet in a modern state to lower inspection costs. Today the desired 160,000 km distance is usually not reached. With increasing market penetration of car-sharing this value seems to be realistic, resulting in 40,000 km per year. The usual distances vary from only about 10 kms in particular for free-floating to 30–60 km for station-based systems [5].

Car-sharing is a growing business segment in Europe. In Germany only 25,400 sharing vehicles exist. This number has increased from 15,000 within 5 years. The number of users has reached 2.5 Mio. A very interesting point is that the share of electric vehicles is more than 20%. This shows the high affinity to new technologies within the users. It can be shown that a single car-sharing vehicle may replace about 10 privately owned cars [5].

1.2 Basic Steps of a Car with Reusable Structures

The reusable car shows a life-cycle differing from a standard vehicle. While many elements are similar, new phases and in particular a circular phase is added. The production of the car and a first utilization phase are similar to the standard process. But after this first use-phase an innovative procedure begins. Already during use, it is possible to gather as much information about the vehicle as possible. This can indicate how to proceed (Fig. 2).

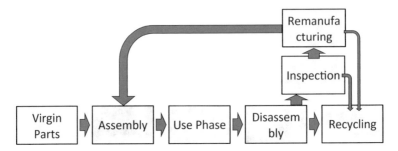

Fig. 2 Basic steps for reuse of structures in cars

The first new step is the detachment of structures. This step is not a novelty, but the result here is not separated materials, but separated healthy reusable structures on one hand and not reusable parts and materials ready for recycling on the other hand. For the recycling, the developed technologies of the two other use-cases are predestined. Here, detachable connections as designed in the discussed car structures can play out their benefits.

It follows the inspection phase. Here, a first visual check is followed by precise measurements of the structure conditions. In an usual case, the precise measurement is not necessary, because the reusable structures were designed for a longer lifespan and several reuse cycles. However, crashes and unforeseen event can lower the mechanical properties and can be a risk. This is a reason why data collection can help to improve the process. Possible technologies are simple measurements of acceleration forces or structural health monitoring.

The remanufacturing phase will be conducted for all structures intended to be reused. To achieve ecologic processes, this must be a very high share. In the remanufacturing step, first the joints are cleaned. Since the structures are made by composites, adhesive connections are preferred. The novel detachable adhesive connections will leave residual adhesive since they are failing cohesively in the adhesive layer. After cleaning, another inspection could take place. The cleaned parts can then be repaired if smaller cracks or surface deflections occurred.

Last step is to bring these parts back into production. This means a logistic effort to bring the structures to the vehicle manufacturer. By tracking the parts, it is possible to retrace the history of any structure. Long-term failures can be detected.

The new procedure begins, when the owner of the car decides, that maintaining the existing vehicle causes more cost than a new vehicle. This can also be induced by new technologies or designs influencing the market. The car sharing provider is always intended to achieve the highest return on his investment. Different business models could be applied to determine the ideal point of re-investment in form of a new vehicle based on the reusable structures.

When the decision is made, the cars first must be recollected. This is a work-intensive logistics process. For station-based car-sharing the vehicles will be located at the stations while for free floating car-sharing the cars must be located and collected. The place for the disassembly, inspection and remanufacturing process is preferably located centralized to reduce logistics and investment costs. Here, it must be noted, that the transport of the cars for longer distances reduces the economic and ecologic assessment of the circular value chain.

2   Disassembly

The disassembly process begins at the centralized remanufacturing facility. The disassembly is conducted similar to classic disassembly. The components worth to be reused as a spare part or using the new reuse procedure must be recovered from the parts for the recycling. The main goal is to separate the vehicle into the different categories. The novel reuse parts are another group beside the spare parts and the different recycling grades.

Usually, the process begins with the add-on parts like doors and front ends. These parts can easily be reused in form of spare parts. It can be necessary to perform a remanufacturing step like dent removal or painting. The processes can be seen in Fig. 3.

In the next steps the reuse-parts can be recovered as well as many other spare parts or recycling material. The battery housing is a bohemian part for battery electric vehicles. The prepared detachable joints are used to recover the battery housing easily. It is assumed that the capacity is usually still high enough for another car life. Due to a high dynamic development in this area, it is not part of this consideration.

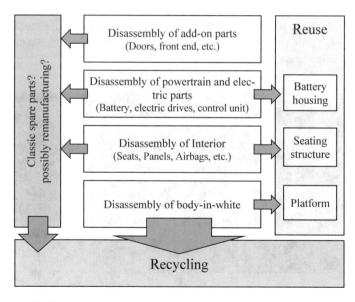

Fig. 3 Schematic disassembly procedure

During the third step the interior is disassembled. Here, the seat can be recovered and goes to the specialized station for reusable seats that will be discussed in Sect. 2.2.

The body-in-white usually is not part of the disassembly process, because all connections are non-detachable. Here, the novel design enables an additional dismantling step.

2.1 Platform Disassembly

The platform is the most important part of the vehicle that consists only of a dismantled body-in-white. To recover the platform several connecting elements must be removed. This is a quite simple process for the numerous bolt connections at the front and the back of the car. They connect the aluminum front-end and back-end to the so-called hat with the passenger cell.

For the rocker panel, the process is more complicated. To meet all crash safety requirements, the aluminum crash structure is connected to the platform with a sealing and structural polyurethane adhesive. A simple technology to cut this connection is to use a wire saw. This is a thin wire with cutting teeth or a specialized geometry to cut through adhesives. These wires are also in use for removing windshields. In an experiment visible in Fig. 4 the disconnection of two carbon fiber reinforced parts was investigated.

It could be shown that cutting through the adhesive is more energy intensive than cutting through a simple sealing adhesive for windshields, but it is possible. Hence, it is highly recommended to use a commercially available CNC-wire-saw. Here, high precision cuts can be performed not influencing the fiber reinforced parts. Using this technology requires a non-curved connection surface. The components can be reused after a following remanufacturing process.

Fig. 4 Disassembly procedure for adhesive joints using the wire saw technology, cutting process (left) and cutting edge after separation (right)

Fig. 5 Disassembly of the
body-in-white

Cutting the connection between the platform and the rocker panel only makes the whole platform ready for reuse (Fig. 5). It is clear that in many cases another step must be performed to disassembly the front-end and the back-end of the car to exchange parts or to achieve another vehicle lifespan for the platform designed for at least seven lifespans. Thus, a last disassembly step is necessary.

Front-end and back-end are connected in a combination of a bolt connection and a sealing structural polyurethane adhesive connection. The connection surface is curved. The wire-saw technology is not applicable here. However, it is possible to separate the connection if the detachable adhesive technology is applied. This means the platform must be put into an oven at the desired temperature to lower the cohesion of the adhesive. After a short time, the platform can be separated mechanically. The use of jigs is recommended here. Theoretically this technology could also be used for the connection at the rocker panel.

2.2 Seating Structure Disassembly

In contrast to the platform described in Sect. 2.1, other or further steps must be observed when dismantling the seating structure.

After removing the seat structure from the vehicle, all moving or mounted parts such as the head restraint or seat belt latch must first be dismantled. Then, it is possible to proceed with the removal of the cover and upholstery from the actual seating structure. In most cases, this requires a specific force, as the cover is usually glued, see Fig. 6.

Fig. 6 Exemplary removal of the cover of an Opel Zafira

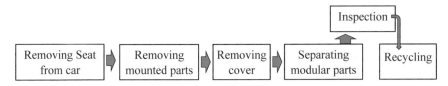

Fig. 7 Basic steps for the disassembly of a seating structure

These steps describe the removal of the visible components that can be replaced each time during a vehicle refurbishment to implement modern new designs in a new life cycle.

Thus, from this point on, the pure seating structure suitable for reuse is present, as described in Chap. 12 (Modular Car Design for Reuse). From here on, the separation of the modular components has to be realized. This is possible by applying for example the technique of non-destructive separation of adhesive bonds as described in Chap. 10 (Composite repair and remanufacturing).

These described individual steps as well as the subsequent steps can be seen in an overview form in Fig. 7.

3   Inspection Process

After the end of the life cycle and the subsequent dismantling of a vehicle with the associated interior, an assessment of the condition of the components and structures designed for reuse is necessary. For reuse, the components must be intact or at least repairable in order to meet the safety requirements for the next life cycle. An inspection process was developed to determine the component status, which consists of various process steps and decisions. The inspection process chain is shown in Fig. 8.

At the centre of the inspection process is non-destructive testing. Prior to this, pre-sorting is carried out by means of a visual inspection. Components are generally

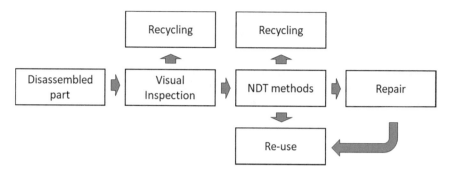

Fig. 8 Process chain for inspection process of reusable composite parts

classified into three categories: direct reuse, reuse after repair, no reuse and continued use in material recycling processes.

3.1   Visual Inspection

The first step in the inspection process chain is the visual inspection. The dismantled components are assessed by trained specialist staff. This can also be done during dismantling. The visual inspection is intended to sort out severely damaged components, which can no longer meet the requirements for the component even through repair.

3.2   Non-destructive Testing

Not all component damage is visible from the outside. Internal damage for which the detection method must be used is also possible. Non-destructive testing methods offer a very good option here. As part of the project, three different non-destructive testing methods were examined: ultrasonic testing, hyperspectral imaging and active thermography. The technical details of the methods and the tests performed are described in Chap. 10.

Complete automation is crucial for the sensible and effective use of non-destructive testing methods. To do this, it is necessary to install the appropriate test equipment, such as sensors on robots that automatically scan the components. Special software programs are required to evaluate the sensor data and detect possible damage. Ideally, the software can be used to decide whether the components can or cannot be reused, or whether they need to be repaired.

3.3 Recycling Strategies

Components that are so badly damaged that they cannot be prepared for a further life cycle even after being repaired should continue to be used in material recycling processes in the interests of sustainable industry. In use cases 1 and 2, mechanical and thermal recycling processes and corresponding processes for manufacturing products using recycling materials were developed.

For non-reusable components from use cases, the thermal processes from use case 2 are particularly suitable due to the use of carbon fibers. However, the use of mechanical processes is also conceivable, especially in the case of very badly damaged components that did not have broken fibers. In any case, the recovery of continuous fibers through the thermal processes is no longer possible here.

4 Remanufacturing Procedure

After a classification of the components has been carried out, which can be transferred into a remanufacturing process, a construction of an overall structure with these reused components is possible. However, specific boundary conditions must be observed for a reuse process. As described in the previous chapter, the components have been tested for their suitability and freedom from damage or damage limitation. Nevertheless, reprocessing must be carried out.

This is necessary for isotropic materials such as metals. But especially this is true for composites, an anisotropic material. This class of material require among other things a special pretreatment. These steps are described in the following sub-chapters.

4.1 Remanufacturing of Adhesive Joints

Adhesive bonds are one of the most important structural connections for composites due to the damage-free and homogeneous joint bond. However, after a separation of the overall structure, adhesive residues are likely to still adhere to the sub-components. These adhesive residues must be removed, and the surfaces prepared before a new bond is made.

This removal of the material is possible by various techniques, for example by grinding, milling or laser processing. Laser and milling can be automated, which enables precise processing. This is recommended for thin adhesive layers and composite materials. Here, the thickness of the layer to be processed can be set very precisely, which ideally means that no fiber layers are employed. Figure 9 shows the results of a series of tests in which various laser parameters were tested. The aim was to obtain an ideal ablation pattern. The lowest box shows a laser processing in

Fig. 9 Test series of laser processing

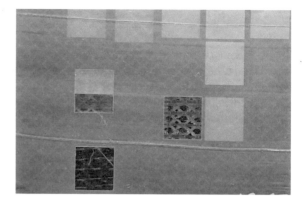

which as much adhesive as possible was achieved with as little damage to the fibers as possible.

A manual and rather rough processing is grinding, which is recommended for thick-walled adhesive layers and metal components, see Fig. 10. Due to the manual guidance, the machining depth can vary depending on the processor and, in the worst-case scenario, damage the fiber layers of composites. However, this process is the economically ideal one when adhesive has to be removed in a layer several millimeters thick. It is applicable for the rocker panel of the platform, where a thick layer of adhesive is necessary to equalize the thermal stresses between the aluminum and the CFRP part.

Fig. 10 Removal of adhesive by grinding

4.2 Repair

To get the remanufactured components into a renewed use, a possible repair is necessary. The non-destructive inspection already carried out and described in the previous section has revealed the extent as well as the positions. This makes it possible to carry out a specifically defined repair.

There are three states of what a selected component for reuse may look like:

1. Components that are severely damaged cannot be reused. These would therefore be sent to a thermal or mechanical recycling process.
2. In the case of very minor damage, only optical defects exist. Here, the structure itself is not damaged. As a repair, for example, an optical reconditioning in the form of a new paint or coating can be carried out.
3. Components with minor damage can be repaired. Specific processes and machining steps are necessary for this.

This third case, minor damage to the structure of a component, is described in detail in Chap. 10 "Composite repair and remanufacturing". In summary it can be said that repairs to fiber composite structures have been carried out in aviation for a long time and are accordingly established. These processes can be transferred to the repair of automotive fiber composite components. A more detailed analysis is explained in the corresponding Chap. 10.

4.3 Reuse of the Structures for New Vehicles

After the described reprocessing of the individual components, it is possible to start a new life cycle. For this purpose, the components must be integrated into the overall structure. In principle, this should be done accordingly to the same instructions or boundary conditions as the initial new build. As a consequence, the same support materials, processes and competencies that already exist can be used.

The more life cycles a component experiences and the more often it is reprocessed as a result, it must undergo a more precise reprocessing of the surface. This is because the reprocessing of an old adhesive layer that has been removed, for example, may damage the fiber layers: The more that is removed, the deeper the fiber layer must be penetrated. As a result, the strength of a composite component is affected, and must be considered in the design process.

Tensile shear tests on bonded specimens showed a good behavior if the component was carefully and well prepared. In contrast, a component that has only been cleaned without further preparation gives low values in the case of rebonding. The numerical values of tensile shear strength corresponding to the types of failure are shown in Fig. 11. The values for the samples without removing the adhesive and the first layer are visible in blue, the values for a good preparation of the surface by milling are shown in red. The values for the initial bonding are shown on the left, and the values for the two repair cases are shown on the right.

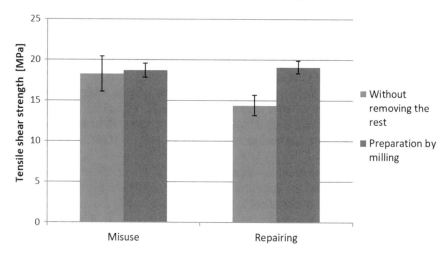

Fig. 11 Numerical results for repairing faulty separated adhesive bonded samples

This figure shows, that a precise reprocessing will allow a wide reuse of composite materials. This usage of composites would greatly reduce pricing pressures in the automotive industry and could therefore be used more widely in large-scale automotive production. As a result, the use of these lightweight materials will lead to a reduction in vehicle mass, resulting in more energy-efficient vehicles.

5 Life-Cycle Assessment

The mission of circular economy is to reduce impacts for the environment. Reducing the waste is the usual objective, but on the other hand the impact can be measured by emissions. The concept is proven by a life-cycle assessment, which is carried out exemplarily for the platform.

The new defined life-cycle of the platform is supposed to lower the overall emissions. The basic idea is to reduce the parts of a car being produced. Electric cars cause low emissions if they are fueled with renewable energies. The production of the car itself is still very energy consuming and causes emissions due to chemical processes.

Not producing a new platform for the next generation of the vehicle lowers the emissions if the inspection and remanufacturing procedure causes lower CO_2 than the reproduction of the platform. On the other hand, a platform lasting for the desired 30 years and more than 1 Mio. km must be designed very robust. In a life-cycle assessment it will be shown if it is possible to lower the emissions and when a break-even point will be reached. Thus, a comparison of the new lightweight platform with a steel-based platform will be made.

5.1 Boundary Conditions

The basis of the examination is a strict set of boundary conditions. These conditions are derived from the purpose of the vehicle in car-sharing. The basic condition is the supposed lifecycle of the vehicle. Here, a standard value is about 180,000 km. This is the distance when the costs for maintenance start to exceed the cost for a new car. Considering that the car is used in public and by people with low experience in driving this car model causing smaller and larger accidents, 180,000 km seems to be a realistic value.

The velocity of the car being used mainly in the city will be about 30 km/h. The energy of the car will be the standard-mix of Germany as the largest country in Europe. The production of the fibers will be in the US using hydrogen power. This is similar to a large CF-producer from central Europe. For the aluminum and steel parts, the up-to-date recycling rates are included. This is lowering the emissions by far. The manufacturing processes and the used media like oil are also included.

The object of comparison is not the whole car, but the platform only. The passenger cell and the suspension can be seen as similar. Even the fuel consumption due to air resistance is equal. Hence, the energy consumption was calculated for the whole vehicle. The sum was divided by weight so that the two different platforms will show slightly different consumptions. The weight difference is 30 kg resulting from the lightweight design for the much more durable new platform.

5.2 Life-Cycle

The life-cycle with the said boundaries is set up in a LCA-tool. Here, the software Gabi was used. Others show similar results. The basis of the examination are databases for materials and processes. Here, the standard-database and in addition the database for reinforced plastics were used.

In Fig. 12 the model for the new CFRP platform is shown. The basic phases of the life-cycle are shown. This picture shows the first life-cycle of the platform. After the use-phase the materials are put to recycling—or to reuse.

The result in absolute numbers for the defined Scenario 2 of 40,000 km per year can be seen in Fig. 13. It shows that the production of the lightweight design platform causes much more emissions during production. The reason is the production of the carbon fibers that must be produced without recycling in a very energy intensive process. During the first use-phase a quite similar slope of emissions can be detected since both platforms show only a medium difference in weight.

After 4 years, the vehicle must be exchanged. This means a new production of the steel-based platform, while the lightweight platform will be remanufactured. This is the point, when both diagrams show an intersection. The new production of the steel platform causes further emissions, while remanufacturing can hardly be seen

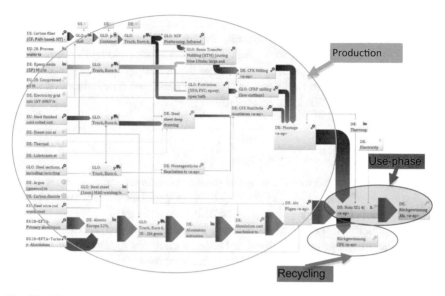

Fig. 12 Life-cycle model of the CFRP platform

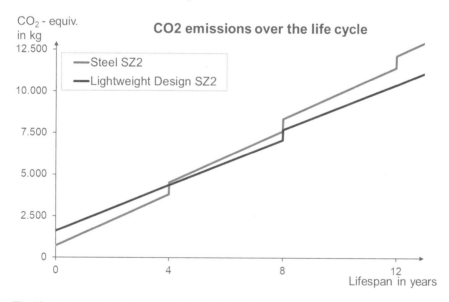

Fig. 13 Emissions of both platforms during the first life-cycles

in this scale. After 8 years, the front-end and the back-end made of aluminum must be exchanged, too. This causes also emissions for the lightweight design platform.

After 12 years, the difference of both diagrams is about 2000 kg CO_2-emissions. This is the amount of the whole production of the lightweight-design platform. For the next 16–18 years, this platform will still be in use and remanufactured saving another about 4000–6000 kg of emissions.

5.2.1 Interpretation

The life-cycle assessment has proven that the new circular design platform will save high amounts of emission during the long lifespan. Similar results are expected for the seating structure. Some points must be discussed upon this.

The production of carbon fibers is a very high CO_2-intensive process. In Fig. 14 the specific emissions are visualized. It becomes clear that lightweight design in a first step is causing more emissions. This is relevant for aluminum, too. The lower emissions during the use-phase can equalize these first emissions. It must be noted that in many cases, lightweight design will not cause a positive effect on the emissions, but on the performance.

Lowering the emission for the production of CFRP is the goal of many other research projects. Here, bio-based precursor materials are used. The resin can also be bio-based. This means the consumption of fossil fuels. Since these materials today are not produced in higher amounts, it must be shown that they cause lower emissions, too.

The batteries of the electric powered car were not under examination. New battery generations with better properties are presented every year. In the near future the batteries will reach a productive level. Hence, it may be possible also to keep the batteries for more than one lifespan. This will reduce the effort for remanufacturing on the one hand and lower the overall emissions of the car by a similar or even higher amount. The reusable platform will be able to support this already today.

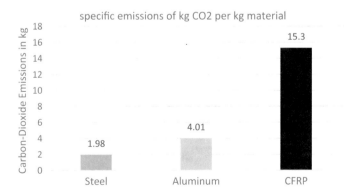

Fig. 14 Specific emissions of the materials

6 Conclusion

The new circular approach for a car-sharing vehicle is shown. The use-case of a sharing vehicle already brings the benefit of lower space consumption in the city and higher utilization per day. In addition, the shared vehicles are owned by the car-sharing provider being able to locate the car and to measure its conditions. Based in this scenario, the basic steps of a reusable platform and a reusable seat are presented. The disassembly, the inspection and repair are the most important phases. New strategies for cutting adhesives, detachable joints, cleaning and inspection are summarized. These technologies lead to the goal to regain as many CFRP platforms and seats as possible to reduce the overall amount of these structures produced.

In a life-cycle-assessment it could be shown that the new circular process pays off in form of much lower emissions. This result was expected and proves the advantages of circular design and remanufacturing instead of producing new parts for the automotive industry.

References

1. Tatsachen und Zahlen, Jahresberichte, Verband der Automobilindustrie (VDA) (2020)
2. TomTom Traffic Index 2019, TomTom International BV (2019)
3. Steierwald, G., Künne, H.-D., Vogt, W.: Stadtverkehrsplanung. Springer, Wiesbaden (2004)
4. Littmeyer, A.: Interview about Carsharing Vehicles, Bundesverband Carsharing (2018)
5. Bundesverband Carsharing: CarSharing in Deutschland 2020 (2020)

Material Library System for Circular Economy: Tangible-Intangible Interaction for Recycled Composite Materials

Alessia Romani⬤, Fabio Prestini, Raffaella Suriano⬤, and Marinella Levi⬤

Abstract Currently the development of new circular materials has brought up the necessity to transfer their knowledge amongst the interested stakeholders for their real exploitation. This chapter aims to illustrate the design of a physical and virtual library system of the FiberEUse project. In particular, this library system wants to foster the development of new applications and value chains through the showcase of the new recycled composite materials and archetypal remanufactured products developed during the project. After the definition of the system concept, specific taxonomies were designed for the physical and virtual parts considering the technical properties and the expressive-sensorial qualities of the new recycled materials and products. A hierarchical organization was then designed to allow both tangible and intangible interactions with the samples, resulting in a coherent experience to explore these new recycled materials. Meanwhile, the physical exhibitors and the library website were developed to collect the physical and virtual samples. At the end, the whole system will be freely accessible through the library website and by booking a visit to the physical part. Thanks to its transdisciplinary nature, this system can stimulate the real exploitation of new value chains and applications.

Keywords Design for sustainability · Circular economy · Material library · Materials experience · Material driven design · Glass fiber reinforced polymers · Carbon fiber reinforced polymers

A. Romani (✉) · F. Prestini · R. Suriano · M. Levi
Department of Chemistry, Materials and Chemical Engineering "Giulio Natta", Politecnico di Milano, Piazza Leonardo da Vinci 32, 20133 Milano, Italy
e-mail: alessia.romani@polimi.it

A. Romani
Department of Design, Politecnico di Milano, Via Durando 38/A, 20158 Milano, Italy

© The Author(s) 2022
M. Colledani and S. Turri (eds.), *Systemic Circular Economy Solutions for Fiber Reinforced Composites*, Digital Innovations in Architecture, Engineering and Construction, https://doi.org/10.1007/978-3-031-22352-5_18

1 Introduction, Motivation and Objectives

1.1 Knowledge Sharing of Circular Materials and Technologies

In recent years, socio-economic and environmental challenges have increasingly emerged and led to the affirmation of new ecologically correct development models, also named sustainable development models [1]. The first definition of sustainable development was given by the United Nations Commission on Environment and Development in 1987 as follows: "sustainable development is development that meets the needs of the present without compromising the ability of future generations to meet their own needs", which should belong to the following sustainability dimensions: social, environmental and economic [2]. For this purpose, in 2015 the member states of the United Nations proposed and began to foster the 2030 Agenda for Sustainable Development. This resulted in the adoption 17 Sustainable Development Goals, which are based on the three sustainability dimensions [3]. Among sustainable development systems, circular economy has been identified in the literature as a paradigm with its own benefits and limitations [4–6]. The implementation of circular economy practices has led to the definition of at least the so-called 10 R-imperatives [7], among which Reuse and Recycling are ones of the most used in literature [8].

However, the use of circular materials is hindered by some barriers, which can be recognized in the identification of new solutions in waste management and materials recycling technologies [9]. More specifically, the development of circular products requires a better sharing of waste- and recycling technology-related knowledge as well as real collaboration between all the potential actors of the new circular product chains (i.e. designers, policymakers, managers, engineers, etc.). In the case of polymer composites, the perceived lack of recyclability or the limited choice between existing recycling technologies could also limit their further development, possibly leading to the progressive elimination from some market segments [10].

Composite materials are heterogeneous, being constituted by several components, which makes their disassembly a fundamental requirement for recycling. It consists of the product dismantling by separation of the material components. Moreover, the recycling processes can be non-destructive or destructive, depending on the final damage level of the components. Shredding is an example of a destructive disassembly process, and it is the most efficient and time-saving technique, even though it limits the possibility of material reuse, remanufacture and recycle due to its high damage level [11]. On the contrary, the material components can be undamaged if a non-destructive disassembly process is carried out. However, the non-destructive disassembly approach is difficult to be implemented, because it poses technical problems, such as strict tool requirements and high operational costs. Usually, a partial disassembly process is employed to sort only the main material components, which can be reused. In addition to the recycling technologies, several consolidated or non-conventional processes can be employed to create new value-chains starting from new recycled polymer composites (i.e. extrusion-based additive manufacturing) [12, 13].

Considering the complexity of circular economy system and the numerous involved players, it is therefore evident that the development of new circular materials, in particular reinforced polymers, and value-added market segments is strictly related to the possibility of showcasing the real technological solutions to all the potential stakeholders. This will encourage the real exploitation of products and materials composed of recycled Glass Fiber Reinforced Polymers (rGFRPs) and recycled Carbon Fiber Reinforced Polymers (rCFRPs).

1.2   Research Goal: Materials and Products Showcase from FiberEUse

This chapter aims to stimulate the scouting of new added-value markets through the showcase of the results achieved within the FiberEUse project. For this reason, a physical and virtual library system was realized including the new recycled materials and the archetypal remanufactured products made with rGFRPs and rCFRPs. The main goal of the library system is to foster the development of new applications and value-added markets by showcasing the main results of the FiberEUse project to all the potential stakeholders that could implement these solutions in new value-chains (i.e. designers, engineers, technicians, industries, and so on). In other words, the purpose of the library system is to encourage the real exploitation of new products and materials made of rGFRPs and rCFRPs by sharing their knowledge through a freely accessible physical and virtual library system.

A taxonomy was deeply studied to classify these parts according to the materials (recyclate fibers and matrix), the manufacturing technology, the process conditions, the surface finishing, the post-processing, and the field of application of the products. Generally speaking, the materials, technologies, and demo-cases developed in the previous experimental activities were considered for inclusion in the library system.

The two main parts of the library system (the physical and the virtual ones) were then created to be available to all relevant European stakeholders involved in the composite sector. Moreover, this library system will be set up and continuously updated. As a matter of fact, the library system is available for all the interested stakeholders according to the different fruition of the two parts, the physical and the virtual parts. The first one can be used, seen, and touched at Politecnico di Milano upon reservation, while the second one can be found at: https://fibereuselibrary.com/.

2 Positioning of the Solution

2.1 Aim of a Material and Product Library

Currently, the real exploitation of new circular materials has been assuming a key role in fostering sustainable models of production and consumption. For this reason, the relevant knowledge related to these materials should be shared to foster the development of new applications. However, their knowledge is not yet well-established amongst the potential stakeholders that could generate new value-chains such as designers, practitioners, and engineers. Furthermore, rGFRPs and rCFRPs are mainly characterized from a technical point of view with little consideration of other important aspects. Indeed, there is a considerable amount of literature on the importance of the material surface finishing for the perception of new products. As a matter of fact, a specific material does not only affect the functional features of a product, but also its expressive-sensorial qualities. While the first ones are strictly related to quantitative data and analysis, the latter ones are more intangible and often measured by means of qualitative methods of investigation [14, 15].

Human perception strongly affects the expressive-sensorial qualities of a material, that mainly derives from the senses, especially from vision and touch. Moreover, perception could subjectively change, according to the specific context or user background [16, 17]. Despite the intrinsic qualitative nature, quantitative technical properties (i.e. physical and chemical properties) play an important role in the perception of a specific material [18]. Therefore, technical properties and expressive-sensorial qualities are equally fundamental for the design of new products, and several studies developed new tools for the material selection by merging these two aspects [19–21].

For these reasons, the attention on the expressive-sensorial qualities is increasingly growing amongst all the stakeholders involved in the product development process. Because of different expertise and level of engagement in the design phases, misunderstandings could be frequent. As a matter of fact, the description of some sensorial aspects with quantifiable physical properties can be extremely challenging and inaccurate. On the contrary, physical samples give the possibility to directly experience expressive-sensorial qualities avoiding the loss of sensorial information [22, 23]. However, some issues should be considered. For instance, managing a large quantity of samples without a well-defined taxonomic classification may lead to confusion.

In this complex scenario, material libraries assume a crucial role. As the term suggests, these libraries collect and showcase different material samples organized according to one or more criteria. In this way, materials can be explored and selected with a strong interdisciplinary approach, taking into consideration both technical properties and expressive-sensorial qualities [17, 24, 25]. On the one hand, material libraries are similar to databases, and help designers and engineers during their activities. On the other hand, they represent a *lingua franca* for the material-oriented professionals, and translate unquantifiable sensorial data without any ambiguity.

For some aspects, material libraries are able to lead materials from the development to their real application [24]. As a matter of fact, they can promote the use of a new material by spreading its knowledge amongst the stakeholder. Thanks to the physical samples, designers can directly understand how to best consider new materials for the design of new products. Especially for recycled materials, this point becomes actually more relevant, since the overall perception of a designed product may be strongly affected by the recycling and remanufacturing processes. For this reason, designers should be aware of the possibilities allowed by the recycled materials thanks to the direct approach that a material library offers.

Recently, a circular material library was designed with this specific purpose. One of its peculiarities is that it includes both material samples and prototypes. Moreover, only recycled materials are showcased in order to demonstrate that virgin ones can be replaced in the existing value-chains [26]. Similarly, the aim of FiberEUse is the demonstration of new value-chains based on the recycle of composite materials. As a consequence, a material and product library system would be a suitable way for the showcase of the complex framework within FiberEuse. As can be imagined, this kind of material library should consider the new recycled materials together with the developed products, giving an exhaustive idea of the real design application.

2.2 Tangible and Intangible Fruition of a Material Library

Although the tangible interaction with the material represents the key point of each material library, this experiential learning method is not always possible or even the most reasonable. Actually, this can be influenced by several economic, social and environmental aspects.

Generally speaking, a material library can be considered as a linking point amongst different stakeholders, i.e. the waste producers and the design practitioners, and it is meant to be consulted more than once during the whole design process and value-chain creation. As a result, several reasons could motivate the use of a material library: from the inspirational seeking in the early stages of design to the material selection during the product development phase. Since a material library is generally placed in a physical space, each interested stakeholder should at least book its use and plan to visit the physical space. However, some external factors may hinder this physical fruition and, consequently, the handling of the samples. As an example, the current situation due to the Covid-19 pandemic has strongly reduced the travelling possibilities of the individuals, hence the in-presence activities have been also affected, including the use of these experiential learning tools [27, 28]. As for other contexts (i.e. education), the knowledge-sharing activities shifted from physical to virtual spaces, and most of the them have been carried out in virtual settings by using different tools, websites and virtual meetings software. Focusing on the material libraries, the design of meaningful intangible experiences therefore represents the main issue to overcome, especially considering the lack of tangible interactions with new materials.

Even if the virtual material tinkering cannot fully replace the tangible fruition of the physical samples, it could be seen as a valid alternative for the early phases of the design process, or even a complementary part of an integrated system that can be used regardless of the specific historical situation. Currently, some consulting companies developed virtual databases that showcase different material samples through pictures and technical datasheets. Moreover, some of them have been connected to the corresponding physical material libraries [29, 30]. This results in a wider experience that allows to preliminary browse the materials and to handling the physical samples at later stages, including both intangible and tangible interactions. To sum up, this opens the way for a paradigm-shift toward the concept of "material library system".

3   Methods and Workflow

3.1   Approach and Workflow

Within this project, the following terms were used to distinguish the different kinds of materials (Fig. 1a):

- Waste: materials from End-of-Life (EoL) products provided by the "waste producers", which means Gamesa Renewable Energy S.A. (GAM), Aernnova Aerospace (AER), and Rivierasca S.p.A. (RIV);
- Recyclates: mechanically and/or thermally recycled composites provided by the "recyclate producers", which means Consiglio Nazionale di Ricerca—Sistemi e Tecnologie Industriali Intelligenti per il Manifatturiero Avanzato (STIIMA-CNR), Tecnalia (TEC), and RIV. They can be constituted by either shredded glass fiber reinforced polymers or recycled carbon fibers;
- Recycled materials: new composite materials reinforced by the recyclates.

The workflow of the experimental activities is resumed in Fig. 1b. First, the partners of the consortium that developed the materials, technologies, and products were contacted to collect all the relevant data for the library system. A hierarchical taxonomy system was then designed for the whole library system according to the provided materials (i.e. technical information about the materials and the remanufacturing process, 3D models, physical samples, pictures, and rendering). Later, specific taxonomic rules were defined according to the different interactions of the stakeholders, hereinafter considered as the main target user of the system, with the physical and virtual parts of the system (i.e. physical fruition of the samples, virtual tour of the products). At the same time, the physical and virtual samples were created according to common guidelines. Then, the physical and virtual exhibitors of the library were designed and developed (physical stands and website) to facilitate the showcase of the samples for the stakeholders. In the end, the link between the physical and virtual parts of the library system was checked and fixed to ensure the coherence of the contents.

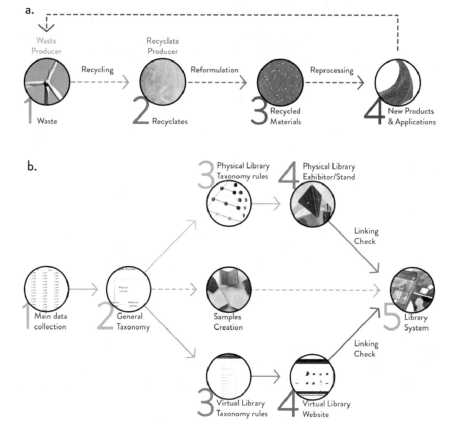

Fig. 1 Main workflows of this work: **a** terminology of the different kinds of materials; and **b** organization of the experimental activities related to the development of the library system

3.2 Physical Library

The Physical library collects the physical samples of the materials and products developed within the project. To better understand the background, a preliminary literature review on the existing material libraries was carried out, especially focusing on the different materials and information showcased. As mentioned before, these libraries are mainly focused on the showcase of different materials rather than products. Furthermore, their taxonomic organization is not always easily recognizable, and each library has its own taxonomy according to the specific samples. Nevertheless, a material library should encourage to see, touch, and compare the physical samples through several aspects, i.e. the process parameters, the visual qualities, the finishing. In most of the cases, the physical samples have similar shapes and dimensions, since the shape itself can strongly affect this experiential comparison. Contrarily, a product library could take advantage of different shapes (and products) because they represent multiple ways to exploit a specific kind of material and/or process. For these reasons, two different paths were followed for the realization of the Physical Material Library and the Physical Product Library. In detail, the Physical Material Library contains flat samples (dimensions: $75 \times 75 \times 6$ mm), while the Physical Product Library is made of cut-off or scaled parts of existing products (dimensions: $80 \times 80 \times 80$ mm max). This dimensional homogeneity allowed to easily manage the spaces of the Physical Library and facilitates their physical handling for comparisons.

The physical material samples were realized by pouring the reinforced polymer in 3D printed molds. The same thermosetting matrix was chosen for the samples to allow the comparison of the recycled fibers. Diverse types and percentages of the recyclates were added to the matrix, and different post-processing was carried out onto the hardened samples. The following ranges for the fiber parameters and post-processing methods were considered:

- Recyclate type: rGF from EoL wind turbine blades (waste producer: GAM; recyclate producer: STIIMA-CNR), rGF from EoL structures (waste and recyclate producer: RIV), rCF from expired prepreg (waste producer: AER; recyclate producers: TEC and STIIMA-CNR);
- Recyclate dimensions: from < 100 μm to 4 mm, long fibers/textiles;
- Recyclate percentage: from 20 to 50% by weight (wt%);
- Post-processing: no post-processing (matte finishing), sanding (matte finishing), gelcoat (glossy finishing), painting (matte and glossy finishing).

The physical product samples were made by the partners of the consortium according to the products from FiberEUse. The samples were cut off to give a dimensional homogeneity of the samples to be handled. A spatial taxonomy based on four different variables was designed to consider the differences related to the specific materials and processes. Different "process units" were created by considering:

- Recyclate type: rGFRPs and rCFRPs;
- Remanufacturing processes: 3D Printing (Liquid Deposition Modelling), Casting, Injection Molding, Lamination, Reuse of components;
- Recyclate dimensions: < 100 μm to 4 mm, long fibers/textiles;
- Recyclate percentages: from 20 to 65 wt%;
- Post-processing: no post-processing, sanding, gelcoat, painting, metallization by PVD (Physical Vapor Deposition);
- Specific process parameters related to some remanufacturing processes: layer height and printing mode (3D Printing), mold finishing (fine-coarse);
- Shape complexity: according to the specific remanufacturing process.

A further literature review was performed to design the taxonomies of the Physical Material Library and Physical Product Library. In particular, the preliminary concept of the latter one is inspired by the spatial color systems (i.e. Munsell color system, the NCS—Natural Colour System®), where the three dimensions represent different parameters, and the spatial position of a specific color is represented by the variation of these three combined parameters. The physical realization of the library was designed by using a CAD software. The custom parts were created by using a desktop 3D printer and laser cutting, and they were then assembled to create the stands of the Physical Library. At the end, a set of QR codes was placed near the physical samples to easily visualize the product and material data.

3.3 Virtual Library

The Virtual Library is a way to virtually collect the samples of the Physical Library, aiming to showcase the samples of the products and materials from FiberEUse. Firstly, a preliminary literature review on the existing virtual material libraries has been carried on and, at the same time, the main tools to design them were searched.

This research brought to three main conclusions. First of all, the core of all the existing library systems is a single database, rearrangeable with some filters. Secondly, the research underlined how much important is to design a User Experience that simplifies the actions that the user needs to find what is looking for. This does not mean that the user must be driven, but that the design of all the needed actions should be precise. Finally, while designing a website, it is particularly important to amaze the user or, at least, to engage him as much as possible to bring him back on the website.

As the number of samples to be showcased in the Virtual Library was substantial, the easiest and fastest way to build a website without coding, i.e. WordPress (https:// wordpress.com/it/), was selected and used. As a matter of fact, this choice allows the integration with WooCommerce (https://woocommerce.com/). This plugin is designed for online shops, but it has a lot of useful tools for complex databases, hence for the Virtual Library. The key feature of WooCommerce is filtering. As previously said, the core element of a virtual database is the presence of filters and WooCommerce allows to easily manage a lot of them.

To manage user interactions to the virtual library, a 360° visualization was added for each product by using a set of nine renderings for each product and a JavaScript developed and shared by 3d web (https://3dweb.nl/). At the same time, the most important part of the work was set, which means the design of the whole experience. For this part, a tool to build quick prototypes and test them with people, called Figma, was used to have first insights of the work done. Finally, surveys were conducted using Google Forms (https://www.google.com/forms/about/) to assess the overall experience related to the use of the Virtual Library.

4 FiberEUse Library System

4.1 From a Library to an Integrated System of Libraries

According to the main goal, a library system was developed to encourage the spread of new applications and value-added markets starting from the results achieved within this project. As mentioned before, the idea of an integrated system of libraries derives from the well-known concept of "material library". From literature, this term refers to an organized collection of samples that allow to showcase, compare, and explore the main technical properties and experiential qualities of one or more materials. Hence, diverse kinds of knowledge can be compared (i.e. mechanical properties, visive and tactile qualities) with a direct experience of the stakeholder that can handle the different samples for an experiential assessment of the potential exploitation. This concept could be used also for the new materials from EoL products developed during the FibeEUse project. However, concrete examples of new applications are generally not part of a material library although their presence may bring additional details on how to best use a specific material. Indeed, their presence could stimulate new different ideas by the users, and, at the same time, they could show some technical features and constraints derived by the material itself and/or the remanufacturing process. The library can also be a point of contact between a future interested customer and the expert who has manufactured the material or the industry that has manufactured the demonstrator. Last but not least, the consultation of a material library can be affected by several economic, social, and environmental factors that do not allow the stakeholders to be physically present or to directly handle the physical samples, i.e. the recent global situation related to the Covid-19 pandemic [27, 28].

For these reasons, an integrated system of libraries was designed and created to merge materials and products on one side, and the physical and virtual fruitions on the other. In this way, distinct levels of fruition can be offered to the stakeholders, reaching a broader scale of potential users. Figure 2a shows the general organization and connection of the library system. According to the different fruition, two parts of the system were developed: (i) a Physical Library linked to the tangible fruition, which means the in-person handling of physical samples; and (ii) a Virtual Library linked to the virtual fruition, which means the remote consultation of virtual samples (i.e. 3D models, pictures, renderings, technical datasheets). In turn, the Physical library is divided into two main parts: (i) a Physical Material Library that contains the physical samples of the new materials from EoL products; and (ii) a Physical Product Library that contains the physical samples of the products developed within FiberEuse. In the same manner, the Virtual library contains the pictures of the new materials and the renderings of the products in the Virtual Material Library and Virtual Product Library, respectively. In addition, the information of the recyclates is shown in the Recyclate Library, as a part of the Virtual Library. Figure 2b represents the various kinds of knowledge that the system gives to the users.

On the one hand, the Physical Library adds the experiential dimension to the showcased materials and products related to the senses and the direct handling of the

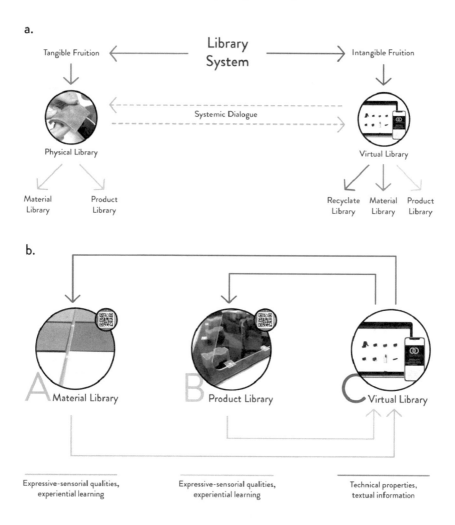

Fig. 2 Library system: **a** general organization and connection of the parts; and **b** the different kinds of knowledge provided by the various parts of the system

samples (Fig. 3a). On the other hand, the Virtual Library gives all the information and quantifiable properties of the samples. This can be possible using a set of QR codes linked to the website of the Virtual Library (Fig. 3b). To sum up, the two libraries, the physical and the virtual ones, can be used separately, but the most complete experience and the highest knowledge transfer can be achieved only through the joint fruition of both parts.

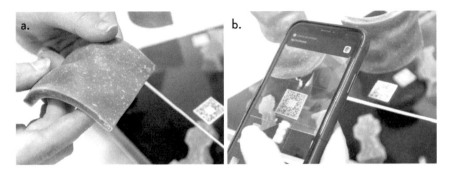

Fig. 3 Showcasing of the library system: **a** direct handling of the Physical Library; and **b** use of the QR codes for the fruition of the Virtual Library

4.2 Physical Library

Physical Material Library. As previously mentioned, the physical material library aims to showcase the recyclates and the recycled materials that the stakeholders can exploit for new applications. For this reason, three variables mainly related to the materials were selected for the taxonomy: recyclate dimension, recyclate percentage, and post-processing.

The resulting samples were organized in different matrixes to let the user make a direct comparison of the samples. To sum up, the Physical Material Library is made of 99 different flat samples divided in three main clusters, as shown in Fig. 4. Further details about the variables and parameters are resumed in Table 1.

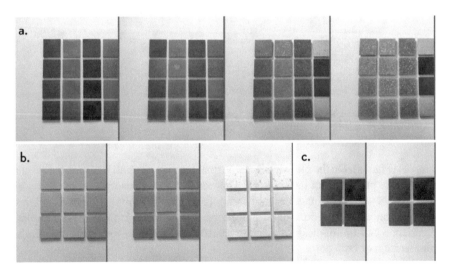

Fig. 4 Physical Material Library: **a** rGFRPs from EoL wind turbine blades; **b** rGFRPs from EoL structures; and **c** rCFRPs from expired prepreg

Table 1 Physical Material Library: the three clusters, the variables and the specific parameters

Fiber	Waste (Cluster)	Variables and parameters
rGFRP	EoL wind turbine blades (64 samples)	1. Recyclate percentage: 20, 30, 40, 50 wt%; 2. Recyclate dimension: 80 μm, 200 μm, 1 mm, 4 mm; 3. Post-processing: no post-processing (matte finishing), sanded (matte finishing), gelcoat (glossy finishing), painting (matte and glossy finishing)
rGFRP	EoL structures (27 samples)	1. Recyclate percentage: 20, 30, 40 wt%; 2. Recyclate dimension: 100 μm, 100 μm with contaminations, 4 mm; 3. Post-processing: no post-processing (matte finishing), sanded (matte finishing), gelcoat (glossy finishing)
rCFRP	Expired prepreg (8 samples)	1. Recyclate percentage: 20, 40 wt%; 2. Recyclate dimension: 100 μm, long fiber textile; 3. Post-processing: no post-processing (matte finishing), gelcoat (glossy finishing)

Physical Product Library. In this case, the physical product library is focused on the showcasing of the technical and experiential possibilities that can be achieved by using the materials shown in the physical product library, especially related to the remanufacturing processes investigated within the FiberEUse project. As a matter of fact, a specific process strongly affects not only the feasibility of a single product but also the expressive-sensorial qualities of the final shape.

For the taxonomic organization, other catalogs and databases were studied to understand the rationale behind complex cataloging systems of samples. Considering the different nature of the processes involved during the project, a unique taxonomy may be not enough to give an exhaustive idea of each process. Hence, a more flexible solution was required, and the starting concept of the taxonomy was built on a system made of different process clusters with the same taxonomic structure based on the most relevant variables of the specific process and/or material.

The basic structure of this concept is a triangle area with three different variables at the vertexes. Each of them can vary along the three sides of the triangle. This variation creates a coordinate system. In the beginning, a number of variations can be defined (i.e. 3, 5, ...), and this number of variations defines the same number of points along the three sides of the triangle, including the vertexes. Then, the segments parallels to the three sides are created by coupling those points, generating a triangular grid. Each intersection represents a sequence of three values, one for each variable, and represents the position of a specific product. A sample can be therefore defined through the combination of the three variables (i.e. recyclate dimensions, recyclate percentages, and post-processing), and each specific variable changes along its corresponding side of the triangle. In addition, the number of the samples, representing a single variable, varies from more pieces along the side to a single object at the vertex, which represents the most complex variation to be achieved with the specific process.

Nevertheless, three variables were not enough to showcase all the possibilities that a specific remanufacturing process can achieve. For this reason, a fourth variable was added, creating a tetrahedral structure. The same concept of the planar taxonomy can be applied to this 3D visualization, creating a four-variable coordinate system, where each sample is described by four values of the variables. In this case, the fourth variable is the same for the entire system, here defined "shape complexity". This variable is an arbitrary value related to the shape and the technical features of a specific product according to the selected process: the more the part is complex to obtain with the process or requires further steps, the more the part will be near to the fourth vertex. This hierarchical spatial taxonomy is resumed here below. Figure 5a represents the triangle unit of the taxonomic organization with the three variables at the vertexes, while the examples of the planar coordinate system and the tetrahedral structure derived from the four-variables system are shown in Fig. 5b and c, respectively.

The Physical Product Library is made of 73 different cut-off or scaled parts of products that are organized according to the hierarchical taxonomy based on four different variables (and previously explained). Two main clusters can be detected according to the fiber type obtained from the EoL waste, which means:

- Glass Fibers (3 tetrahedral structures shown in Fig. 6a): "Liquid Deposition Modelling—3D printing" structure, "Casting" structure, "Lamination and Injection Molding" structure;
- Carbon Fibers (2 tetrahedral structures shown in Fig. 6b): "Casting and Injection Molding" structure, "Reuse" structure.

Further details related to the variables and the parameters are resumed in Table 2.

4.3 Virtual Library

Virtual Product Library. The Virtual Product Library mainly contains the samples of the Physical Product Library, but it differs from the corresponding physical part for the different interaction. First of all, the samples are not organized following the same specific taxonomy of the Physical Product Library. Since the fruition of the two parts (the physical and the virtual) changes a lot, the samples of the Virtual Product Library are arranged in a cluster.

A series of filters, the "Virtual library organizers", was then designed based on the taxonomy of the Physical Library, considering the different contexts of use related to the virtual fruition. By changing them, the cluster of samples changes and so the library will display only that section of products identified through the filtering actions selected by the user. They can be grouped into the following three main categories: (i) Manufacturing processes; (ii) Recyclates; and (iii) Recycled materials. Further details about the filters of these categories are described in Table 3.

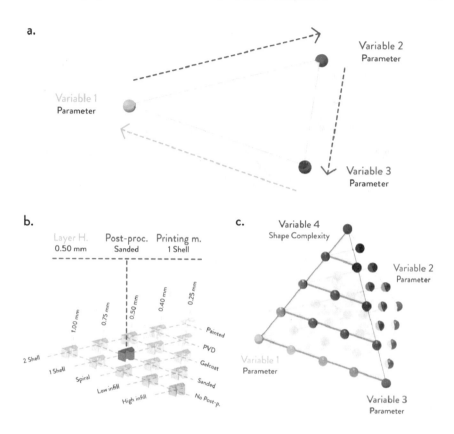

Fig. 5 Physical Product Library spatial taxonomy: **a** basic structure of the spatial taxonomy with 3 variables; **b** example of the planar coordinate system (3D printing); and **c** basic structure of the spatial taxonomy with 4 variables

To simplify the user's experience in finding what he is looking for, three distinct colors were selected to identify the different kinds of recyclate with an immediate visual inspection in addition to the filters. In this way, the user can easily recognize which recyclates were employed to develop the products and the corresponding waste and recyclate producers thanks to the identification of the colors. Accordingly, a common visual identity was designed for each product page. For this reason, mostly realistic renderings on a white background were used. These renderings are also useful for a 360° visualization of the products. This allows to better visualize the products and, at the same time, to engage the user during the fruition. Furthermore, each product has a data chart with all the information about the production, the material, and the recyclate. Figure 7 a shows a preview of the webpage with the filters and clusters of the Virtual Product Library, while an example of the 360° visualization of the products is visible in Fig. 7 b.

Fig. 6 Physical Product Library: **a** GF structures (from left to right: "liquid deposition modelling—3D printing, casting, lamination and injection molding); and **b** CF structures (from left to right: casting and injection molding, reuse of components)

Virtual Material Library and Recyclate Library. The Virtual Material Library was designed to showcase the variations related to the recycled material itself, in particular to the concentration of the recyclates types: rCFs and shGFRPs (shredded glass fiber reinforced polymers, rGFRPs). By changing the recyclate size and percentage and trying different finishes, the recycled material changes not only for its specific technical properties but also considering its expressive-sensorial qualities. Although this experiential investigation can be done through the Physical Material Library, it is interesting to show this aspect in a section of the website for a complete overview of the recyclate materials. As mentioned before, the same matrix material (epoxy resin) was chosen to better compare the recyclates.

In this section, there are only two filters for the classification of the sample pictures: rCF and shGFRP fibers. The main data related to the recycled materials are displayed through the name of the specific sample. As an example, the image below (Fig. 8) shows the name of one of the samples: "shGFRP 50% 4 mm—from mold". In this case, "shGFRP" indicates the recyclate type, "50%" the weight percent of recyclate, "4 mm" the recyclate size (fibers length), and "from mold" the finishing (in this case, no post-processing).

Table 2 Physical Product Library: the five tetrahedral structures, the variables and the specific parameters

Fiber	Cluster	Variables and parameters
rGFRP	Liquid deposition modelling 3D printing (35 samples)	1. Layer height: 0.2, 0.4, 0.5, 0.75, 1 mm; 2. Printing mode: two shells, one shell, spiral, low infill, high infill; 3. Post-processing: no post-proc. (matte finishing), sanded (matte finishing), gelcoat (glossy finishing), PVD (metallic glossy finishing), painting (glossy finishing); 4. Shape complexity: Five different 3D models from the base to the top of the tetrahedral structure
rGFRP	Casting (20 samples)	1. Recyclate percentage: < 35, 40, 45, > 50 wt%; 2. Recyclate dimensions: 80 μm, 1 mm, 2 mm, 4 mm; 3. Post-processing: no post-proc. (fine matte finishing), no post-proc. (rough matte finishing), sanded/polished (glossy finishing), PVD (metallic glossy finishing); 4. Shape complexity: 11 different 3D models from the base to the top of the tetrahedral structure
rGFRP	Lamination and injection molding (4 samples)	1. Recyclate percentage: 25, 30 wt%; 2. Recyclate dimension: 1, 5 mm; 3. Finishing: fine, rough; 4. Shape complexity: 4 different 3D models from the base to the top of the tetrahedral structure
rCFRP	Casting and injection molding (10 samples)	1. Recyclate percentage: 20, 30, 40 wt%; 2. Recyclate dimension: < 100 μm, > 100 μm, 1 mm; 3. Post-processing: no post-proc. (fine matte finishing), no post-proc. (rough matte finishing), sanded/polished (matte finishing); 4. Shape complexity: 4 different 3D models from the base to the top of the tetrahedral structure

(continued)

Table 2 (continued)

Fiber	Cluster	Variables and parameters
CFRP	Reuse (4 samples)	1. Fiber percentage: 50, 65 wt%; 2. Process: Infusion, pultrusion; 3. Post-processing: no post-proc. (matte finishing), gelcoat (glossy finishing); 4. Shape complexity: 4 different 3D models from the base to the top of the tetrahedral structure

Table 3 Virtual Product Library: the filters categories, namely the "Virtual library organizers"

Filter category	Filters
Manufacturing processes	1. Re-Manufacturing Processes: additive manufacturing, closed mold pressing, compression molding, continuous lamination, indirect manufacturing, infusion, injection molding, open mold casting, pultrusion, spray lay-up; 2. Producers: the consortium partners
Recyclates	1. Recyclate type: shGFRP or rGFRP, rCF, CFRP; 2. Recyclate producers STIIMA-CNR, TEC, RIV; 3. Recycling process: mechanical shredding, pyrolysis, reuse; 4. Waste producers: AER, GAM, RIV
Recycled materials	1. Recyclate size: 80 μm, 100 μm, 200 μm, 1 mm, 2 mm, 4 mm, 5 mm, 6 mm, long fiber textile; 2. Composite matrix: acrylic, epoxy, polyamide, polyester, polypropylene, polyurethane, orthophtalic unsaturated polyester; 3. Recyclate percentage: 20, 25, 30, 35, 40, 45, 50, 55, 65 wt%; 4. Finishing: not required, gelcoat, painted, sanded, PVD

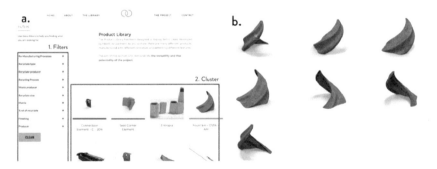

Fig. 7 Virtual Product Library: **a** preview of the webpage with the filters and clusters; and **b** example of the 360° visualization of the products

HOME   ABOUT   THE PROJECT   THE PROJECT   CONTACT

shGFRP
50% – 4mm
–from mold
shGFRP

Fig. 8 Virtual Material Library: main page with the info from the Physical Material Library

Generally speaking, each recycled material of the library system is composed of a matrix and a type of recyclate. Each recyclate has its own features, and therefore it was important to better clarify them in a separate page. This page, named Recyclate Library, is linked to the Virtual Material Library and shows the technical information needed to understand the composition of each sample.

Virtual Library Website. The other pages of the website are meant to summarize the FiberEUse project and give an overview to arouse the user's curiosity. In general, the developed website was well received either by partners of FiberEUse and general users. Both the User Interface and the User Experience were designed to ease the user's navigation through the products through a neat and clean interface. This should allow its fruition from a large range of users. As a matter of fact, the Virtual Library has turned out to be interesting for a wide range of users: from naïves to materials-oriented professionals such as technicians, designers, engineers. In this case, the extensive work performed to summarize the contents (data and renderings), the tests and surveys done for each step of the process, and the collaboration between all the partners played a crucial role in determining the success of the Virtual Library and, consequently, of the whole library system.

5 Conclusions and Future Research Perspectives

This chapter aimed to illustrate the integrated library system developed within FiberEuse project. This material and product library system was designed to show-case the main output of the project. In short, it consists of a physical and a virtual part that are interconnected to communicate, and it contains the recycled materials and the archetypal remanufactured products made with rGFRPs and rCFRPs.

First, a brief theoretical perspective on the concepts of material libraries and library system was provided. After the collection of the data from the consortium partners, a preliminary concept of the library system was defined considering the physical and virtual fruition of materials and products. In detail, a hierarchical organization of the virtual and physical contents was designed to combine the physical and virtual fruition for a complete experience with the samples. Specific taxonomies were then defined considering the specific divisions of the library system (i.e. Physical Material Library, Physical Product Library, Virtual Material Library, Virtual Product Library, Recyclate Library). The virtual and physical samples were firstly produced and collected by Politecnico di Milano and the consortium partners and then placed in custom exhibitors, which means the physical stands and the library website. Finally, the coherence of the link between the physical and virtual parts was checked to ensure the right interconnection of the whole system.

Through this library system, new applications and value-added markets can be investigated to exploit the main results of the FiberEUse project. This tool can be seen as a way to facilitate the real exploitation of new products and materials made of rGFRPs and rCFRPs through the sharing of knowledge that it allows. As a matter of fact, the system will be freely accessible for the stakeholders through the library website and booking the consultation of the physical samples. Also, this system will be updated by the consortium partners with new products and/or materials, fostering the generation of new value-chains linked to the composite market.

In detail, the classified sample collection of the library system comprehends: (i) 99 material samples of three different kinds of recyclates in the Physical Material Library; (ii) 73 different samples of 28 products in the Physical Product Library; (iii) 8 recyclate samples from different EoL composites in the Recyclate Library; (iv) 99 material samples of three different kinds of recyclates in the Virtual Material Library; and (v) 89 different samples of 33 products in the Virtual Product Library. To conclude, this system can be considered a valid tool to foster the development of new real applications and follow-ups thanks to its transdisciplinary nature, merging technical and expressive-sensorial aspects with tangible and intangible materiality.

References

1. Seiffert, M.E.B., Loch, C.: Systemic thinking in environmental management: support for sustainable development. J. Clean. Prod. **13**, 1197–1202 (2005)
2. United Nations: Report of the World Commission on Environment and Development. Our Common Future. http://www.un-documents.net/wced-ocf.htm. Accessed 16 Oct 2021
3. Hauschild, M.Z., Kara, S., Røpke, I.: Absolute sustainability: challenges to life cycle engineering. CIRP Ann. **69**, 533–553 (2020)
4. Andersen, M.S.: An introductory note on the environmental economics of the circular economy. Sustain. Sci. **2**, 133–140 (2007)
5. Murray, A., Skene, K., Haynes, K.: The circular economy: an interdisciplinary exploration of the concept and application in a global context. J. Bus. Ethics. **140**, 369–380 (2017)

6. Geissdoerfer, M., Savaget, P., Bocken, N.M.P., Hultink, E.J.: The circular economy—a new sustainability paradigm? J. Clean. Prod. **143**, 757–768 (2017)
7. Campbell-Johnston, K., Vermeulen, W.J.V., Reike, D., Brullot, S.: The circular economy and cascading: towards a framework. Resour. Conserv. Recycl. **7**, 100038 (2020)
8. Panchal, R., Singh, A., Diwan, H.: Does circular economy performance lead to sustainable development?—a systematic literature review. J. Environ. Manag. **293**, 112811 (2021)
9. Salmenperä, H., Pitkänen, K., Kautto, P., Saikku, L.: Critical factors for enhancing the circular economy in waste management. J. Clean. Prod. **280**, 124339 (2021)
10. Pickering, S.J.: Recycling technologies for thermoset composite materials-current status. Compos. Part A: Appl. Sci. Manuf. **37**, 1206–1215 (2006)
11. Oliveux, G., Dandy, L.O., Leeke, G.A.: Current status of recycling of fibre reinforced polymers: review of technologies, reuse and resulting properties. Progr. Mat. Sci. **72**, 61–99 (2015)
12. Romani, A., Rognoli, V., Levi, M.: Design, materials, and extrusion-based additive manufacturing in circular economy contexts: from waste to new products. Sustainability. **13**, 7269 (2021)
13. Adams, R.D., Collins, A., Cooper, D., Wingfield-Digby, M., Watts-Farmer, A., Laurence, A., Patel, K., Stevens, M., Watkins, R.: Recycling of reinforced plastics. Appl. Compos. Mater. **21**, 263–284 (2014)
14. Crippa, G., Rognoli, V., Levi, M.: Materials and emotions: a study on the relations between materials and emotions in industrial products. In: Brassett, J., McDonnell, J., Malpass, M. (eds.) Proceedings of 8th International Design and Emotion Conference, p. 10, London, United Kingdom (2012)
15. Veelaert, L., Du Bois, E., Moons, I., Karana, E.: Experiential characterization of materials in product design: a literature review. Mat. Des. **190**, 108543 (2020)
16. Zuo, H.: The selection of materials to match human sensory adaptation and aesthetic expectation in industrial design. Metu J. Fac. Archit. **27**, 301–319 (2010)
17. Miodownik, M.A.: Toward designing new sensoaesthetic materials. Pure Appl. Chem. **79**, 1635–1641 (2007)
18. Whitaker, T.A., Simões-Franklin, C., Newell, F.N.: Vision and touch: Independent or integrated systems for the perception of texture? Brain Res. **1242**, 59–72 (2008)
19. Camere, S., Karana, E.: Experiential characterization of materials: toward a toolkit. Presented at the Design Research Society Conference 2018 June 28 (2018)
20. Veelaert, L., Ragaert, K., Hubo, S., Van Kets, K., Du Bois, E.: Bridging design and engineering in terms of materials selection. In: International Polymers & Moulds Innovations Conference 2016, pp. 319–326. Ghent, Belgium (2016)
21. Pedgley, O., Rognoli, V., Karana, E.: Materials experience as a foundation for materials and design education. Int. J. Technol. Des. Educ. **26**, 613–630 (2016)
22. Wilkes, S., Wongsriruksa, S., Howes, P., Gamester, R., Witchel, H., Conreen, M., Laughlin, Z., Miodownik, M.: Design tools for interdisciplinary translation of material experiences. Mat. Des. **90**, 1228–1237 (2016)
23. Karana, E., Barati, B., Rognoli, V.: Material Driven Design (MDD): a method to design for material experiences. Int. J. Des. **9**, 20 (2015)
24. Wilkes, S.E., Miodownik, M.A.: Materials library collections as tools for interdisciplinary research. Interdiscip. Sci. Rev. **43**, 3–23 (2018)
25. Wilkes, S.: Materials libraries as vehicles for knowledge transfer. Anthropol. Matt. **13**, 12 (2011)
26. Virtanen, M., Manskinen, K., Eerola, S.: Circular material library. An innovative tool to design circular economy. Des. J. **20**, S1611–S1619 (2017)
27. Iranmanesh, A., Onur, Z.: Mandatory virtual design studio for all: exploring the transformations of architectural education amidst the global pandemic. Int. J. Art Des. Educ. **40**, 251–267 (2021)
28. Komarzyńska-Świeściak, E., Adams, B., Thomas, L.: Transition from physical design studio to emergency virtual design studio. Available teaching and learning methods and tools—a case study. Buildings **11**, 312 (2021)

29. Material ConneXion—Every idea has a material solution. ®, https://materialconnexion.com/. Accessed 18 Ott 2021
30. Materioteca Matters—Makte. https://www.makte.it/it/. Accessed 18 Ott 2021

New Business Models and Logistical Considerations for Composites Re-use

Giacomo Copani⊙, Maryam Mirpourian⊙, Nikoletta Trivyza⊙,
Athanasios Rentizelas⊙, Winifred Ijomah⊙, Sarah Oswald⊙,
and Stefan Siegl⊙

Abstract The growing use of composites in various industries such as aerospace, automotive and wind turbine has increased environmental concerns regarding their waste disposal methods. Deploying circular economy practices to reuse composites could play a crucial role in the future. In this regard, this chapter addresses the development and implementation of new business models for composites re-use, as fundamental enabler for the industrial exploitation and diffusion of technological and methodological innovations developed in the FiberEUse project. Seven products were chosen as representatives for composites reuse application in four industrial sectors: sanitary, sports equipment, furniture and automotive. Re-use business models are presented describing their value proposition, with particular reference to the provision of advanced product-service bundles, the revenue models (including schemes such as leasing), as well as new supply chain configurations entailing new partnership between producers and recyclers to access post-use composites to re-use. Given the importance of reverse supply networks, the potential reverse logistics pathways for mechanical recycling of Glass Fiber Reinforced Plastic (GFRP), thermal recycling of Carbon Fiber Reinforced Plastic (CFRP) and remanufacturing of CF composites waste in Europe for 2020 and 2050 have been investigated. We concluded that the optimal reverse logistics network needs to be decentralized in more than one country in Europe. Therefore, it is suggested that policy makers address regulation to allow

G. Copani (✉) · M. Mirpourian
Institute of Intelligent Industrial Technologies and Systems for Advanced Manufacturing,
National Research Council of Italy, Via A. Corti, 12, 20133 Milan, Italy
e-mail: giacomo.copani@stiima.cnr.it

N. Trivyza
Department of Naval Architecture, Ocean and Marine Engineering, Maritime Safety Research
Centre, University of Strathclyde, 100 Montrose Street, Glasgow G4 0LZ, UK

S. Oswald · S. Siegl
Saubermacher Dienstleistungs AG, Hans-Roth-Straße 1, 8080 Feldkirchen Bei Graz, Austria

A. Rentizelas · W. Ijomah
Design, Manufacturing and Engineering Management Department, University of Strathclyde, 75
Montrose Street, Glasgow G1 1XJ, UK

M. Colledani and S. Turri (eds.), *Systemic Circular Economy Solutions for Fiber
Reinforced Composites*, Digital Innovations in Architecture, Engineering
and Construction, https://doi.org/10.1007/978-3-031-22352-5_19

the transportation of waste between European countries to facilitate the development of recycling networks for composites reuse.

Keywords Business model innovation · Circular business model · Composites reuse · Product-service system · Reverse logistics · Optimization · Recycling

1 Introduction

1.1 What Is Circular Business Model Innovation?

Circular business models refer to models that articulate the logic of how organizations create, offer and deliver value to its broad range of stakeholders while, at the same time, minimizing ecological and social costs [1]. It is a conceptual logic for value creation based on utilizing the economic value retained in products after use in the production of new offerings [2]. However, there are diverse views on what circular business models are. For example, business models can focus on utilization of intermediary outputs that have no further use in the creation of activities that are monetized in the form of either cost reductions or revenue streams [3]. Business models could also focus on the use of 'presources' over the use of resources in the process of creating, delivering and capturing value [4] or they can focus on market value of sustainable remanufacturing [5]. Inherent in all these definitions is value creation using pre-existing products [6]. Implicit in these definitions of circular business models are business processes that support sustainable manufacturing strategies in the circular economy.

With increasing global population and limited scarce natural resources, the linear economic model whereby raw materials are utilized in the manufacture process and disposed of is unsustainable. There is need to consider circular industrial business models. The linear economic model is giving way to the circular economy model focuses on careful alignment and management of resource flows across the value chain by integrating reverse logistics, design innovation, collaborative ecosystem, and business model innovation [7]. This is an innovative area that is receiving attention as governments push to cut carbon emissions to combat global warming and meet their climate change targets [8]. Circular business promotes sustainability by allowing companies to generate maximum return from given resources and reach zero waste targets. It also contributes to greater customer satisfaction through service innovation that reduces negative impact on the environment [9].

High investment costs and risks, deficiency in take back programs, and lack of awareness of environmental concerns are among factors that stand as barriers to implementation of circular business models [10]. Circular business models represent solutions to move towards zero waste, improving environmental impacts and increasing economic profit [11]. However, issues such as cultural barriers and risk of cannibalization could serve as barriers to the implementation of circular business models [12].

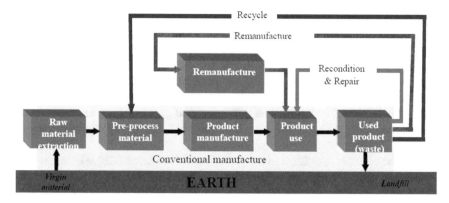

Fig. 1 Remanufacturing concept and significance [13]

There is increasing demand for sustainable manufacturing that takes responsibility in conserving natural resources and limiting negative impact on the environment. This has given rise to the evolution of the circular economy that investigates how raw materials can be managed in a sustainable and ethical manner. Circular economy could help in reducing harmful emissions to the environment on the one hand, while at the same time reduce utilization of energy. As suggested in Fig. 1, sustainable manufacturing strategies that add value in the circular economy in the life cycle of the product include recycling, recondition and repair, reuse and remanufacture. At every point in the life cycle of a product, virgin raw materials energy and resources are being utilized while emissions and waste are released into the land, sea and air [13].

Underpinning the circular economy are circular business models that take cognizance of the contextual environment in which the sustainable manufacturing is taking place. Life Cycle Analysis could be useful if comparing and contrasting the advantages of different circular business models. It could be argued that the type of business model may be indicated by the type of sustainable manufacturing activity undertaken. However, this opens up debates on how and when to apply specific circular business models (i.e. for remanufacturing instead than recycling).

1.2 Circular Business Model Trends

Business model for sustainability deals with creating competitive advantage through superior customer value that contributes to a sustainable development for companies and society [14]. This is distinguished from Business Models for Circularity (BMCs) that explicitly link business models to the product life cycle [15].

In the literature various business model strategies are suggested. Firstly, those that provide performance (through the capability of satisfying user needs without needing to own products), secondly, those that extend product value (exploiting residual value of products from manufacturers, customers and back to manufacturers), thirdly, those that focus on delivering long life of products supported by design for durability and repair, fourthly, those that encourage sufficiency, such as those that actively reduce end user consumption through principle such as upgradability and service, fifthly, those that extend resource value, such as exploiting residual value of recourses through collection of "wasted materials" and turning them into new forms of value and sixthly industrial symbiosis that utilize residual outputs from one process and a resource for another process. However, they may not fit into neatly separate strategies [1].

With the emergence of Circular Business Model Innovation (CBMI), there is need for development of tools that support these processes. There is need for standardization of terminology in what is meant by circular business models [16]. From the literature it could range from different sustainable remanufacturing strategies to the business activities that underpin sustainable manufacturing. For example, conceptualizing sustainable business models into 5 categories including, Coordinating circular value chains through data, Circular product design, Use, Reuse, Share and repair, Collection and reverse logistics, Sorting and pre-processing [17]. These could be argued to be concepts that underpin sustainable business models and not business models in their selves.

The need to better conceptualize circular business models demonstrates the variety in definitions and operationalization of circular business processes [6, 18]. An ambitious attempt to categories circular business models delineates them into those that exploit resource supplies, resource recovery, product service systems and open innovations [10]. While product service systems are often highlighted as a potential enabler for new circular business models, not all business models based on product service systems have superior impact on the environment and efficient use of resources [19].

In contrast, circular business models may be conceptualized into those grounded on pay per use concepts (where products are rented or leased instead of owned), product life extension (such as reuse, repair, reset) and resource value extensions (such as recycling materials) [20]. There is a risk of mixing up the sustainable manufacturing strategies that contribute to the circular economy from the business models that underpin these strategies [4]. This buttresses the need for research that provides clear standardization of terminology and nomenclature in the circular business model landscape.

1.3 State of the Art of Current Research

Business Model Canvas (BMC) can be used for circular business models as a conceptual framework. It is suggested that there is an urgent need for research in business models that take cognizance of issues such as customer relationships, key partners, value proposition, costs and revenue into consideration [21]. These will include research on how risk of cannibalization are managed and how policy makers can be enablers of circular business models. There is need for research that investigates how policy influences the operationalization of circular business models in creating value for customers [22]. Research is needed on how circular business models inform circular economy as well as the development of sustainability theories that inform current challenges to circular business models [14]. There is also need for research that tests the hypothesis of effectiveness of different circular business models which could include business to business and business to customer which evaluates reverse logistics and viability of operationalization of different sustainability strategies [18].

1.4 Need for Circular Business Models for the Case of Composites Re-use

Composite materials made of glass and carbon fibers are extensively used in various industries such as aerospace, automotive and sport equipment, due to their durability and superior mechanical properties, light weight and high corrosion resistance [23]. By 2020, the yearly production of Carbon Fiber (CF) composites is expected to be more than 140,000 tons [24]. Also, the market of glass fiber reinforced plastic (GFRP) is huge and is expected to rise at a Compound Annual Growth Rate (CAGR) of 6.4% from 2017 to 2022 [25]. As a consequence of such rising demand, the waste management of composites is crucial. However, at the moment the common waste treatment of composites is landfilling [26]. On the other hand, there are serious environmental concerns on landfill disposal of composites, which have raised serious concerns among policy makers [27]. For instance, EU legislation is making the composites landfilling unviable in most EU countries [28].

As the future recycling of composites will be at the top in the agenda of many countries, manufacturers in several sectors shift towards implementing a circular economy business model, which encourages the use of recycled composites. This way, the new businesses will be more economical and contribute to environmental legislations [26].

FiberEUse project aims at diversion of waste from landfill through recycling and reuse of composites and development of innovative circular business models. In the following sections the circular business models of composites reuse in the sanitary, ski equipment, furniture and automotive industries are presented.

2 Rational of Business Model Research Approach

As the purpose of this study is to develop innovative circular business models based on composites reuse, the state of the art of business model innovation in the context of circular economy was firstly investigated to derive insights for the development of novel value propositions, innovative value network architectures and revenue models.

Fig. 2 Methodology of developing composites reuse business models

Both academia and industry sources have been considered. In particular, 29 publications on composites waste management and business model have been taken in to account. The search keywords were chosen as following: *business model innovation, circular business model, composites reuse, product-service system.*

The proposal of innovative circular business models referred to the specific cases of products chosen from sectors where composites are key-materials, such as the sanitary, sports, furniture and automotive industry.

To develop a novel business model, it is significant to study the current market conditions, envisaged trends and customers' needs [29]. Focus group research methodology was selected to derive such market insights. Two focus groups were organized involving automotive industry experts from Spain and Germany, sanitary sector experts from Italy, ski industry experts from Austria, and furniture industry experts from Italy. The combined insights from these workshops as well as literature review supported the formulation of realistic novel business models. Graphical representation of the methods used is presented in Fig. 2.

3 New Circular Business Models for Composites Reuse

In this paragraph, a new business models for each selected industry is proposed.

3.1 New Sustainable Business Models for Products in Sanitary and Sport Equipment

Sanitary and sports equipment are among the industries that can benefit from using recycled composites. Composites are an appropriate alternative to metals and are widely used in industrial and sports equipment sectors [30]. As some examples of using GF composites in sanitary products, we can mention shower trays and bathtubs. Ski is an example of sports product using massively GF composite materials.

Sanitary products. The global demand for sanitary products is increasing. As an example, the market of shower trays is expected to grow by 4.9% from 2018 to 2025 [31].

The new business model for shower tray and bathtub is focused on offering a product made of recycled composites via mechanical recycling (Table 1). The model suggests a manufacturer offers a product which is built using scraps generated from the production process, that are usually lost. In this new paradigm, a manufacturer could replace at least 40% of virgin GFRP with post-industrial GFRP waste (the percentage is calculated based on average performance/sustainability considerations). In case in-house produced scraps are not sufficient to produce new products, manufacturers can use external waste materials to meet the minimum level of scraps they need to reach such a percentage (main available scrap source is the construction industry). The intrinsic benefit of using in-house or external scraps implies securing the value chain for raw material (cost saving) as well as assuring the quality of secondary materials (non-mixed fractions of waste) are met.

From the value proposition point of view, the new business model aims at offering products that have lower environmental impact, while keeping the price and performance at the same level of products made out of virgin material. Thus, this is mainly the case of a product-oriented type of business model. However, the offering of greener products provides also the opportunity to provide a set of services that increase the value for customers and that can also decrease eventual customers' acceptance barriers. In particular, in order to address concerns about products performance, maintenance packages and enhanced performance guarantee, which include product substitution in case of failures, can be offered in bundle with the product (with no or limited cost increase). Another type of possible advanced service consists in the product upgrade after a certain use period, coupled with an efficient take-back system. This service might be particularly interesting for segments such as hotel, Spa, sport club, etc., which may plan in this way renovation periods in the frame of long-term supply and service contracts.

Regarding the revenue stream, the selling price of products such as shower tray and bath tub made out of recycled material need to be set at least the same as

Table 1 Summary of the new business models of recycled products in the sanitary and ski industry

Business model dimensions	
Value proposition	• Similar performance compared to the original product made of virgin material
	• Lower environmental impact
	• Offering maintenance/repair/upgrade services
Value chain architecture	• Partnership with waste management companies or manufacturers selling scraps (if the in-house production scrap for recycling is not enough)
Value capture	• Revenue from: sale of products with the same price
	• Cost is less than the original product built from virgin material

that of products realized with virgin material. Customers may be reluctant to spend more money for a partly recycled product, especially until when a generally-diffused consumers' green culture will not be established. With respect to cost structure, the increase in cost due to the recycling process will be compensated by the use of the in-house scraps.

Since the value proposition emphasizes green product, it is crucial to properly communicate in order to let customers know the green value embedded in the products. Eco-marketing strategy could be deployed to highlight the company's effort in decreasing its ecological footprint through more sustainable production.

Key-assets of manufacturers to implement this business model are on the one hand the mastery of the mechanical recycling process and, on the other, the establishment of a green brand image. Scrap collection, mechanical recycling and customized after sale services such as offering maintenance and repair packages are the key activities of manufacturer to create and deliver value.

From the value architecture point of view, if the internal waste to build a recycled product is not enough, it is necessary that manufacturers establish new partnerships with waste management companies or with other composites manufacturers willing to sell their GFRP production scraps.

Ski. By having 43% of the global ski market, Europe holds the largest market share of this industry. America and Asia, with 34% and 20% of the market respectively, are the other largest markets for the ski industry [32]. Major customer of ski industry are ski rental companies.

Regarding the ski's End-of-Life (EoL), a systematic take back system to collect the product does not exist. Moreover, since ski is made out of different types of materials, disassembly is a very challenging task. As a consequence, currently, the product at its EoL goes to landfill or incineration after removal of aluminum.

The new business model of ski is based on offering a greener product via replacement of 10% of virgin GFRP (via mechanical recycling) with in-house scraps from the production process. Such a limited percentage of recycled material is due to the need to guarantee extreme mechanical performance to the products.

The business model proposed for skis is the same as sanitary products (Table 1). The manufacturer sells a greener product with similar price and performance, while offering additional value-added service to ski renting companies. The service includes enhanced warranty to customers during extended periods. In this way, the customer uncertainty regarding the recycled product's performance will be reduced.

In terms of cost structure, the product's cost is lower than the original one.

The key assets of manufacturer are the mechanical recycling technologies and green brand image.

From a value architecture point of view, there will be less need to recur to external scraps supplier with respect to the sanitary case, since the target percentage GFRP post-industrial waste is lower and compatible with internal production scraps of ski-making companies.

3.2 Leasing of Composite-Made Components in Furniture Industry

Green public procurement (GPP) is one of the promising tools in moving towards a more resource efficient economy [33]. GPP brings environmental factors into the core of decisions that are related to purchasing public goods and services [34]. The expenditures of public organizations that use GPP for goods and services account nearly the 14% of the EU's Gross Domestic Products (GDP). Such a large demand for green products acts as a catalyst on green innovations and largely contributes to circular production and sustainable consumption [35].

One of the areas that GPP can widely be used is in the office furniture industry. Currently, there is a serious environmental concern regarding the inefficient life cycle management of office furniture. These concerns are mainly related to the high volume of waste at the end of products life cycle [36]. In Europe, around 1.2 million tons of office furniture are discarded every year [33]. Therefore, new business models that suggest reusing the furniture materials instead of disposing them can generate significant environmental and economic gains in comparison to conventional business models. In order to show how recycled materials can be used in this sector, we consider the case of office desks built from recycled composites. Desks are composed of two main parts: a table top made of GFRP and metal pillars. The two parts can be easily disassembled and re-assembled, for example for on-site substitution and renewal of table top or for re-use in other products.

In the new business model, we consider a business partnership between a furniture supplier and a GFRP-made tabletops supplier (Table 2). The furniture producer offers an eco-friendly office desk, with similar performance and aesthetic features compared to those products made out of virgin material. He is responsible for installation of desks at customers' site and take-back service. The tabletop supplier is specialized in the production of green GFRP materials by include a certain percentage of end-of-life composites in the production of parts through mechanical recycling technologies integrated in its plant. In the case of tabletops, about the 50% of virgin GFRP can be replaced by recycled composites.

The proposed case falls under the category of a usage/access oriented business model where the product ownership remains with the furniture producer and customer pays for using the product rather than owning it [37]. In other words, this business model is based on leasing the product instead of selling it. In reference to circular economy concept, leasing would guarantee a constant return flow of old office desks from customers to furniture producer. After recollecting the used products, the furniture producer disassembles the desks and send their tabletops to the manufacturer. This way the dismantled tabletops will be recycled and re-enter the production cycle.

A long-term partnership between tabletop manufacturer and furniture producer is crucial for the successful implementation of the new circular business model. Part manufacturer's capacity in recycling materials, enables the furniture producer to offer a lease concept centered on "green products" and that incorporates continuous renovation services to the customers. On the other hand, the leasing model guarantees

Table 2 Summary of the new business models of recycled office desk in the furniture industry

Business model dimensions	
Value proposition	• Similar performance
	• Lower environmental impact
	• Non-ownership model
	• Maintenance/repair/upgrade services
Value chain architecture	• Partnership with logistics providers
	• Link with government/public bodies
Value capture	• Revenue from leasing of products
	• Cost is less than the original product

the tabletop manufacturer to have a constant and planned return flow of parts to be recycled.

In the new financial scheme, the furniture producer offers a 4–7 years lease contract to public organizations and by the time that lease expiries, the furniture producer would take back the old product and provide customers with new green office desks. In addition, the furniture producer can offer regular maintenance checks and ergonomic desk adjustments to address customers' concerns on product performance.

From an economic point of view, the furniture producer will experience postponed incomes, which affect the revenue model. As a result, furniture producer need to precisely reflect this into the leasing fee proposed to customers. Therefore, adoption of advanced marketing and financial approaches are crucial for the economic sustainability of this business models. Overall, the increase in the expenses need to be compensated by the cost saving associated with reuse of materials and by market increase due to the new attractive value proposition.

After 4–7 years that the lease expires, most of the metal pillars might still be in good conditions. These metal parts can easily be integrated into the manufacturing of new office desks and the manufacturer can decrease its related production costs. Therefore, although composites reuse is the main area of focus in this business model, reusing the metal pillars of the office desks can be another attractive opportunity to consider for efficiency and sustainability, as well as to strengthen the "green image".

Using recycled composites in the manufacturing process complies with regulations related to environmental protection. For instance, Italy has introduced an environmental legislation that requires a mandatory GPP requirement for all the Italian public offices. [38]. Therefore, business risks associated with the novelty of circular model can be mitigated through company's green marketing strategies. Eco-labeling, as part of the green marketing strategy (company's interface with customers) can be applied to highlight the use of recycled composites in the product. This would help customers to be aware of the green aspect of the product and accept the new business model as an added value. From a customer perspective, green element embedded in a product can be a key driver towards accepting the new business model.

From a value chain network building blocks perspective, new partnership with logistics companies for collection of old office desks might be considered, in case the furniture supplier does not manage logistics internally. Finally, it is important to consider building a strong partnership with government and public bodies. Such a partnership will contribute to the evolution of legislation/directives that entail incentives to increase the sustainability of GPP practices.

3.3 Greener Structural Components in Automotive Industry

Automotive industry is going through a radical transformation. There are signs which indicate that new technology and circular business models will be the leading trends of the future [39]. The use of lightweight parts made of composite materials in vehicles is another leading trend. One of the major drivers to use lightweight material stems from the fact that heavyweight components used in Internal Combustion Engine (ICE) cars are one of the sources of pollution. Therefore, the lightweight and remarkable fatigue behavior of composite materials are the reasons, which the automotive industry is moving towards using more composites [40]. Such a trend will be even more marked in the future due to the transition towards electric vehicles, in which the significant weight of battery pack requires reduction of weight of other components.

In the automotive industry, GF composites and carbon fiber reinforced polymer (CFRP) are used in the structural components of cars [41]. Regarding the EoL waste of CFRP, the common scenario is landfilling or incineration [42]. In order to address the EoL waste problem of composites, three case studies of car components made out of recycled composites reclaimed from thermal recycling are discussed in this section.

The first vehicle structural component is cowl top, which is built from 20 to 30% rGF. The manufacturer uses the rGF derived from EoL of wind turbine blades reclaimed from thermal recycling. Wind turbines have a lifecycle around 20–25 years. Therefore, the removal and dismantling of wind turbines in near future will lead to a substantial amount of composites waste [43]. The second vehicle structural component is clutch pedal that is built from 20 to 40% thermally recycled CF (rCF). Finally, the third vehicle component is a front-end carrier that is built from 15 to 25% rCF thermally recovered from the production waste of the aeronautic sector. CFRP is widely used in the aeronautic industry [41].

In the new business model of cowl top, a tier 1 manufacturer provides eco-friendly products with lower price, while keeping the aesthetic values and performance at the same level of those products made out of virgin material.

Regarding the clutch pedal and front end carrier, the tier 1 manufacturer offers green products with the enhanced performance, while the product's price and aesthetic value will be similar to the original products. Products' technical performances are improved since the new product is lighter, thanks to the low density of the rCF.

The new business models are characterized as product-oriented (Table 3). The revenue is generated through selling greener performing products. In order to address possible concerns about the performance of parts manufactured with post-consumer material, extended guarantee will be offered with direct substitution of eventual defective products as a product-oriented service. The targeted customers are Original Equipment Manufacturers (OEMs), in particular those with a high environmental profile. The evolution of regulation in the automotive sector will favor the diffusion of this business model, since OEMs will be forced to increase the rate of recycled materials in automotive parts. Regarding the cost structure, the product cost can be either lower or higher than the product made out of virgin material. Given the proportion of recycled material used in a product and the price of rCF, the product final cost may vary.

The key asset of the tier 1 manufacturer includes long term contracts with suppliers of rCF/rGF and logistics providers. Through such contracts, the manufacturer is assured of a steady supply of a certain volume of materials with specific quality requirements. Green brand image is another key resource of manufacturer.

From a technical point of view, the manufacturing process is the same as the one from virgin composites (GF/CF). Therefore, key activities of the manufacturer include keeping the production cycle time equivalent to the original product and making sure that the reliability and quality of the recycled materials fulfill the technological specification of the final products.

From the value chain architecture point of view, new partnerships with recyclers and logistics providers need to be established. Another key partner includes government and public bodies. Such a partnership allows government to update legislation/directives and offer new incentives that promotes the circular economy and EoL waste management based on the new business models.

Table 3 Summary of the new business models of recycled vehicle's components in the automotive industry

Business model dimensions	
Value proposition	• Similar/enhanced performance
	• Lower environmental impact
	• Similar/lower price
	• Enhanced guarantee
Value chain architecture	• Partnership with recyclers (rGF/rCF suppliers)
	• Partnership with logistics providers
	• Link with government/public bodies
Value capture	• Revenues from sale of products with the same/lower price
	• Cost can either be higher or lower, depending on the percentage of the recycled material and price of rCF used in products

3.4 Selling of Remanufactured Components in Automotive from EoL Carbon Fiber Reinforced Plastic

Composite materials are an appropriate alternative to metals [30] due to their strength, lightweight and long life cycle [40]. In comparison to metal the deterioration of composites is quite slow. Therefore, at the EoL of a car, technical performance of CF composite structural components is close to new ones. This means that a great proportion of composite components can be reused in several car generations. Based on this consideration, a new business model is proposed, where the manufacturer adopts lightweight CFRP car platform for an electric vehicle, that can be re-used at the end of life, as an alternative to a steel one. The platform will be made using both virgin and recycled composites, mainly drawn from the automotive industry.

In this novel business model, the tier 1 manufacturer offers a greener product with the same quality and reliability, while the technical performance and aesthetic values are enhanced and the product life cycle is extended through product reuse in several car generations. In other words, the new product has the same stiffness and can tolerate the same fatigue, while it is 40% lighter and has a better design compared to the steel one. In addition, the product life cycle is extended through quick and non-destructive disassembly. After the disassembly, the product will be tested to identify whether it can be reused or not. In case the product cannot be reused, it will be sent to recyclers.

The new business model is a product-oriented business model, with the inclusion of End-Of-Life related services (Table 4). The tier 1 manufacturer establishes long-term contracts with the OEM for the supply of CFRP platform, guaranteeing their re-use in vehicles other than the first one at their end-of-life within a fixed time horizon. Thus, recovered platform can be re-used either in vehicles of new production (in a futuristic scenario), in fleets of vehicles sold to leasing/renting companies, or as spare parts for used or refurbished vehicles. In the contractual relationship, OEM returns to tier 1 manufacturer the platform dismantled from End-Of-Life vehicles. The latter tests them, eventually carries out repair and adaptation operations, and provides back to the OEM platforms to be installed in other vehicles with the same performance guarantee than the initial ones. Supplied platforms can come from his stock of re-usable platforms (after testing and necessary repair/adaptation operations) or can be platforms of new production. Thus, tier 1 manufacturer becomes the responsible of the circular economy of platform for the OEM, guaranteeing performance in-use in multiple use cycles.

A promising market scenario for this business model is the one of vehicles fleets leased to car sharing/renting companies, private or B2B customers. In particular, it is projected that by year 2030, the sharing paradigm will reshape the automotive industry. In Europe, car sharing will stand for almost 35% of road trips [39]. For fleet vehicles that are used by customers occasionally or in the frame of full-service contract limited in time, the main requirement is efficiency-in-use and cost saving, while the non-ownership of the product lowers possible acceptance barriers related to re-used parts. Furthermore, contractually fixed leasing periods with certain return

Table 4 Summary of the new business models of remanufactured light weight platform in the automotive industry

Business model dimensions	
Value proposition	• Enhanced performance
	• Lower environmental impact
	• Enhanced design
	• Reusable products via several lifecycle
Value chain architecture	• Partnership with raw material supplier and recyclers
	• Partnership car sharing and fleet management companies
	• Partnership with logistics providers
	• Partnership with government/public bodies
Value capture	• Revenue from selling a product and offering reuse/repair services during several product life time
	• Cost is higher in the first life cycle, however it decreases in the next life cycles of products

time of parts, allow the exact definition of use cycles time, which is a fundamental variable both for long-term contracting and operations planning.

From the cost structure point of view, in the first diffusion phase of this business model, the first production cost of CFRP platform is significantly higher than the current traditional platform cost (it will decrease when scale economies will be realized through the massive adoption of this concept by OEMs). However, the possible re-use of the part in 4–5 use cycles will more than compensate the increased initial cost and will provide economic advantage in the long term. However, this implies additional risk and postponed revenues.

In terms of key assets, the platform manufacturer should rely on detachable joining technologies enabling the disassembly of the lightweight platform quickly and without any destruction. Other key resources are the non-destructive test and repair technologies. When the platform is dismantled, special technologies are needed to specify if the product can be reused or not. In addition, in case of product's small damages, the manufacturer repairs the products via innovative repair technologies. The green brand image and the long term contract with recyclers and raw material suppliers are other key resources of the manufacturer. Finally, special agreements need to be established with OEM/fleet management companies/car sharing. Such agreements serve as a key resources, which guarantees a strong take back system to collect a lightweight platform at vehicle's EoL.

From value chain architecture point of view, it is important to develop a partnership with raw material (CF) supplier, recyclers, logistics providers and fleet management / car sharing companies' and taxis. Another key partner includes public bodies, so that they can update regulations based on the new business model to increase the circular economy practices in the automotive industry.

4 Reverse Logistics

4.1 Current Status of Logistics for Composite Materials Circular Business Models

It is identified both in the academia and industry that the introduction of reverse logistics is essential to the transition towards a circular economy business model [44]. It is considered one of the main modifications for the companies' circular economy business models [45]. During the reverse logistics management on a circular business model it is highly important to investigate the potential of each return product and decide regarding the reuse, recycle, remanufacture of the material, component and product itself. In addition, another necessary issue is to understand the sources and the frequencies of the returns, as well as their impending benefits [46]. This includes concerns regarding the returns' quality and their quantification, especially in the future. In this section, potential reverse logistics pathways for composite materials are presented, along with the quantification of the sources.

It is argued that one of the first issues on the reverse network design in general is the location-allocation problem, hence there is a rising interest on the reverse flows and optimal facilities identification [47]. Extensive collection of papers can be found in the literature that discuss the deterministic optimization of closed loop and reverse supply chains [48] or optimization under uncertainty [49]. In Table 5 a review of reverse logistics circular economy methods is presented, including the optimization method, the objectives and the application sector.

It is evident from Table 5 that in the existing literature there are no studies focusing in the reverse supply chain optimization of reinforced composite parts. This can be explained by the fact that the reverse supply chain for composite parts and materials does not currently exist, apart from some exceptions (e.g. the carbon fiber recycling sector). Models for the automotive reverse supply chain have also been proposed [52], however the focus was not on the reinforced composite parts of the EoL vehicles.

There are few cases that considered reinforced composite products. A study developed an optimization model to identify suitable locations in UK to process the waste from wind blades, however the study focused only on the material flow form the wind farms to the recycling facilities [57]. Another study introduced an optimization model for the design of an aerospace CFRP waste management supply chain in France, including various recycling technologies [58]. However, the reverse supply chain proposed did not include the outbound flow of the recycled fiber to the customers. This gap in the literature has to do with the fact that there is uncertainty and challenges regarding the reinforced composite parts recycling, as well as the end market of the recycled composites.

In the following section, potential, novel reverse logistics pathways for composite materials are proposed.

Table 5 Review of reverse logistics circular economy methods

Authors	Sector	Method	Objectives	Level of decision	Type of supply chain	Reuse pathway
[50]	WEEE	MILP	Cost	strategic	reverse supply chain	Recycle
[51]	Sand	MILP	Cost	strategic	reverse supply chain	Recycle
[52]	Vehicle	MILP	Cost	strategic	reverse supply chain	Remanufacture/repair/reuse/recycle
[53]	Electronic equipment	MILP	Cost	strategic	reverse supply chain	Remanufacture
[54]	Fashion	MILP	Profit	strategic	closed loop	Recycle
[55]	Refrigerators	MILP	energy used and residual waste and cost	strategic	closed loop	Recycle
[56]	Sand	MILP	Revenues	strategic	reverse supply chain	Recycle

4.2 Novel Reverse Logistics Pathways for Composite Materials

Proposed Reverse Logistics Pathways. Three promising reverse logistics pathways for composite materials are proposed, and are presented in Fig. 3. These pathways were identified within the FiberEUse project.

The three pathways recommend different treatment for the FRP waste: mechanical and thermal treatment, as well as remanufacturing. For Glass Fiber (GF) material the mechanical treatment option was considered as the most reasonable way forward due to the low value of virgin material and the high volumes of waste available, in combination with the availability of potential end use applications. The wind sector was selected as a representative example of waste supplier of EoL GFRP, due to the fact that wind blades consist more than 82% of GF and it is estimated that by 2050 there will be more than 300,000 tons of waste wind blades in Europe [59]. The Sheet Molding Compound (SMC) and Bulk Molding Compound (BMC) manufacturers are assumed the most promising customers of the recycled material according to the literature [60] and experts, where the shredded fibers are used as fillers in new thermoset polymers. In that respect the reverse supply network has a cross-sectorial approach since the SMC and BMC material is employed in various end applications such as transportation, electronics and building. Furthermore, the SMC/BMC material has an increasing demand with approximately 287,000 tons produced in 2018 and an expected 1–2% annual growth [61].

Regarding the Carbon Fiber material, thermal treatment was assumed, due to the expensive processing required that, according to the literature and experiments, it

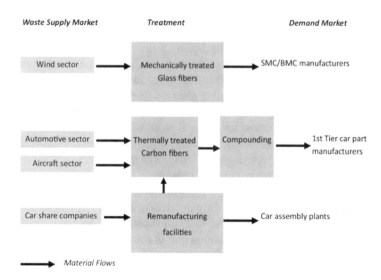

Fig. 3 Proposed reverse logistics pathways

is not a feasible option for a low value material such as Glass Fibers, and the fact that thermal treatment has the potential to maintain the original fiber properties to a higher extent than mechanical treatment, which enhances the potential for high-value applications and circularity of the rCF. In addition, the production of carbon fibers is energy-intensive [62], which has as a result the high price of the virgin material. Therefore, the possibility of recycling the CFRP waste has both economic and environmental incentives. Two sectors have been assumed to provide material for the thermal treatment, the automotive and aircraft sector, since CFRP material use has great potential in those two sectors due to its lightweight properties [63]. However, the latter sector, has the greatest amount of material with better mechanical properties. In total it is forecasted that from both sectors, there will be more than 130,000 tons of CFRP available in Europe by 2050. A potential pathway, after the rCF are retrieved is to create pellets that can be used for injection molding in the automotive sector. For this reason, the rCF are sent to chemical facilities (compounders) and after to the end customer, the 1st Tier car part manufacturers, where the demand of CFRP in Europe is forecasted for 2050 to be more than 200,000 tons [61].

For the final reverse supply chain pathway, remanufacturing of FRP parts is considered. As mentioned earlier in the business model section, the focus is on an innovative and high value product from the automotive, the reusable CF car frames (platform) from car share companies. Car sharing is an alternative for sustainable transportation and it supports towards a low emission mobility [64]. It is estimated that Europe has currently almost a 60,000 cars fleet [65] and an increase is expected. In addition, the use of the CF platform for the shared cars is promising due to the lightweight properties of the material, it improves the fuel consumption, which is highly important for the car users. Secondly, these cars have a high usage (almost 12,000 miles per year) and they are usually dismissed after three years of service, however, the platform of the car is usually in a good condition and only the engine or the body of the car actually needs replacement. Therefore, the EoL cars are assumed to be sent to a new entity, the remanufacturing facility, where they are dismantled and the car platform is inspected. Some frames are then remanufactured depending on their condition, otherwise they are sent to thermal treating facilities. The remanufactured car frames are returned to the car assembly plants, where they are reused for new car production.

For the first pathway with the mechanical treatment, four logistical scenarios were identified according to the wind sector and waste management experts that are presented in Fig. 4, as already presented in Chap. 3 of this Book.

In all four scenarios the main process, shredding, is performed in a dedicated processing facility, however in scenario 2 a pre-shredding process is performed in the wind farm location. It is challenging to fully shred the wind turbine blades at the wind farm currently, as pre-shredding at the wind farms is controlled by many regulations concerning the prevention of dust formation and in some countries (as Germany) it is almost impossible. Scenarios 1 and 3 are quite alike, with only difference in the final size of the blades at the wind farm. Scenario 1 is similar with the currently adopted procedure, when the blades are cut and transported for landfilling. These two scenarios showcase the trade-offs between the more cost-efficient cutting in

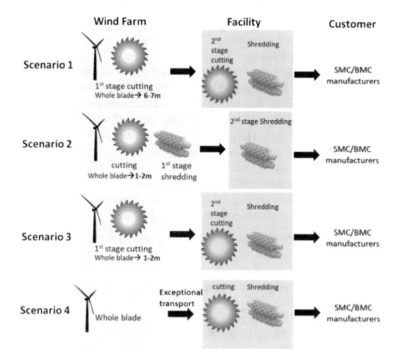

Fig. 4 Scenarios for the wind blades reverse supply network

the facility and the lower transportation cost of scenario 3, with the more efficient transport density of the blades. In the final scenario, the blade is transported as it is to the facility with exceptional transport, which has a much higher cost compared to the other alternatives.

A sensitivity analysis was performed for the four scenarios against the most influential parameters in order to identify the most interesting ones from a cost and practical perspectives. The four scenarios were investigated for the following variables, which are the most impactful:

- **Number of blades in a decommissioned wind farm**: 3, 6, 12, 18, 30 and 40
- **Distance wind farm-plant**: 20, 100 and 1000 km
- **Particles final output dimension**: 2, 4, 6, 8, 10 mm.

The scenarios that were optimal in greatest percentage of combinations of the above factors are scenarios 1 and 2, as it is depicted in Table 6. Therefore, these two reverse supply networks are considered the most promising.

Table 6 Sensitivity analysis of scenarios: number of cases of optimal cost per scenario

	Scenario 1	Scenario 2	Scenario 3	Scenario 4
	53%	37%	0%	10%

The first two proposed reverse logistics pathways of Fig. 3 are optimized for Europe and the latter one for the case of United Kingdom, due to data availability. A location-allocation problem was followed with the inputs and outputs of Fig. 5. A cost optimization model was developed, because the focus was to maximize the potential for viability of the proposed reverse supply chains. At the current stage where the cost is one of the key barriers, focusing on configurations that reduce the carbon footprint, while potentially increasing the cost would be less relevant. In addition, the proposed networks that introduce a circular pathway are considered more sustainable than the current end-of-life option, which is the landfilling or the incineration. In future work, the model could be adapted to include environmental objectives as a multi-objective optimization problem.

Some indicative results regarding the optimal configurations and their economics as well as environmental performance are presented in the following section.

Results from Optimal Networks. The model developed for the reverse logistics network cost optimization was employed to find the optimal facility location and material flows for the three previously presented reverse supply chains for composite materials. The optimal networks are presented in Table 7 for the proposed supply networks for both 2020 and 2050.

For the mechanically treated glass fiber it can be observed in Table 7 that the network proposed from the optimization will need to change from 2020 to 2050 for both scenarios. Only the facilities in Germany and Spain remain the same and facilities in UK and Poland are introduced. The optimal reverse supply chains have a decentralized nature and similar locations of the optimal facilities for both scenarios and for 2020 as well as 2050. This can be inferred as an approximate indication of where the facilities should be initially located, whichever treating process is selected and these locations will remain optimal while the system would be expanding over time. It should be noted first that the facilities are located near the SMC/BMC manufactures and second that the largest facility is in Germany, which has the highest supply and demand of material.

For the second reverse logistics pathway (CFRP), it is observed that the proposed network is semi-decentralized, with only two facilities for 2020 since the amount of waste material is quite low and four facilities for 2050. The two proposed networks are quite different, with only similarity the facility in the Netherlands. Generally, in both networks the facilities are located in central Europe.

Fig. 5 Reverse logistics network optimization

Table 7 Optimal facilities location for each reverse logistics pathway

Reverse logistics pathway	Scenarios	Year	Location
Mechanical treated glass fibers	*1*	2020	Spain Germany Switzerland
		2050	Poland UK Spain Germany
	2	2020	Spain UK Italy Germany
		2050	Spain UK Poland Germany
Thermal treatment of carbon fibers		2020	Austria Netherlands
		2050	Switzerland France Netherlands Sweden
Inspection, repair, remanufacture of CFRP	*Optimistic*	2020	1 × South East UK
	Pessimistic		1 × South East UK
	Optimistic	2050	2 × South East UK 1 × South West UK 2 × Central West UK 3 × North East UK
	Pessimistic		2 × South East UK 1 × Central West UK

In the final reverse logistics pathway (reusable CF car platforms), for 2020 both scenarios indicate as optimal a centralized facility near the hot spot of CFRP EoL products, due to the low amount of material. It is observed that both the optimistic and pessimistic scenario propose the same facility location. On the other hand, the optimal network for 2050 differs and becomes very decentralized. This is highlighted on the optimistic scenario, where 8 facilities are proposed, due to the higher amount of CFRP EoL product.

Finally, it can be inferred for the two first reverse supply networks that the optimal supply network structure identified is not centralized and the facilities receive material in most cases from more than one country, as well as they supply material in customers in other countries. Therefore, the optimal networks showcase the need of intercountry transportation of both waste and recycled material.

In Fig. 6 the breakdown of the cost of each process is presented for all the optimal reverse logistics pathways. For the first pathway, it is evident that the transportation cost for scenario 1 is much higher compared to scenario 2. This is due to the fact that in the former case there is no pre-shredding of the blades in situ; therefore, the blades are transported into pieces, which is more costly. The most prominent way to transport one blade in pieces is with two trucks with two containers each. On the other hand, it is inferred that the processing cost in situ for scenario 2, represents almost 50% of the total cost, whereas for scenario 1 it is only 4–6% because in that case the only process required in situ is the cutting of the blades in large pieces.

For the second reverse supply network, it is observed that the compounding cost constitutes the greatest percentage on the total costs. It is evident that in 2050 the pyrolysis cost percentage is decreasing; this is due to the benefits of the economy of scale. So, the material increases and therefore, the processing costs increase proportionally.

For the final reverse logistics pathway, it is evident that the processing constitutes the greatest part of the costs with more than 85%. For both 2020 and 2050 it is observed that the processing cost decreases, when the scenario becomes pessimistic thus the percentage of material that is sent for recycling and is not remanufactured

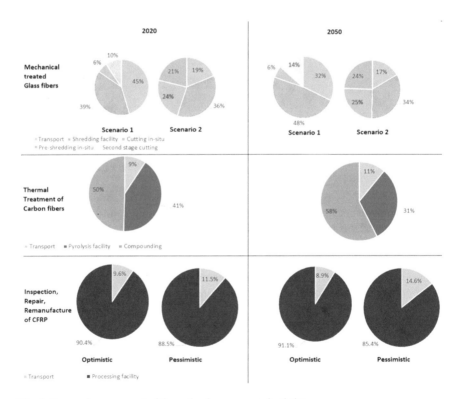

Fig. 6 Economic assessment of the optimal reverse supply chains

increases. On the other hand, the transportation cost increases, since more material is transported to the recycling facility.

In Fig. 7, the percentage of emissions per stage on the total emissions is displayed. In the first case, the emissions from the shredding process correspond to more than 70% of the total emissions, with highest emissions for scenario 1 due to the highest electricity consumption in the facility. The carbon emissions from transportation are higher on scenario 1 compared to scenario 2 with almost 25% of the total emissions, due to the inefficient blade transportation process, the higher amount of material transported and generally the higher transportation effort involved.

In the second investigated reverse logistics pathway, the carbon emissions from pyrolysis constitute more than 42% of the total emissions, whereas the compounding around 37% for 2020. It is observed that in 2050 the emissions from compounding consist a higher percentage than those of pyrolysis, this is due to the fact that the carbon emissions per kWh per country differ. According to the data, Sweden, Norway, France and Switzerland have a very low carbon emissions coefficient compared to Austria and Netherlands. Finally, in both cases the contribution of the transportation to the emissions is very low.

For the final reverse logistics pathway, it is observed that the greatest contribution derives from the remanufacturing facilities, where mostly electricity is used. This

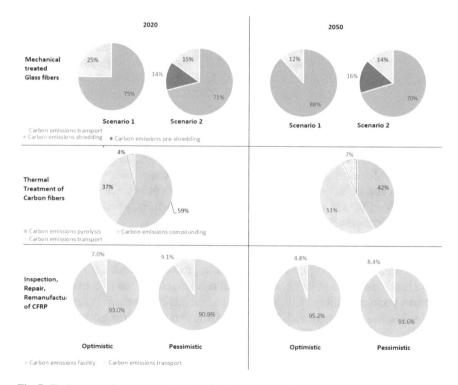

Fig. 7 Environmental assessment of the optimal reverse supply chains

percentage is higher on the optimistic scenario, due to the greater amount of CFRP EoL products that are processed. On the other hand, it is evident that in 2050 optimistic scenario the percentage of the emissions from the transportation is decreased compared to 2020. This decrease derives from the fact that the facilities on these scenarios in 2050 are more decentralized therefore the distances travelled per EoL CFRP product are reduced.

4.3 Logistics Challenges in Closing the Loop of Fiber-Reinforced Composites

A barrier in closing the loop for fiber-reinforced composites is the current lack of some links in the reverse logistics/supply chain, that are needed for the reverse flow of the material to happen. For example, in the mechanically treated GF logistical pathway examined, there is currently no mechanical processing capacity available to undertake this task. Despite the fact that the technology required exists, with modifications of commercially available heavy-duty shredders, there is currently very limited capacity available as there is no market for the end product yet. This is enhanced by the fact that in many countries e.g. Austria landfills currently accept the EoL wind blades in small pieces. In a similar way, pyrolysis capacity for CF currently exists at the scale of the current CF waste production levels, but this will need to be significantly enhanced in the future, since the anticipated amounts of related waste will increase drastically. These supply chain links, essentially the material processors, could be created by a market pull, either as new entities/startups specializing in offering these services, or by existing players in relevant areas that could exploit the new business opportunities arising, for example waste management and recycling companies.

Additional transport obstacles arise when it comes to cross border transportation. At a cross border transport of waste, it has to be ensured if a notification in the origin country as well as in the receiving country is necessary. Shipments of waste on the "Green List" for closing the loop do not require notification. Since fiber-reinforced composites are not included in the "Green List", notifications are needed. It was also observed that FRP are generally materials with low density, which is not ideal for achieving logistical efficiencies.

Another logistics-related barrier in the emergence of the proposed circular economy pathways relates to the geographical spread of the waste material availability. Glass and carbon fibers are included in products that are used widespread, and in most cases a recycling system would need to recover the EoL material from too many remote locations. This leads to logistical inefficiencies, as (a) the materials involved have low density and do not make efficient use of transportation vehicle capacities and (b) they require complex routing decisions and long transportation distances, as the material is not currently aggregated at more centralized locations. Recycling systems generally benefit from economies of scale at the processing facility level, while the geographical spread of material availability favors more decentralized systems to reduce logistical costs. This is a trade-off between designing centralized and decentralized systems that is difficult to resolve. Ultimately, some of the EoL material that is available in very remote locations leads to much higher costs overall for the system; hence, if systems were not forced to collect the material from everywhere, or if the remote material logistical operations were subsidized in some way, the overall circular economy pathways economics would be much more favorable.

It was also identified that increasing the transported material density significantly reduces the overall system cost. In more detail, the scenarios for the wind blades reverse supply network show that the transportation cost in scenario 1 is much more expensive, especially the inbound due to the inefficient transport of the blade pieces as well as the transportation density difference of the cut and pre-shredded blades. This leads to low utilization of the transportation capacity when transporting large blade pieces. Therefore, this issue would require further research and significant attention, in order to reduce the costs.

Ultimately, a current major barrier in the emergence of the proposed circular economy pathways for fiber-reinforced composite materials is the low cost of the virgin materials, especially in the case of GF. For any circular economy pathway to be successful in practice, a clear prerequisite is to be able to provide the recycled/reused material at a price equal or lower to the virgin material currently used in the specific end processes.

4.4 Managerial and Policy Implications of Novel Reverse Logistics Pathways

The three logistical pathways investigated are quite different from each other; therefore, generic conclusions are difficult to draw. A conclusion valid for both pathways 1 and 2 is that the optimal supply network structure that leads to minimum cost contains a handful of processing facilities at the European level. Hence, the material flows are not country-specific and in many cases the proposed facilities source waste from many countries or may serve end users in a different country. This means that if the aim is to minimize the recycled end product cost, movement of waste and recycled material between countries should be allowed and any existing barriers removed, as very few European countries in isolation would have enough supply and demand to be able to support feasible circular economy pathways. For a successful circular economy approach, especially in the GF case investigated where the feasibility is marginal against the virgin material competitors, a change of perspective from 'waste' to 'valuable raw materials' should be adopted. For the CF case, the economics are more favorable due to the much higher cost of the competitor virgin CF materials; however, even in this case the supply network proposed indicates large facilities around Europe with a waste material 'catchment area' spanning many countries each. Ultimately, standardization of legislation on movement of waste material, when these are for recycling and reusing within a circular economy approach, would be of great importance to facilitate the development of economically viable recycling systems.

Another important lever in the hands of policy makers is the disincentives for alternative waste disposal pathways, such as landfilling or energy recovery. Currently, there is a large variation in policies and costs of disposal of GF and CF, with many European countries having very cheap landfilling, whereas others have already banned the landfilling of these materials. Therefore, if policy makers ensure that the alternative options that divert the material from the circular economy become more expensive, or even cease to exist, they would automatically support the development of such recycling networks for reuse of the FRP.

It is also interesting to note that when one designs a system to process/recycle all the available waste material of one type, this comes at a higher cost compared to a system that only processes the profitable quantities. This of course leaves a question on what would be done with the remaining amounts of FRP waste. It is apparent that some locations will incur a higher cost to provide the waste material for processing, therefore incentives from the policy makers' perspective should be considered for the reverse supply system to include this more 'expensive' waste.

Another key point for consideration of managers and policy makers alike is that the trade-off between larger facilities for ensuring economies of scale, versus the need for more decentralized network with more and smaller facilities to reduce the logistical costs, has no obvious solution. This is apparent from the results of the various pathways and scenarios investigated, where the optimal network structure is different for most of them. Therefore, modelling tools can support a better understanding of these trade-offs and allow investigation of the impact of alternative scenarios, policies or forecasts.

It was also observed that FRP are generally materials with low density, which is not ideal for achieving logistical efficiencies. Despite the fact that there was not one approach identified as best, in most cases increasing the transported material density significantly reduces the overall system cost. Therefore, this issue would require further investigation and significant attention, in order to bring the costs down. Still, one would need to consider the whole network before focusing only on the logistics efficiency in a part of the network, as in our work it has been shown that increasing the transportation efficiency in logistical pathway 1 by pre-shredding at the wind farms ultimately may come at a disproportional cost due to the cost of additional operations in remote locations.

Logistical pathway 3 was quite different in scope than 1 and 2, as it investigated the potential for remanufacturing and reuse of a CF-based automotive part, a novel car frame designed specifically for car sharing applications. Despite the numerous assumptions that had to be made, due to the novelty of this idea, it was identified that even at the country level there would be multiple facilities required in the future, if the demand grows significantly. This shows that the remanufacturing network would be much more decentralized at European level compared to the pathways 1 and 2, and indicated the need for these remanufacturing facilities to be located primarily close to the end users and secondarily close to EoL car availability hotspots. It also shows that transportation costs for reusing whole parts are even more important than in the first two scenarios, ultimately favoring more local reverse supply chain networks.

5 Conclusion

In this chapter, new potential business opportunities of composites reuse were presented. Case studies of products made out of recycled composite materials in the sanitary, sports equipment, furniture and the automotive industry were selected to propose innovative circular business models. The new business models of products made from secondary composites in the sanitary, ski and automotive industries were all focused on selling a greener product together with product-related services. The innovative business model of recycled office desk, in the furniture sector, was based on leasing (non-ownership). Finally, the proposed business model for designing and remanufacturing of lightweight platform built from FRP was based in a customer–supplier long-term supply relationship where the supplier sells greener and more performing products that are periodically taken back and re-supplied at favorable conditions for new production.

In addition, pathways for implementation of reverse logistics, which leads to viable recycling and reuse of composites in Europe, were also proposed. Three possible logistical pathways to reuse composites in Europe were investigated as following: mechanical recycling of GFRP derived from EoL of wind turbine blades; thermal recycling of CFRP derived from aeronautic waste; remanufacturing of CFRP component (reusable CFRP platform) in the automotive industry. The optimal supply network for the first two pathways, which leads to cost reduction, indicates that the waste material flow should be semi-decentralized and that an open multi-country approach should be pursued. The logistics pathway for the latter case suggested that the remanufacturing network needs to be more decentralized in Europe in comparison with the former pathways. In addition, it was defined that even if remanufacturing is carried out in one country, several recycling facilities are required in the future.

Future research is needed at business model level to carefully investigate sustainability and risks, considering the high number of possible business model implementation options and scenarios. Based on techno-economic results, policy research is also needed to elaborate evidence-based policy guidelines to support the diffusion of circular business models in this sector.

References

1. Hoffman, F., Jokinen, T., Marwede, M.: Circular business models. European Union Development Fund (2017). https://sustainabilityguide.eu/methods/circular-business-models. Last accessed 01 Dec 2020

2. Linder, M., Williander, M.: Circular business model innovation: inherent uncertainties. Bus. Strateg. Environ. **26**, 182–196 (2015)
3. Roos, G.: Business model innovation to create and capture resource value in future circular material chains. Resources **3**, 248–274 (2014)
4. Den Hollander, M., Bakker, C.: Mind the gap exploiter: Circular business models for product lifetime extension. In: Proceedings of the Electronics Goes Green 2016, pp. 1–8, Berlin, Germany (2016)
5. Vogtlander, J.G., Scheepens, A.E., Bocken, N.M.P., Peck, D.: Combined analyses of costs, market value and eco-costs in circular business models: eco-efficient value creation in remanufacturing. J. Remanuf. **7**(1), 1–17 (2017)
6. Nußholz, J.: Circular business models: defining a concept and framing an emerging research field. Sustainability **9**(10), 1810 (2017)
7. Goyal, S., Esposito, M., Kapoor, A.: Circular economy business models in developing economies: lessons from India on reduce, recycle, and reuse paradigms. Thunderbird Int. Bus. Rev. **60**(5), 729–740 (2018)
8. Tse, T., Esposito, M., Soufani, K.: How businesses can support a circular economy. Harvard Business Review (2016). https://hbr.org/2016/02/how-businesses-cansupport-a-circular-eco nomy. Last accessed 19 Nov 2020
9. Upadhyay, A., Akter, S., Adams, L., Kumar, V., Varma, N.: Investigating "circular business models" in the manufacturing and service sectors. J. Manuf. Technol. Manag. **30**(3), 590–606 (2019)
10. Chen, C.W.: Improving circular economy business models: opportunities for business and innovation: a new framework for businesses to create a truly circular economy. Johnson Matthey Technol. Rev. **64**(1), 48–58 (2020)
11. Oghazi, P., Mostaghel, R.: Circular business model challenges and lessons learned—an industrial perspective. Sustainability **10**(3), 739 (2018)
12. RISE—Research Institutes of Sweden, Ict, and Viktoria.: Circular business model innovation: inherent uncertainties. Bus. Strat. Environ. **26**(2), 182–196 (2017)
13. Ijomah, W.L., McMahon, C.A., Hammond, G.P., Newman, S.T.: Development of robust design-for-remanufacturing guidelines to further the aims of sustainable development. Int. J. Prod. Res. **45**(18), 4513–4536 (2007)
14. Fraccascia, L., Giannoccaro, I., Agarwal, A., Hansen, E.G.: Business models for the circular economy: opportunities and challenges. Bus. Strateg. Environ. **28**(2), 430–432 (2019)
15. Hansen, E.G., Große-Dunker, F., Reichwald, R.: Sustainability innovation cube. A framework to evaluate sustainability-oriented innovations. Int. J. Inno. Manage. **13**, 683–713 (2009)
16. Bocken, N., Strupeit, L., Whalen, K., Nußholz, J.: A review and evaluation of circular business model innovation tools. Sustainability **11**(8), 2210 (2019)
17. Shahbazi, K.: 10 circular business model examples, https://www.boardofinnovation.com/blog/circular-business-model-examples. Last accessed 19 Nov 2020
18. Lewandowski, M.: Designing the business models for circular economy—towards the conceptual framework. Sustainability **8**(1) (2016)
19. Pieroni, M.P.P., McAloone, T.C., Pigosso, D.C.A.: Configuring new business models for circular economy through product–service systems. Sustainability **11**(13), 3727 (2019)
20. Whalen, C.J., Whalen, K.A.: Circular economy business models: a critical examination. J. Econ. Issues **54**(3), 628–643 (2020)
21. Reim, W., Parida, V., Sjödin, D.R.: Circular business models for the bio-economy: a review and new directions for future research. Sustainability **11**(9), 2558 (2019)
22. OECD.: Business models for the circular economy: opportunities and challenges from a policy perspective. OECD Publishing, Paris (2018)
23. Naqvi, S.R., Prabhakara, H.M., Bramer, E.A., Dierkes, W., Akkerman, R., Brem, G.: A critical review on recycling of end-of-life carbon fibre/glass fibre reinforced composites waste using pyrolysis towards a circular economy. Resour. Conserv. Recycl. **136**, 118–129 (2018)
24. Meng, F., Olivetti, E.A., Zhao, Y., Chang, J.C., Pickering, S.J., McKechnie, J.: Comparing life cycle energy and global warming potential of carbon fiber composite recycling technologies and waste management options. ACS Sustain. Chem. Eng. **6**, 9854–9865 (2018)

25. GFRP Composite Market. https://www.marketsandmarkets.com/Market-Reports/glass-fiber-reinforced-plastic-composites-market-142751329.html#:~:text=%5B160%20Pages%20Report%5D%20The%20global,6.4%25%20from%202017%20to%202022. Last accessed 12 Nov 2020
26. Rybicka, J., Tiwari, A., Leeke, G.A.: Technology readiness level assessment of composites recycling technologies. J. Clean. Prod. **112**(1), 1001–1012 (2016)
27. Jacob, A.: Composites can be recycled. Reinforced Plastic **55**(3), 45–46 (2011)
28. Pickering, S.J.: Recycling technologies for thermoset composite materials—current status. Compos. A Appl. Sci. Manuf. **37**(8), 1206–1215 (2006)
29. McGrath, R.G.: Business models: a discovery driven approach. Long Range Plan **43**(2–3), 247–261 (2010)
30. Oliveux, G., Dandy, L.O., Leeke, G.A.: Current status of recycling of fibre reinforced polymers: review of technologies, reuse and resulting properties. Prog. Mater Sci. **72**, 61–99 (2015)
31. Global Shower Trays Market. https://www.qyresearch.com/index/detail/1087978/global-shower-trays-market. Last accessed 03 Dec 2020
32. Hudson, S.: Snow Business a study of international Ski industry, 1st edn. Cassell, London (2000)
33. Parikka-Alhola, K.: Promoting environmentally sound furniture by green public procurement. Ecol. Econ. **68**(1–2), 472–485 (2008)
34. European Environment Agency.: https://www.eea.europa.eu/help/glossary/eea-glossary/green-procurement. Last accessed 23 March 2021
35. European Commission, ICLEI and PPA: Buying Green! A Handbook on Green Public Procurement, 3rd ed. Publications Office of the European Union, Luxembourg (2016). https://ec.europa.eu/environment/gpp/pdf/Buying-Green-Handbook-3rd-Edition.pdf
36. Besch, K.: Product-service systems for office furniture: barriers and opportunities on the European market. J. Clean. Prod. **13**, 1083–1094 (2005)
37. Tukker, A.: Eight types of product–service system: eight ways to sustainability? Experiences from SusProNet. Bus. Strategy Environ. **13**, 246–260 (2004)
38. Forrest, A., Hilton, M., Ballinger, A., Whittaker, D.: Circular economy in the furniture sector. Arditi, S (ed.) European Environmental Bureau (EEB) (2017)
39. Kuhnert, F., Stürmer, C.: Five trends transforming the automotive industry. https://www.pwc.at/de/publikationen/branchen-und-wirtschaftsstudien/eascy-five-trends-transforming-the-automotive-industry_2018.pdf. Last accessed 16 Jan 2018
40. Kumar, S., Bharj, R.S.: Emerging composite material use in current electric vehicle: a review. In: International Proceeding of Conference on Composite Materials: Manufacturing, Experimental Techniques, Modeling and Simulation, pp. 27946–27954. Material Today Proceedings, Jalandhar (2018)
41. Koniuszewska, A.G., Kaczmar, J.W.: Application of polymer based composite materials in transportation. Progr. Rubber, Plastics Recycl. Technol. **32**(1), 1–24 (2016)
42. Hagnell, M.K., Åkermo, M.: The economic and mechanical potential of closed loop material usage and recycling of fiber-reinforced composite materials. J. Clean. Prod. **223**, 957–968 (2019)
43. Larsen, K.: Recycling wind turbine blades. Renew. Energy Focus **9**(7), 70–73 (2009)
44. Werning, J.P., Spinler, S.: Transition to circular economy on firm level: Barrier identification and prioritization along the value chain. J. Clean. Prod. **245**, 118609 (2020)
45. Urbinati, A., Chiaroni, D., Chiesa, V.: Towards a new taxonomy of circular economy business models. J. Clean. Prod. **168**, 487–498 (2017)
46. Özkir, V., Başligil, H.: Modelling product-recovery processes in closed-loop supply-chain network design. Int. J. Prod. Res. **50**(8), 2218–2233 (2012)
47. Difrancesco, R.M., Huchzermeier, A.: Closed-loop supply chains: a guide to theory and practice. Int. J. Logistics Res. Appl. **19**(5), 443–464 (2016)
48. Van Engeland, J., Beliën, J., De Boeck, L., De Jaeger, S.: Literature review: Strategic network optimization models in waste reverse supply chains. Omega **91**, 102012 (2020)

49. Govindan, K., Fattahi, M., Keyvanshokooh, E.: Supply chain network design under uncertainty: a comprehensive review and future research directions. Eur. J. Oper. Res. **263**(1), 108–141 (2017)
50. Achillas, C., Vlachokostas, C., Aidonis, D., Moussiopoulos, N., Iakovou, E., Banias, G.: Optimizing reverse logistics network to support policy-making in the case of electrical and electronic equipment. Waste Manage. **30**(12), 2592–2600 (2010)
51. Barros, A.I., Dekker, R., Scholten, V.: A two-level network for recycling sand: a case study. Eur. J. Oper. Res. **110**(2), 199–214 (1998)
52. Cruz-Rivera, R., Ertel, J.: Reverse logistics network design for the collection of end-of-life vehicles in Mexico. Eur. J. Oper. Res. **196**(3), 930–939 (2009)
53. Jayaraman, V., Guide Jr, V.D.R., Srivastava, R.: A closed-loop logistics model for remanufacturing. J. Oper. Res. Soc. **50**(5), 497–508 (1998)
54. Kim, J., Do Chung, B., Kang, Y., Jeong, B.: Robust optimization model for closed-loop supply chain planning under reverse logistics flow and demand uncertainty. J. Clean. Prod. **196**, 1314–1328 (2018)
55. Krikke, H., Bloemhof-Ruwaard, J., Van Wassenhove, L.N.: Concurrent product and closed-loop supply chain design with an application to refrigerators. Int. J. Prod. Res. **41**(16), 3689–3719 (2003)
56. Listeş, O., Dekker, R: A stochastic approach to a case study for product recovery network design. Eur. J. Operational Res. **160**(1), 268–287 (2005)
57. Sultan, A.A.M., Mativenga, P.T., Lou, E.: Managing supply chain complexity: foresight for wind turbine composite waste. Procedia CIRP **69**, 938–943 (2018)
58. Vo Dong, A., Azzaro-Pantel, C., Boix, M.: A multi-period optimization approach for deployment and optimal design of an aerospace CFRP waste management supply chain. Waste Manage. 95, 201–216 (2019)
59. Lichtenegger, G., Rentizelas, A., Trivyza, N., Siegl, S.: Offshore and onshore wind turbine blade waste material forecast at a regional level in Europe until 2050. Waste Manage. **106**, 120–131 (2020)
60. Mamanpush, S.H., Li, H., Englund, K., Tabatabaei, A.T.: Recycled wind turbine blades as a feedstock for second generation composites. Waste Manage. **76**, 708–714 (2018)
61. Witten, E., Mathes, V., Sauer, M., Kuhnel, M.: Composites Market Report 2018 (2018)
62. Song, Y.S., Youn, J.R., Gutowski, T.G.: Life cycle energy analysis of fiber-reinforced composites. Compos. A Appl. Sci. Manuf. **40**(8), 1257–1265 (2009)
63. Li, X., Bai, R., McKechnie, J.: Environmental and financial performance of mechanical recycling of carbon fibre reinforced polymers and comparison with conventional disposal routes. J. Clean. Prod. **127**, 451–460 (2016)
64. Goldman, T., Gorham, R.: Sustainable urban transport: four innovative directions. Technol Society **28**(1–2), 261–273 (2006)
65. Phillips, S.: Car sharing market analysis and growth. https://movmi.net/carsharing-market-growth. Last accessed 15Jan 2020

Economic and Risk Assessment of New Circular Economy Business Models

Winifred Ijomah⊕, Nikoletta L. Trivyza⊕, Andrea Tuni⊕,
Athanasios Rentizelas⊕, Fiona Gutteridge, and Volker Mathes

Abstract Circular economy business models are key enablers of the circular economy. However, they must also be economically viable to materialize in reality, since profitability is a major business driver. Assessment of the economic potential of circular economy business model is subjected to significant uncertainties and to a range of risks, due to their novel nature. This chapter firstly discusses the economic assessment of the circular business models specifically for composites, based on five business model cases from various sectors, focusing on the identification of the major causes of uncertainty and then performing sampling-based sensitivity analysis to investigate the viability of the circular economy business models under uncertainty. The findings show that the proposed circular economy business models are not always more profitable than the existing models. Thus, in some instances new market stimulus would need to be identified and implemented to increase attractiveness of the proposed solutions. As a consequence, the chapter identifies and prioritizes risk factors for new circular economy business models for composites. The risk analysis uses input from industry experts and the literature to come up with a key risk factors list relevant to circular economy business models for composite materials. The risk analysis concludes that the key risk factors are both from the demand/market and supply side.

Keywords Circular economy · Business models · Economic assessment · Risk assessment

W. Ijomah (✉) · A. Tuni · A. Rentizelas · F. Gutteridge
Design, Manufacturing and Engineering Management Department, University of Strathclyde, 75 Montrose Street, Glasgow G1 1XJ, UK
e-mail: w.l.ijomah@strath.ac.uk

N. L. Trivyza
Department of Naval Architecture, Ocean and Marine Engineering, University of Strathclyde, Maritime Safety Research Centre, 100 Montrose Street, Glasgow G4 0LZ, UK

V. Mathes
AVK—Federation of Reinforced Plastics, Am Hauptbahnhof 10, 60329 Frankfurt Am Main, Germany

© The Author(s) 2022
M. Colledani and S. Turri (eds.), *Systemic Circular Economy Solutions for Fiber Reinforced Composites*, Digital Innovations in Architecture, Engineering and Construction, https://doi.org/10.1007/978-3-031-22352-5_20

1   Introduction

Accelerating costs of materials, waste disposal, energy, environmental penalties coupled with the increased expectations of customers and investors for environmental conscious manufacturing have integrated environmental issues into the cost equation. Therefore, sustainable manufacturing is becoming a critical business need. Circular economy business models are modes of business operation designed to maximize the life of products and materials that are kept in use to extract the maximum value from them. Such models are a key plank of the circular economy and are imperative in the drive towards sustainable manufacturing. A one size fits all approach cannot be taken in the design of effective circular economy business models. Each business scenario must be evaluated to obtain full understanding of its nature and from there which potential circular economy models could be viable. Circular economy business models must be designed to ensure their applicability, robustness and profitability. Thus, each prospective circular economy model must be analyzed to identify the circumstances under which it would be profitable, the risks involved, their gravity and potential mitigating actions. This chapter describes the strategies that can be used to enhance the potentiality of circular economy business models. The initial part of the chapter describes the economic assessment of the business models, the identification of the major causes of uncertainty via information from the literature and experts, and the sampling-based sensitivity analysis to investigate the viability of the business models under uncertainty. Following from there it describes the identification of circular economy business model risk factors (including their extent of impact and likelihood of occurrence) and their prioritization to form a bank of knowledge that can be used to develop mitigation strategies for critical risks. Major results and limitations of the work are described with potential future work to use the key research outcomes.

2   Economic Assessment

In the literature it is evident that one of the methodologies followed to support the development of new Circular economy business models is the Business Model Canvas (BMC) [1] and frameworks have been proposed to adapt the traditional BMC to the circular models [2, 3]. One of the main parts of the BMC and greatest strategies on a new business model is the value capture, which is expressed with the cost structure and revenue stream [4]. When introducing a circular economy business model various issues need to be considered amongst technological and regulatory. These challenges have an impact on the business model starting from the cost structure and revenue streams and then influencing the rest of the BMC blocks [5].

In that respect, it is defined that 'the business model's key role is to incorporate the circular economy principles into a design or redesign of business activities and partnerships and to create a cost and revenue structure, which is compatible both with sustainability and with profitability' [6]. Companies in order to enhance their sustainability and circularity are required to impose changes on the way they generate value [7]. Existing studies indicate that the new circular economy business models can potentially be more effective and efficient compared to the traditional ones [8]. Therefore, when introducing new circular economy business models, a critical step is the cost structure model.

Another important issue when considering circular economy business models is the dynamic environment and the uncertainty [9]. The uncertainty can be related with various factors in new circular economy business models, since the products proposed as well as the technologies and materials are novel and innovative.

In this chapter, the focus is on assessing the economic potential of new circular economy business models for composite products under significant uncertainty due to the data accuracy and availability. As a result, a sensitivity analysis approach has been adopted in order to consider this uncertainty in the assessment.

2.1 Sensitivity Analysis on Circular Economy Business Models of Composites

Sensitivity analysis investigates how the output is affected by the changes of the input parameters and in many cases it is used for quantification of the system's uncertainty [10]. As a result, a sensitivity analysis should be performed with assumptions on the uncertain parameters that differ from the data employed in the primary analysis [11]. Performing a sensitivity analysis supports in determining which parameters are key drivers for the final outcome and increase the level of confidence in the outputs.

A sensitivity analysis is required in order to map the region of the input parameters variability space that has significant impact on the output. A sampling based sensitivity analysis approach follows the steps according to [12]:

1. Input sampling
2. Model evaluation
3. Post processing.

For the input sampling either a one-at-a-time or all-at-a-time approach can be adopted. In the former, one input parameter is varied and the others are considered fixed, whereas in the latter the inputs are simultaneously varied. The one-at-a-time approach is easier to apply and visualize, however it is not possible to identify the interactions among the parameters. Therefore, in this work the all-at-a-time approach is employed. The process followed for the economic assessment is presented in Fig. 1.

Fig. 1 Sensitivity analysis for new business models

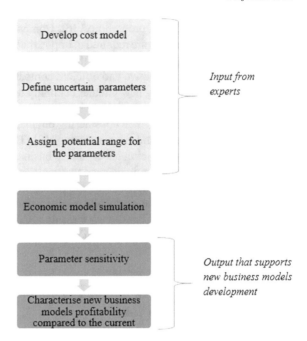

2.2 Results from the Sensitivity Analysis

The economic assessment was carried out for the proposed circular economy business models of composites. Due to confidentiality issues it is not possible to provide the exact measures used for the assessment.

The first business model considered is about sanitary ware products such as bathtubs and shower trays. The new business model proposes as a baseline scenario replacing the roving, which consists of 40% glass fibers with recycled production waste. As a result, the cost of raw material is reduced and there is no cost for 40% of the roving material. However, there is a cost for the recycling process. The parameters considered for the economic assessment are displayed in Table 1.

The uncertain parameters considered for the economic assessment are the 'percentage of material replaced (roving)' and the 'roving price with recycled glass fibers'. The outputs of the economic assessment are presented in the following figures.

Table 1 Parameters considered in the economic assessment of bathtubs

Parameter	Bathtub
Raw materials cost without roving (€/unit)	24.91
Amount of waste (kg/unit)	2.87
Disposal cost (€/kg)	0.21
Labor cost per unit (€/unit)	13.61

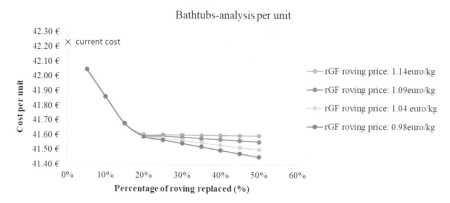

Fig. 2 Bathtubs cost per unit

In Fig. 2, the cost per unit of bathtub is presented for different percentages of roving material replaced and recycled glass fiber prices. In addition, the current cost with zero percentage replaced is displayed in the figure. It is inferred that the new circular economy solution proposed manages to reduce the unit cost for different percentages of material replaced and recycled fibers price. It is derived from the figure that the roving cost has a great impact on the unit costs and as a result the replacement of the material positively influences the total cost. The changes in the percentage of material replaced lead to changes in the unit cost. It is observed that replacing more than the current baseline value of 40% material makes the product even more profitable. The highest % replacement of the roving with the recycled waste and the lowest recycled fibers price leads to savings of approximately 2% per unit. As a result, the proposed baseline solution with 40% replacement of roving material for the bathtubs reduces significantly the annual and unit cost, and increasing this percentage further can lead to even higher savings.

A subsequent economic assessment is performed for the circular economy business model proposed for skis, as an example of a composite product from the sports sector. A percentage of the virgin glass fibers is replaced by recycled fibers. In addition, it is assumed that the amount of PU polymer used is reduced due to the use of recycled fibers. The parameters considered for the economic assessment are displayed in Table 2.

The uncertain parameter considered for the economic assessment is the cost of the recycled fibers and the cost range is considered to be up to 20% lower than the virgin price. The output of the economic assessment is presented in the following figure along with the current cost.

In Fig. 3 the cost of each unit for varying recycled fiber cost is shown. It is evident that the solution proposed is more competitive than the current one due to the fact that it uses recycled fibers and less PU polymer that has higher cost compared to the recycled fibers. In addition, there is a linear relationship between the recycled fibers price and the skis' cost. However, regardless of the change of recycled fibers' price, the proposed solution is still slightly more competitive.

Table 2 Parameters considered in the economic assessment of ski

Parameter	
Cost of material per unit (without fibers) (€/pair)	28
PU polymer quantity used in one unit (kg/pair)	1.48
PU polymer system price (€/kg)	2.05
Recycled fibers quantity per unit (kg/pair)	0.10
Virgin fibers quantity per unit (kg/pair)	0.75
Amount of waste (kg/pair)	0.32
Disposal cost (€/kg)	0.20
Volume of production (sqm/year)	20,500
Depreciation cost of machinery (€/pair)	7.50
Labor cost per unit (€/pair)	8.00

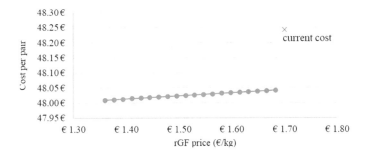

Fig. 3 Skis unit cost

The next business case investigated is from the automotive sector (a car pedal bracket) and the parameters considered for the economic assessment are displayed in Table 3.

Table 3 Parameters considered in the economic assessment of pedal bracket

Parameter	
Target cost per part (€)	2.66
Cost of raw materials (€ per part)	1.76
Approximate weight for each part (kg)	0.51
Process cost (€/part)	0.5
Volume of production per year (parts)	200,000
Labor cost per part (€/part)	0.4
Investment needed for changes in the current process (€)	100,000
Pellet production (€/kg)	0.8
Polyamide polymer price (€/kg)	2.4
Polypropylene polymer price (€/kg)	1.7

In the new business model, the glass fibers are replaced with recycled carbon fibers. In the original case, the percentage of glass fibers used in the raw material is 40% in the pedal bracket. However, in the proposed solution the recycled carbon fibers will replace glass fibers in the range of 20–40% for the pedal. The rest of the part's material is polymer, therefore the changes on the carbon fiber percentage affect the respective percentage of polymer. In addition, the type of polymer is also an uncertain factor and either polypropylene or polyamide is considered. It should be noted that there is an additional cost due to the pellet production, because pellets are required for the injection process. The recycled carbon fibers price is uncertain but the part price target must be the same as before (2.66 €). An investment is needed for introducing a new material, which is divided per annual volume and accounted for over 6 years. Therefore, the uncertain factors are the percentage of the recycled carbon fibers and its price, the type of polymer and the polymer percentage which is directly estimated from the carbon fibers percentage.

The economic assessment was performed separately for the two polymers. In Fig. 4 the economic analysis for polypropylene is presented, and in Fig. 5 for the polyamide. The part cost for different values of the other two uncertain factors, the recycled carbon fibers price and the percentage of carbon fibers is estimated. The current cost is highlighted in the figures indicating a benchmark for the potential combinations of carbon fibers percentage and price.

It is evident from Fig. 4 that the proposed solution is more cost efficient than the current alternative for different recycled carbon fiber prices provided that they range between 1 €/kg and 3.5 €/kg for all the carbon fibers replacement percentages. On the other hand, for 20% replacement of carbon fibers the proposed solution is more cost efficient than the current. It is worth highlighting that the price of the proposed pedal bracket is lower than the current despite the same percentage of fibers replacement due to the lower assumed recycled fibers price.

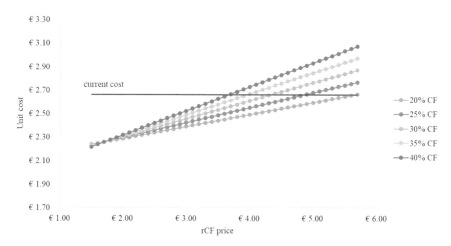

Fig. 4 Pedal bracket cost with polypropylene

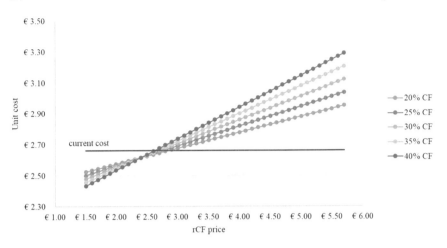

Fig. 5 Pedal bracket cost with polyamide

The lowest price for the part can be 2.25€, which is 15% less than the current solution. It is observed that when the recycled carbon fibers price is less than the polypropylene then the alternatives that have greatest percentage of carbon fibers (40%) are more cost efficient. However, when the price of recycled carbon fiber increases then the solutions with lowest percentage of carbon fibers have lower cost. It can also be inferred that with lower prices of carbon fibers the differences on the cost between 20 and 40% of carbon fibers are less significant than when the carbon fibers price increases.

In comparison with Fig. 4, in Fig. 5 the majority of the alternative scenarios have a higher cost than the current price, specifically, when the recycled carbon fibers price is higher than 2.5€/kg. It is evident that when the carbon fibers price is less than 2.4€/kg, which is the polyamide price, then the best solutions with the lowest cost are the ones with 40% of carbon fibers.

The next business model investigated considers roofs used in construction. The parameters considered for the economic assessment are displayed in Table 4.

In the new business model, the glass fibers material is completely replaced with recycled material. This leads to eliminating landfilling waste cost compared to the current situation. The price of the recycled fibers is different from the virgin fibers. It is assumed that the recycled glass fibers price range will be up to 20% lower that the virgin price. In addition, extra investment is required for the changes from the current process, which is divided per annual volume and accounted for six years. The uncertain parameter considered for the economic assessment is the cost of the recycled fibers.

In Fig. 6 the sensitivity analysis of the unit cost is presented for the uncertain parameters. It is evident that the current cost is much higher than the cost of the proposed solution. The proposed solution offers a decrease on the cost within the ranges of 2–4% lower than the current cost. Therefore, it is evident that the proposed

Table 4 Parameters considered in the economic assessment of roof used in construction

Parameter	
Fibers quantity per unit (kg/sqm)	0.36
UP polymer system price (€/kg)	1.5
UP polymer system quantity per unit (kg/sqm)	1.09
Cost of internal waste reprocessing (€/kg)	0.12
Amount of waste (kg/sqm)	0.11165
Disposal cost (€/kg)	0.235
Volume of production (sqm/year)	540,480
Depreciation cost of machinery (€/sqm)	0.12
Labor cost per unit (€/sqm)	0.71
Investment needed for changes in the current process (€)	1,500,000

solution even with recycled fibers price only 1% lower than the virgin, it is still more cost efficient. The recycled fibers price has a great impact on the total and unit cost.

The final business model investigated is a reusable car platform for electric vehicles. In this case, a reusable part is investigated, rather than recycled composite material use. The parameters considered for the economic assessment are displayed in Table 5.

In the new business model, the replacement of the steel car platform (current case) with a lightweight one made of carbon fibers (proposed case) is proposed. The lightweight platform has a higher initial cost but leads to lower fuel consumption. The electricity cost during the operational phase is assumed to be 0.30 € per kWh.

An economic assessment is performed to investigate the economic viability of the solution proposed. In Fig. 7, the breakeven point of the lightweight platform is presented. It is evident that the lightweight platform does not provide an economic advantage against the traditional steel platform before the 21st year of life. However, considering the life cycle of the platform is 30 years, the proposed solution is viable.

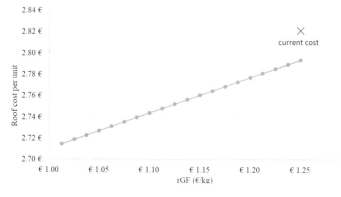

Fig. 6 Roofs for construction unit cost

Table 5 Parameters considered in the economic assessment of the car platform

Parameter	Car platform	
	Proposed	Current
Initial platform cost (€/platform)	1670	505
Remanufacturing cost every 4 years (€/platform)	508	505
Daily energy consumption (kWh/day)	1.87	2.27
Selling volume in each year (number of units)	10,000	10,000
Tooling cost (the investment is divided per annual volume for 10 year)	34 M€	5 M€

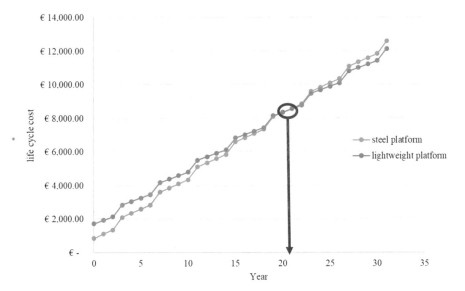

Fig. 7 Life cycle cost of car platform

2.3 *Economic Assessment Key Findings*

The analysis demonstrates that the proposed business models can be classified into two categories, based on the profitability of the proposed circular business model compared to the existing business model, as presented in Table 6.

Table 6 Classification of products profitability

Products strongly profitable	Products profitable under certain conditions
Bath tub	Pedal bracket
Roofs used in construction	Car platform
Skis	

The new business models for skis, roofs and bath tubs is always more profitable than the existing business model, including all potential scenarios evaluated through the sensitivity analysis. There is therefore a strong case to move from the existing business model to the proposed one. It is not required to add extra services for these proposed business model to increase profitability. For the other products evaluated, the proposed business case is more profitable than the existing business model only under specific conditions, among those evaluated through the sensitivity analysis. Additional in-depth analysis is required to evaluate the inclusion of extra services to enhance the market attractiveness of the proposed solution.

2.4   Limitations of Research

In this work the economic assessment of the proposed circular economy business models was performed. The most uncertain parameters of the business models were identified, and a sensitivity analysis was performed in order to investigate the viability of the proposed models in an uncertain environment. The most uncertain parameters considered are the recycled fibers price and the percentage of recycled fibers used in the final product.

However, the analysis was limited due to some assumptions. First, the virgin fibers price was considered fixed, when in reality it depends on the market and there are variations on the price. Another parameter that was assumed fixed is the cost of disposing of the waste, which varies depending on the regulations. Another consideration that was not taken into account is the market availability of material, since the recycled fiber is a novel material. This could lead to production delays or the need for standardizing arrangements with different suppliers. Furthermore, the use of an innovative material may require regulatory incentives to stimulate the demand for such materials and if the incentives are not placed or less than expected it might have an impact on the business model [13–16].

All of the aforementioned parameters can have an impact on the proposed circular economy business models and they should be further investigated.

3   Risk Assessment

The economic assessment of new circular economy business models carried out in Sect. 2, illustrated their expected economic performance, taking into consideration specific uncertain parameters, such as the recycled fibers price and the percentage of recycled material included in the final products. However, as detailed in Sect. 2.3, the new circular economy business models are subjected to a number of additional indirect uncertainties and risks, arising within the competitive environment, including supply dynamics, market dynamics and the effect of regulations. All these factors can have a significant impact on the new business models and need to be carefully evaluated as they may ultimately affect the economic viability of such business models. Therefore, this section aims to identify key risk factors for new circular economy business models and prioritize them in order to investigate suitable risk management strategies for critical risk factors in the future.

Figure 8 visually summarizes the risk assessment process followed to identify and prioritize such risk factors. A literature review at the intersection of circular economy, business models and risk assessment areas led to the identification of a set of risk factors, which were validated by experts, who have extensive experience either in the domain of circular economy and new business models or specifically in industries adopting glass fibers and carbon fibers. The experts belong to a variety of industries, thus adequately representing the various actors of reverse supply chains, such as waste material suppliers, waste management organizations, materials' recycling companies and manufacturers of circular economy products, as well as additional stakeholders from academia and relevant research centers (Fig. 9a). The most represented industries are 'Automotive' and 'Research and Development' industries with four experts each being part of the sample, whereas the most common position held by experts is the 'Project Manager/Leader' position within research and development area (Fig. 9b). The experts average 10.61 years of experience within the organization they currently belong to and 13.42 years of experience in the industry they belong to.

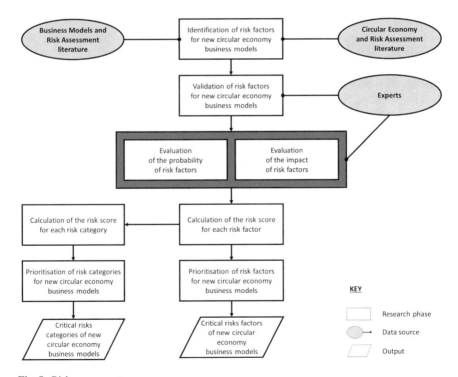

Fig. 8 Risk assessment process

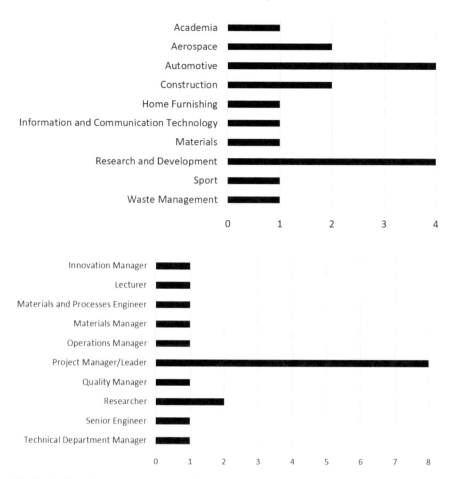

Fig. 9 Profiles of experts: **a** industry and **b** role

The validation stage led to a list of 24 risk factors. These risk factors were classified into six categories and their probability and impact were qualitatively evaluated by experts adopting a linguistic scale. The outcomes of such evaluation were then transformed into relevant quantitative values, allowing the calculation of the risk score per risk factor and per risk category. The risk assessment ultimately identified the critical risk factors that need to be prioritized for new circular economy business models, in order to identify suitable risk mitigation strategies or alternative risk response strategies in the future.

3.1 Circular Economy Risk Factors

Circular Economy (CE) aims to realize closed material flows in the whole economy [17], thus reducing the pressure on the environment [18] and creating a restorative and regenerative economic system [19]. Several aspects related to the circular economy in action, such as CE practices [18], critical success factors [20], drivers [18] and barriers [18] have been analyzed, however risk factors associated with circular economy have not been exhaustively addressed.

Green supply chain risks from a CE perspective were evaluated using a fuzzy-AHP method [16], distinguishing between organizational risks, control system risks, supply risks, demand risks and market risks and adopting a strictly supply chain perspective. Similarly, a multi-criteria decision-making framework specifically for supply chain risks under a CE context has been developed, taking into consideration environmental risks, such as biological and natural disaster risks, logistical and infrastructural risks, management and operational risks as well as risks associated to the macro-political environment such as public policy and institutional risks and security risks [15]. The limitations of current regulations were also identified as a factor currently preventing a wider implementation of CE, along with perceived lower quality of CE products by customers and high up-front investment costs [18]. At the same time, additional supply, demand and organizational aspects were identified as both barriers and potential threats to the development of CE business models [18].

3.2 New Business Models Risk Factors

CE business models, also referred to as circular business models, are a subset of sustainability-oriented business models, in which "the conceptual logic for value creation is based on utilizing economic value retained in products after use in the production of new offerings" [21], thus putting all three pillars of sustainability at the center of the value proposition [22]. CE business models are innovative business models, as they implement a new conceptual logic to create, deliver and capture value [23].

New business models are characterized by inherent uncertainties and risks due to their "complexity, modularity and integrative nature" [13]. Risks can endanger the "profitability and sustainability of the business model or even company goals" [23]. Both risks internal within the organization and external to the organization exist, spanning in areas such as customer, value proposition offer, infrastructure [13]. These risks, combined with the overall uncertainties in the competitive environment due to political and regulatory risks, can negatively impact the financial viability of new business models [13]. The introduction of environmental regulations particularly, as highlighted by the energy sector, can determine disruptive changes to the industry [14, 24, 25]. Specific business models may become obsolete as incentives and/or taxation exemptions are introduced or new regulations define actors who are allowed to compete in the market due to environmental criteria [14]. Regulatory risks have a direct effect, impacting primarily on the revenue structure of business models, but also their cost structure [14, 25]. However, there is a cascading effect on other critical elements of business models with additional risks, such as increased capital costs [25], increased operations costs such as production costs and maintenance costs [24, 25], technological risks [24], know-how and human resources risks [24] and even market risks [24]. These risks require to be accurately evaluated to inform decisions about the development of new business models [13].

3.3 Summary of Risk Factors for New Circular Economy Business Models

The combination of circular economy risk factors and new business model risk factors led to the identification of a list of 61 risk factors obtained from the literature. Experts were asked to identify relevant risk factors to be forwarded to the evaluation stage as well as to identify further risk factors, which are applicable to the context of new circular economy business models using recycled glass fiber and carbon fiber and were not explicitly mentioned by the literature. Several risk factors were excluded by the experts, as they were deemed not relevant for the new circular economy business models, while other risk factors were merged due to their similarity in scope. Finally, a 'COVID-19 Pandemic' risk factor was added to the list, following the input from the experts, thus leading to a finalized list containing 24 risk factors, which are displayed in Table 7. The 24 risk factors are clustered in six risk categories, namely market (M), supply (S), finance (F), political and regulatory (PR), technical (T) and other (O).

3.4 Risk Factors Evaluation

Risk Score Calculation. The 24 risk factors were evaluated by the panel of experts, who were asked to provide their judgement on the probability and the impact of each risk factor using a linguistic scale. Probability is defined as the likelihood that a particular risk will occur during a specific time frame [14, 15], whereas impact is defined as the severity of the financial effect should the risk occur within the specified time frame [14, 15]. The time frame for the risk factor evaluation was defined as five years, in line with [14], meaning that the timeframe for evaluation of the risk factors is up to 2025. Moreover, the five-years' time frame is consistent with the timeline of future regulations expected to become operative in the European Union, which will further tighten the limits concerning disposal of carbon and fiber glass, hence introducing a disruptive regulatory change in the industry [26]. The probability and

Table 7 Summary of risk factors

Risk category	Risk factor	Risk factor description	Source
Market	M1: Consumer perception (quality vs. new products)	Resistance of end-users towards CE products due to customer perception about inferior quality compared to new products, resulting in low public acceptance and low market penetration	[18, 24]
Market	M2: Customer perception (ownership)	Resistance of end-users towards CE products due to preference of customers for ownership rather than access to service resulting in low public acceptance of products (applicable only to non-ownership business models)	[18]
Market	M3: Market forecast	Demand risk, namely the uncertainty about the market size (applicable only to non-ownership business models)	[16]
Market	M4: Economic cycle	Risk in the business environment/market due to adverse economic cycle and/or uncertain financial stability, resulting in low demand for products at the global level	[13, 15]
Supply	S1: Supply availability	Limited and/or not timely availability of recycled materials to support demand of CE final products	[18]
Supply	S2: Logistics	Risk associated to changes in the supply network and transportation determining an increase of the logistic costs	[15, 16]
Supply	S3: Supplier quality	Supplier quality risk, resulting in quality criteria for the input materials not being achieved	[16]

(continued)

Table 7 (continued)

Risk category	Risk factor	Risk factor description	Source
Supply	S4: Take-back system	Lack of structured take-back systems, including the lack of accurate information about the tracking of materials in the reverse supply chain	[18]
Finance	F1: Lifecycle revenues	Lifecycle risk and uncertainty factors related to a possible uncalculated change of revenues due to price volatility of the final product resulting in lower than expected revenues	[13, 16, 24]
Finance	F2: Capital costs	Risk associated to high up-front investment costs and capital costs required to create and deliver the value proposition of the innovative business models, including costs for production plants and inventory costs	[13, 14, 18, 24, 25]
Finance	F3: Recycled material costs	Unpredicted increase of recycled material cost, affecting the economic viability of using recycled materials	[14, 18, 25]
Finance	F4: Virgin material costs	Unpredicted decrease of virgin materials cost, affecting the economic viability of using recycled materials	[18]
Finance	F5: Production costs	Increased production costs, including increased costs for energy and maintenance within the production plant	[18]
Finance	F6: Financial resources	Factors impacting the capability to finance the CE business model, including access to finance	[13]

(continued)

Table 7 (continued)

Risk category	Risk factor	Risk factor description	Source
Political and regulatory	PR1: Regulatory standards	Lack of standards for CE products, potentially affecting compatibility, quality and sustainable branding	[14, 18]
Political and regulatory	PR2: Legal and regulatory	Commitment of regulatory and legal circumstances, including the ineffectiveness of existing laws and/or their insufficient implementation	[13–15, 18, 25]
Political and regulatory	PR3: Public policy and institutional	Risk arising from adverse changes in policy support schemes or regulations, including economic incentives to shift from linear economy to circular economy, and/or uncertainty regarding changes in government policies	[15, 18, 24]
Political and Regulatory	PR4: Taxation	Risk arising from adverse changes in the taxation regulations, including declining or eliminated tax advantages for green products, and/or uncertainty regarding changes to the fiscal policies	[13–16]
Political and regulatory	PR5: Non-ownership business model	Legal issues emerging for non-ownership business models as service providers cannot legally retain owner-ship of a sold product (applicable only to non-ownership business models)	[18]
Political and regulatory	PR6: Intellectual property	Risks associated to the drainage of IP or know-how, including sensitive data on the organization's partners	[13]

(continued)

Table 7 (continued)

Risk category	Risk factor	Risk factor description	Source
Technical	T1: Human resources	Risks related to the lack of qualified human resources required to realize the CE business model	[13, 18]
Technical	T2: Quality	Risks related to the quality of the final CE products, such as gaps in expected vs. delivered performance, durability and functionality throughout lifecycle of the CE product;	[13, 15, 18]
Technical	T3: Technology	All risk and uncertainty factors that are linked to the use of technologies that are new or still in a premature state, highly complex, or for which the company lacks experience, potentially leading to lower than expected technological efficiency	[13, 14, 18, 24]
Other	O1: COVID-19 pandemic	Risk arising from the COVID-19 pandemic, including major supply chain disruptions, supplier failure and/or customer solvency	NA

impact were evaluated using a five-level linguistic scale and a quantitative score was allocated to each level of the linguistic scale as per Table 8.

Following the evaluation of the risk probability and impact by each expert, the linguistic scale values were transformed into the relevant quantitative scores and,

Table 8 Five-level linguistic scale used for the evaluation

Linguistic scale	Quantitative score
Very low	1
Low	2
Medium	3
High	4
Very high	5

subsequently, averaged to generate an aggregated probability and an aggregated impact per risk factor.

These were functional to calculate the risk score of each risk factor, which was obtained by multiplying the aggregated probability and the aggregated impact of each risk factor.

Finally, the obtained risk scores were normalized. The min–max method was used to normalize the risk score nRS_k of the k-th risk factor to a value between 0 and 1 [27, 28], according to Eq. (1), where the normalized risk score of the k-th risk factor was calculated as:

$$nRS_k = \frac{RS_k - RS_{min}}{RS_{max} - RS_{min}} \qquad (1)$$

Probability of Risk Factors. The aggregated probability of risk factors displays an average value equal to 2.84, with values in the range 2.06–3.67, as illustrated in Fig. 2. "S4: Take-back system" was identified as the most probable risk factor according to respondents (Table 9). Experts showed concerns about the behavior of reverse supply chains, which can ultimately hamper the possibility to have a reliable supply through new circular economy business models, as well as on the relevant information available to track materials in the reverse supply chains, a critical element to plan the production. Additionally, "PR1: Regulatory standards" and "PR2: Legal and regulatory" risk factors were also deemed as having a very high probability, both displaying an aggregated probability score of 3.29. Subsequently, four risk factors belonging to the 'Market' risk category occupy the following positions in the ranking (4th–7th), which determines 'Market' risk category to be considered the one displaying the highest probability according to experts, with a probability score equal to 3.21 (Table 10).

Impact of Risk Factors. The aggregated impact of risk factors displays an average value equal to 3.10, with values in the range of 2.41–3.61 (Fig. 10), thus showcasing higher average value and reduced variation compared to probability values. "T2: Quality" was identified as the potentially most impactful risk factor according to experts (Table 9). Should products obtained through new circular economy business model not match the quality requirements of comparable products produced using virgin raw material, this would ultimately have a disruptive impact on businesses with financial consequences potentially spanning across a long-term timespan. While quality testing of products obtained through circular economy business models is widely available, the risk is particularly relevant taking a lifecycle perspective, with potential impacts affecting the durability and functionality of products in the long-term. "M2: Customer perception (ownership)" and "T3: Technology" follow as the most impactful risk factors, displaying scores of 3.50 and 3.41 respectively. The impact of M2 would be financially significant as it would disrupt the transition from traditional linear business model to new circular economy business model, which are often based on product-service systems to reduce the uncertainty associated with the returned products [29, 30] and therefore do not necessarily envisage the ownership

Table 9 Risk factors: scores and rankings for probability, impact and normalized risk

Risk category	Probability		Impact		Normalized risk	
	Score	Rank	Score	Rank	Score	Rank
M1	3.28	4	3.35	4	0.89	2
M2	3.13	7	3.50	2	0.88	3
M3	3.25	5	3.25	6	0.83	5
M4	3.19	6	3.20	10	0.77	7
S1	2.93	12	3.33	5	0.71	10
S2	2.82	13	3.24	8	0.62	14
S3	2.65	16	3.06	15	0.46	15
S4	3.67	1	3.20	10	1.00	1
F1	2.63	17	3.06	14	0.45	16
F2	3.06	9	3.25	6	0.74	9
F3	3.00	10	3.13	13	0.66	13
F4	2.44	20	2.78	21	0.27	20
F5	2.24	22	2.78	21	0.18	23
F6	2.71	14	2.82	20	0.40	18
PR1	3.29	2	3.06	17	0.75	8
PR2	3.29	3	3.21	9	0.83	6
PR3	2.29	21	3.06	15	0.30	19
PR4	2.60	18	3.00	18	0.42	17
PR5	2.57	19	2.62	23	0.26	21
PR6	2.06	24	2.41	24	0.00	24
T1	2.24	22	2.94	19	0.24	22
T2	2.67	15	3.61	1	0.69	11
T3	3.12	8	3.41	3	0.84	4
O1	3.00	10	3.18	12	0.67	12

Table 10 Risk categories: scores and rankings for probability, impact and normalized risk

Risk factor	Probability		Impact		Normalized risk	
	Score	Rank	Score	Rank	Score	Rank
Market (M)	3.21	1	3.33	1	0.84	1
Supply (S)	3.02	2	3.32	2	0.80	2
Finance (F)	2.68	5	3.25	3	0.73	3
Political and regulatory (PR)	2.68	4	3.21	4	0.64	5
Technical (T)	2.67	6	3.21	4	0.70	4
Other (O)	3.00	3	3.14	6	0.63	6

of the product by the customer. Should the customer still prefer the ownership of products which are in use for traditional linear business models, this would ultimately lower the demand for CE products. On the other hand, the financial implications of T3 are mostly linked to the research and development phase required to achieve the level of technological maturity to guarantee a technological efficiency that is competitive with products being produced using virgin raw materials.

The risk matrix (Fig. 10) visually plots the risk factors according to their probability and impact scores. Overall, nine risk factors fall within the upper-right quadrant, which is typically labelled as the top-risks quadrant [14]. These are: "M1: Consumer perception (quality vs. new products)", "M2: Customer perception (ownership)", "M3: Market forecast", "M4: Economic cycle", "S4: Take-back system", "F2: Capital costs", "PR1: Regulatory standards", "PR2: Legal and regulatory" and "T3: Technology". Furthermore, two additional areas of focus are the *black swans* area and the *white swans* area [14]. The former is characterized by probability of occurrence below the medium-value threshold but a high financial impact, whereas the latter features a high probability of occurrence with a financial impact below the medium-value threshold. Among the evaluated risk factors, only "T2: Quality" falls within such areas, namely within the *black swans* area, thus still being a risk factor decision makers should focus on due to its potential significant financial impact [14].

Total Risk of Risk Factors. Combining the probability and impact of risk factors, the overall risk score of each risk factor was calculated. Normalized risk scores are displayed in Table 9. "S4: Take-back system" was identified as the factor having the highest risk, largely due to its probability score, thus being the reference maximum score for other risk factors. "M1: Consumer perception (quality vs. new products)" and "M2: Customer perception (ownership)" follow next in the ranking with normalized risk scores of 0.89 and 0.88 respectively, demonstrating that experts are still

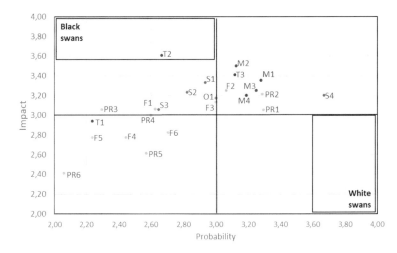

Fig. 10 Risk matrix by risk factor

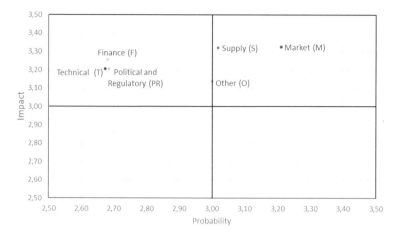

Fig. 11 Risk matrix by risk category

concerned regarding the acceptance of CE products by customers, both in the case of innovative business models based on product-service systems and in the case of simple material substitution, moving from virgin fibers to recycled fibers within similar business models.

Risk Categories. Moving from individual risk factors, to wider risk categories, results show that risk categories do not show significant differences in terms of impact scores, with all categories showing values in a close range (3.14–3.33), with 'Market' risk category showing the highest impact score (Table 10).

'Market' risk category displays also the highest probability score, however more significant differences in terms of probability scores exist, with values ranging from 2.67 to 3.21 (Table 10).

The combination of probability and impact scores per risk category are visible in the risk matrix presented in Fig. 11, which highlights that only 'Market' risk category and 'Supply' risk category fall within the top risks quadrant, meaning that such areas need to be prioritized to identify suitable risk management strategies. This was also confirmed by the normalized risk scores displayed in Table 10, with 'Market' risk category and 'Supply' risk category showing an average normalized risk score of 0.84 and 0.80 respectively, thus being at the top of the ranking.

3.5 Risk Assessment Key Findings

This work aimed to identify and prioritize risk factors for new circular economy business models using recycled glass fibers and carbon fibers. Reviewing the literature at the intersection of the fields of risk assessment, circular economy and business models, a list of risk factors was obtained and validated by experts, leading to the identification of 24 key risk factors, spanning across six risk categories. The probability and impact of such risk factors were evaluated by 18 experts, leading to the definition of the risk score per risk factor and the prioritization of risk factors. Major risk factors for new circular economy business models lie both on the demand side, i.e. the market, and on the supply side. Regarding the market, experts highlighted risks related to the acceptance of CE products by customers and the uncertainties regarding the definition of the market size for innovative products. On the supply side, experts highlighted the risk of low efficiency of take-back systems as part of the reverse supply chains. Finally, technical aspects, such as quality of CE products and technological challenges associated with using recycled materials, were considered potentially among the most impactful risk factors for the new business models, however their overall risk was lowered by their relatively lower probability of occurrence.

3.6   Limitations and Future Research Directions

The research identified and evaluated risk factors for new circular economy business models, leading to a prioritization of risk factors, however further research is required to identify the most suitable risk management strategies to tackle such risks, such as risk prevention to reduce the probability of occurrence of the risk [31–33], risk mitigation to reduce the impact of the risk in the case of occurrence [24, 31–34], risk avoidance [24, 32, 34], risk transfer to different parties [24, 32], risk monitoring [34] or risk assumption [32]. Different strategies may be appropriate for different risk factors taking into account the efforts required and budget constraints [13, 35] and a more detailed quantification of risk factors belonging to the top risks' quadrant is required: these could involve predicting the monetary evaluation of the financial impact should a risk occur and a more accurate evaluation of the probability of such risk factors, potentially adopting probabilistic methods. With this respect, the scenario analysis method and bow-tie analysis are potential future methods to identify the most suitable risk management responses.

4   Conclusion

Circular economy business models are key enablers of the circular economy and are designed to maximize the length of time products and materials are kept in use such that maximum value is extracted from them. Such models must be carefully designed to reflect the reality and needs of the prospective use environment in order to ensure their usefulness and usability for prospective stakeholders. They must also be economically viable since profitability is a major business driver. This chapter has described the steps that can be taken to increase the potentiality of composite circular economy business models. The work described include the economic assessment of the business models, the identification of the major causes of uncertainty via information from the literature and expert informants and sampling-based sensitivity analysis to investigate the viability of the business models under uncertainty. Following from that the identification and prioritization of risk factors for new circular economy business models for composites were described. In the case of economic assessment of the new business models the major findings was that the proposed new business models were not always more profitable than the existing models. Thus, in some instances new market stimulus would need to be identified and implemented to increase attractiveness of the proposed solution.

The risk analysis concluded that the key new circular economy business model risk factors were from the demand/market and supply side. On the demand side, the risks are lack of consumers acceptance of CE products and the uncertainties regarding the size of the market for innovative products. On the supply side, the key risks are low efficiency of take-back systems that feed the reverse supply chains. Technical aspects, such as the quality of circular economy products and the challenges of working with recycled materials were among the major risk factors but their lower probability of occurrence significantly reduced their gravity. The research limitations include the assumptions of the analysis including the fixed costs for virgin fibers and disposal when in reality both would vary with regulations and market considerations. Also, some issues were not considered, for example availability of recycled fibers was not considered even though its novelty could lead to shortages. These limitations could impact the potential of the circular economy models thus require further analysis. Further research is required to identify the most suitable risk management strategies to tackle the risks by reducing their impact and the potential of their occurrence.

References

1. Rosa, P., Sassanelli, C., Terzi, S.: Towards circular business models: a systematic literature review on classification frameworks and archetypes. J. Clean. Prod. **236**, 117696 (2019). https://doi.org/10.1016/j.jclepro.2019.117696
2. Urbinati, A., Chiaroni, D., Chiesa, V.: Towards a new taxonomy of circular economy business models. J. Clean. Prod. **168**, 487–498 (2017). https://doi.org/10.1016/j.jclepro.2017.09.047
3. Lewandowski, M.: Designing the business models for circular economy—towards the conceptual framework. Sustainability **8**, 43 (2016). https://doi.org/10.3390/su8010043
4. Vegter, D., van Hillegersberg, J., Olthaar, M.: Supply chains in circular business models: processes and performance objectives. Resour. Conserv. Recycl. **162**, 105046 (2020). https://doi.org/10.1016/j.resconrec.2020.105046
5. Daou, A., Mallat, C., Chammas, G., Cerantola, N., Kayed, S., Saliba, N.A.: The Ecocanvas as a business model canvas for a circular economy. J. Clean. Prod. **258**, 120938 (2020). https://doi.org/10.1016/j.jclepro.2020.120938
6. Geissdoerfer, M., Pieroni, M.P.P., Pigosso, D.C.A., Soufani, K.: Circular business models: a review. J. Clean. Prod. **277**, 123741 (2020). https://doi.org/10.1016/j.jclepro.2020.123741
7. Pieroni, M.P.P., McAloone, T.C., Pigosso, D.C.A.: Business model innovation for circular economy and sustainability: a review of approaches. J. Clean. Prod. **215**, 198–216 (2019). https://doi.org/10.1016/j.jclepro.2019.01.036

8. Fehrer, J.A., Wieland, H.: A systemic logic for circular business models. J. Bus. Res. (2020). https://doi.org/10.1016/j.jbusres.2020.02.010
9. Atabaki, M.S., Mohammadi, M., Naderi, B.: New robust optimization models for closed-loop supply chain of durable products: towards a circular economy. Comput. Ind. Eng. **146**, 106520 (2020). https://doi.org/10.1016/j.cie.2020.106520
10. Alam, M., Abedi, V., Bassaganya-Riera, J., Wendelsdorf, K., Bisset, K., Deng, X., et al.: Agent-based modeling and high performance computing. Elsevier Inc (2016). https://doi.org/10.1016/B978-0-12-803697-6.00006-0
11. Chin, R., Lee, B.: Analysis of data. In: Principles and Practice of Clinical Trial Medicine. Academic Press (2008)
12. Pianosi, F., Beven, K., Freer, J., Hall, J.W., Rougier, J., Stephenson, D.B., et al.: Sensitivity analysis of environmental models: a systematic review with practical workflow. Environ. Model Softw. **79**, 214–232 (2016). https://doi.org/10.1016/j.envsoft.2016.02.008
13. Brillinger, A.S., Els, C., Schäfer, B., Bender, B.: Business model risk and uncertainty factors: toward building and maintaining profitable and sustainable business models. Bus. Horiz. **63**, 121–130 (2020). https://doi.org/10.1016/j.bushor.2019.09.009
14. Leisen, R., Steffen, B., Weber, C.: Regulatory risk and the resilience of new sustainable business models in the energy sector. J. Clean. Prod. **219**, 865–878 (2019). https://doi.org/10.1016/j.jclepro.2019.01.330
15. Yazdani, M., Gonzalez, E.D.R.S., Chatterjee, P.: A multi-criteria decision-making framework for agriculture supply chain risk management under a circular economy context. Manag. Decis. (2019). https://doi.org/10.1108/MD-10-2018-1088
16. Yang, Z., Li, J.: Assessment of green supply chain risk based on circular economy. In: IEEE 17th International Conference Industrial Engineering and Engineering Management, Xiamen, China. IEEE, pp. 1276–1280 (2020)
17. Geng, Y., Doberstein, B.: Developing the circular economy in China: challenges and opportunities for achieving "leapfrog development." Int. J. Sustain. Dev. World Ecol. **15**, 231–239 (2008). https://doi.org/10.3843/SusDev.15.3:6
18. Govindan, K., Hasanagic, M.: A systematic review on drivers, barriers, and practices towards circular economy: a supply chain perspective. Int. J. Prod. Res. **56**, 278–311 (2018). https://doi.org/10.1080/00207543.2017.1402141
19. Lieder, M., Rashid, A.: Towards circular economy implementation: a comprehensive review in context of manufacturing industry. J. Clean Prod. **115**, 36–51 (2016). https://doi.org/10.1016/j.jclepro.2015.12.042
20. Khan, S., Maqbool, A., Haleem, A., Khan, M.I.: Analyzing critical success factors for a successful transition towards circular economy through DANP approach. Manag. Environ. Qual. An Int. J. **31**, 505–529 (2020)
21. Linder, M., Williander, M.: Circular business model innovation: inherent uncertainties. Bus. Strateg. Environ. **26**, 182–196 (2017). https://doi.org/10.1002/bse.1906
22. Boons, F., Lüdeke-Freund, F.: Business models for sustainable innovation: state-of-the-art and steps towards a research agenda. J. Clean Prod. **45**, 9–19 (2013). https://doi.org/10.1016/j.jclepro.2012.07.007
23. Brillinger, A.S.: Mapping business model risk factors. Int. J. Innov. Manag. **22**, 1–29 (2018). https://doi.org/10.1142/S1363919618400054
24. Gatzert, N., Kosub, T.: Risks and risk management of renewable energy projects: the case of onshore and offshore wind parks. Renew. Sustain. Energy Rev. **60**, 982–998 (2016). https://doi.org/10.1016/j.rser.2016.01.103
25. Gross, R., Blyth, W., Heptonstall, P.: Risks, revenues and investment in electricity generation: why policy needs to look beyond costs. Energy Econ. **32**, 796–804 (2010). https://doi.org/10.1016/j.eneco.2009.09.017
26. WindEurope. Accelerating Wind Turbine Blade Circularity (2020)
27. Bai, C., Sarkis, J.: Determining and applying sustainable supplier key performance indicators. Supply Chain Manag. **19**, 275 (2014). https://doi.org/10.1108/SCM-12-2013-0441

28. Wang, J.J., Jing, Y.Y., Zhang, C.F., Zhao, J.H.: Review on multi-criteria decision analysis aid in sustainable energy decision-making. Renew. Sustain. Energy Rev. **13**, 2263–2278 (2009). https://doi.org/10.1016/j.rser.2009.06.021
29. Tukker, A.: Product services for a resource-efficient and circular economy—a review. J. Clean Prod. **97**, 76–91 (2015). https://doi.org/10.1016/j.jclepro.2013.11.049
30. Singhal, D., Tripathy, S., Jena, S.K.: Remanufacturing for the circular economy: Study and evaluation of critical factors. Resour. Conserv. Recycl. **156** (2020). https://doi.org/10.1016/j.resconrec.2020.104681
31. Aqlan, F., Lam, S.S.: A fuzzy-based integrated framework for supply chain risk assessment. Int. J. Prod. Econ. **161**, 54–63 (2015). https://doi.org/10.1016/j.ijpe.2014.11.013
32. Ali, A., Warren, D., Mathiassen, L.: Cloud-based business services innovation: A risk management model. Int J Inf Manage **37**, 639–649 (2017). https://doi.org/10.1016/j.ijinfomgt.2017.05.008
33. Dong, Q., Cooper, O.: An orders-of-magnitude AHP supply chain risk assessment framework. Int. J. Prod. Econ. **182**, 144–156 (2016). https://doi.org/10.1016/j.ijpe.2016.08.021
34. Qazi, A., Dickson, A., Quigley, J., Gaudenzi, B.: Supply chain risk network management: a Bayesian belief network and expected utility based approach for managing supply chain risks. Int. J. Prod. Econ. **196**, 24–42 (2018). https://doi.org/10.1016/j.ijpe.2017.11.008
35. Venkatesh, V.G., Rathi, S., Patwa, S.: Analysis on supply chain risks in Indian apparel retail chains and proposal of risk prioritization model using interpretive structural modeling. J. Retail Consum. Serv. **26**, 153–167 (2015). https://doi.org/10.1016/j.jretconser.2015.06.001

Impact of Policy Actions on the Deployment of the Circular Value-Chain for Composites

Marco Diani(ID)**, Abdelrahman H. Abdalla**(ID)**, Claudio Luis de Melo Pereira, and Marcello Colledani**(ID)

Abstract The effect of legislation on composites recycling can be both a driving force, such as in the case of End-of-Life Vehicle legislation, that is making it mandatory to reuse the materials used in vehicle manufacturing, or a boundary, increasing the burden to manufacturers reusing composite materials. As a consequence, a deep study on the impact of policies on the reuse of composites is fundamental to promote those actions boosting the deployment of circular value-chains. In this Chapter, a model based on System Dynamics theory, representing the entire industrial environment of composite materials, has been developed, leading to a prioritization of most impacting legislations, providing conclusions and recommendations derived from data.

Keywords System dynamics · Causal loop diagram · Stock and flow map · Legislation · Policies · Policy actions · Prioritization

1 Introduction

One of the major issues of the implementation of Circular Economy in the composites sector is the comparison of the cost savings brought by these procedures and the additional costs entangled by their setup. Commonly, the savings are not sufficient to compensate the expenditures in an acceptable time interval for stakeholders in their current mindsets, leading to underinvestment.

The enhancement of the economic viability of FRP's de-manufacturing processes may require the introduction of new methods and technologies in such activities. However, this action may find resistance among stakeholders, for example operators and managers accustomed to the usual practices and procedures. They would have to change their behaviors adapting to the new circumstances, not to mention eventual training efforts, which would represent additional costs to the enterprise. If

M. Diani (✉) · A. H. Abdalla · C. L. de Melo Pereira · M. Colledani
Department of Mechanical Engineering, Politecnico di Milano, Milan, Italy
e-mail: marco.diani@polimi.it

© The Author(s) 2022
M. Colledani and S. Turri (eds.), *Systemic Circular Economy Solutions for Fiber Reinforced Composites*, Digital Innovations in Architecture, Engineering and Construction, https://doi.org/10.1007/978-3-031-22352-5_21

445

high enough, this reluctance can eventually bar the company's adoption of circular practices, especially if shared by decision-makers.

Additionally, technological development requires investments, a sensitive matter already discussed, and the related boundaries and limitations are still unclear. Financing is indeed an obstacle since sources of funds usually base their lending decisions on risks and returns, characteristics that are not the allure of FRP Circular Economy business models, and there is no alignment across the sector regarding the search for funds and pricing methodologies. Another issue that might arise halting the development of CE solutions in the industry regards the compatibility of proprietary systems among different players along the supply chain. It is evident, from the FiberEUse project, that integration and information exchange in the supply chain could boost circular practices on the market.

In addition, there may be limits in respect to the market penetration and applications that could hamper the implementation of circular chains for composites. Concerning the co-processing of GFRP waste in cement kilns, there is a limit around 10% of the fuel input not to compromise cement's properties, particularly because of E-glass fibers boron content. Moreover, the amount of powder to add in compounds as reinforcement or filler is curbed according to the requirements for FRP final properties and not to disturb fiber-matrix adhesion. Consequently, recycled composites may not be suitable substitutes to virgin fiber reinforced plastics in all their applications under the allegation of unsafety, especially in those with the most demanding mechanical properties' standards. The argument of unsafety referring to rFRP correlates with a current belief in society, which belittles recycled materials, conceiving them as a class of products of inferior quality. Although alarming for Circular Economy business model evolution since recycled products may face resistance in their uptake, there are also present trends of increased environmental responsibility, which boost the development of circular solutions, opposing the belief.

Governance aspects may represent further barriers to the implementation of Circular Economy systems for handling composite materials. The success of de-manufacturing supply chains processing FRP waste may require the association of several players from different sectors; nevertheless, the dispersion of stakeholders across many industries might bring coordination challenges and result in misalignments. Hence, establishing communication mechanisms between players involved in composites' recycling is foremost. Stakeholders' appeals regarding composites policy hold a lack of priority in legislators' agendas even though plastics are at the spotlight of discussions, a scenario that discourages agents from engaging in composite de-manufacturing activities.

In Europe, composite collection and de-manufacturing activities have no specific regulation, yet there are general legislations on waste handling that must be followed by stakeholders operating in the industry within the block's territory. For example, we see changes in legislations regarding the cost of landfilling (which can be quite cheap in some countries in Europe and beyond) which constitutes a barrier towards widescale adoption of CE. The main standard currently in place is the European Directive on Waste (2008/98/EC) that provides fundamental concepts and definitions regarding the management of waste flows. It offers definitions for waste, discerning it from by-products, and for processes related to its processing, as recycling and recovery. However, it poorly embraces remanufacturing activities and does not go in depth on technical aspects with the provision of standards and metrics.

Additional frameworks that affect Circular Economy business models are the Directive on End-of-Life Vehicles (2005/53/EC) and the Extended Producer Responsibility (EPR) Legislation (2002/96/EC). While the first imposes recycling requirements in weight fractions for vehicles reaching their EoL state, the second obliges producers to offer customers return possibilities for the products upon end of use so they enter pathways compliant to the legislation for that type of good. Both regulations include important stakeholders in the products' EoL handling and decision-making; they also define targets and timelines based on items' type, not on composition, but still lack specifications on the extent of stakeholders' obligations.

Landfill Directive (1999/31/EC) regulates the different types of landfills available, determines the waste flows than can be landfilled as an EoL option and establishes a tax for this action. It defines landfilling as the least desirable option for goods, but in the case of non-hazardous composites, it still allows it to occur. Notably, a few countries have already forbidden this practice for EoL FRP, for instance Germany, and others are expected to adhere to that decision; there are further legislation packages under discussion that will impose extra restrictions on landfilling in general.

In terms of supervising the movement of waste flows within the European Union (EU), there is the Waste Shipment Regulation (2006/1013/EC) and its amendment (2014/660/EU), which enforce a need of notification of competent authorities and their approval before the movements of waste imported by, exported by or in transit through EU member states. Regarding transboundary shipments, legislation is even stricter and establishes that all the countries crossed by the route must be notified about the movement. These terms contribute to an increase in the complexity of collection activities, hence to the overall complexity in respect to the organization of Circular Economy business models. This aspect is particularly relevant to the case of composites, in which waste movements are necessary to achieve higher volumes needed to compensate the low margins.

The above-mentioned directives are eventually complemented by country- and region-specific rules, as previously mentioned, with varying level of enforcement. These complementary rulings vary across countries and regions according to the specific circumstances within their territories. Consequently, there is a misalignment between regional regulations concerning FRP, yielding intricacies and inconsistency, which imply stakeholders in different locations must comply with divergent standards. Once more, the complexity related to the establishment of composites de-manufacturing supply chains increases, since these would likely contain players spread over different regions thus subject to disparate rules, to which the system would have to concurrently comply.

There are aspects still lacking regulation that if organized within a framework could aid the development of FRP Circular Economy business models. For example, a directive on composite materials waste management would be helpful, as it would define the practices to be adopted to handle waste at the time of their generation, possibly after the creation of standards for residues and offering waste generators information on such materials' pathways (for example Material Passports). This could lead to higher availability of flows to de-manufacturing systems and better sorting, increasing their efficiency, as well as educate people on opportunities arising from waste, thus changing their perceptions about EoL materials and about the products they originate too.

2 State of the Art

System Dynamics (SD) is a field of study initially developed by Forrester [1] during the 1950's at Massachusetts Institute of Technology, addressing the investigation of complex, non-linear, dynamic systems by means of formal mathematical modeling and computer simulations. It acknowledges that real world problems arise in consequence of the dynamics of the system in which they are embedded, and when trying to solve them people often are misled by their mental models to wrong inferences about these dynamics, regardless of the simplicity of the system [2].

Because of the limitations of their mental models, people's actions do not take into consideration all the possible outcomes, and the efforts applied to solve specific problems frequently create unpredicted side effects. In turn, those side effects generate new problems in the near future, hence, the instruments applied as response to an issue can be the cause of new issues ahead. System Dynamics is presented as a methodology to overcome these limitations and enhance the comprehension of the system and its dynamics prior to decision-making [2].

Sterman [2] points that complex systems' behavior in which the system's response to an intervention prevails over the interference, denoted policy resistance, is caused by dynamic complexity, described as the counterintuitive response of such systems in reaction to the agents' interactions. Additionally, he presents the characteristics of systems culminating in dynamic complexity, which are:

(a) Constantly changing: change is always happening inside systems, yet in different speeds, and the distinct time scales can interact as well;
(b) Tight coupling: all elements inside a system are interconnected even though they may not seem, either among themselves or with the environment, and they all strongly interact;
(c) Feedback governance: the tight couplings result in subjects' actions feeding back to themselves, since their decisions change the state of the environment, making others react, which produces further disturbances that will in turn shape the first group's next decisions; feedbacks are the source of dynamics;
(d) Non-linearity: cause and effect usually are not proportional, local behavior seldom is applicable to distant regions of the system and the system's components can have boundaries;
(e) History-dependence: the present state of the system depends on the previous actions taken, and these can be irreversible;

(f) Self-organization: the internal structure of the system gives birth to its dynamics, in which the feedback structures determine the behavior following any disturbance;

(g) Adaptation: agent's decision criteria and capabilities evolve over time, some being replicated, and others extinguished, and the evolution is not necessarily for the best;

(h) Trade-off characterized: long-run and short-run responses are different, and usually short-run and long-run benefits occur in antagonism;

(i) Counterintuitive: cause and effect are distant in time and space, and frequently attention is not directed to root causes but rather to gleaming evidence close to the problem;

(j) Policy resistant: system's complexity is too great for mental models to handle; thus they are oversimplified by people's minds, resulting in possibly threatening obviosities.

The reasons behind erroneous decision making in complex, dynamic systems is the misinterpretation of causal relations, commonly built from heuristics incapable of coping with the main sources of dynamic complexity, namely feedbacks, stocks and flows, and time delays, the principal components of SD thinking [2].

The development of a successful System Dynamics model involves following certain steps during the process. Initially, there is the Problem Articulation step, in which the issue to be assessed is identified along with the reasons behind its characterization as a problem and the key variables and concepts affecting it. Additionally, there is the definition of the time horizon to consider for the analysis, and the collection of the system's historical behavior, searching for insights regarding its dynamics.

The following step encompasses the formulation of Dynamic Hypothesis, initial assumptions for explaining the undesired behavior, which should be focused on the system's elements themselves rather than blaming the erratic pattern on exogenous factors. Moreover, there is also the mapping of the system's causal structure grounded on the generated hypotheses and additional available information. In this stage, the diagrams representing the system emerge, thus, there is the definition of the model's boundaries and the representation of subsystems, as well as the development of Causal Loop diagrams, describing the mental models and feedback structure, and Stock and Flow maps, which further detail the functioning of the system, apart from other tools.

In sequence, there is the formulation of the Simulation Model, which specifies the structure and decision rules adopted by agents, estimates parameters, behavioral relationships and initial conditions, and tests the model built for purpose and boundary adequacy. Then, additional testing is performed in an ulterior step, this time focusing on the reproduction of reference modes, robustness and sensitivity.

Finally, in the policy design and evaluation stage there is the specification of the possible scenarios to be faced and policies to be implemented, along with the conduction of sensitivity analysis, hypothetical cases assessment and policy interaction effects observation. Although it seems a cascaded process, modeling is iterative.

Therefore, downstream steps may generate the need of upstream changes in the model, a loop fed by additional knowledge and information about the system.

System Dynamics models find innumerous applications in the real world and are used in many different contexts. Nassehi and Colledani [3] point out SD is particularly good to model and assess long-term policies and strategies, and their effects on production, which is largely verified by the amount of studies available having this as finality, and they apply it together with agent-based techniques for the study of remanufacturing under Circular Economy scenarios. Scholz-Reiter et al. [4] use the technique to model an autonomously controlled shop floor in comparison to discrete-event simulation, finding out SD does not require much programming effort to implement autonomous control strategies in the model and offers a description of the logistic processes with high-level of aggregation. In their study, Trailer and Garsson [5] use System Dynamics to analyze public policy impacts on new ventures' growth rates, directly inserting into the model parameters representing the policy effects and varying their values for testing different scenarios. Sterman [2, 6] presents a series of practical applications of SD theory in occasions such as vehicle leasing, epidemics spreading, and technology adoption, among many others.

Regarding elements within the Circular Economy's perspective, Wang et al. [7] apply the theory to assess the impacts of subsidy policies on recycling and remanufacturing of auto parts in Chinese territory, offering a bunch of examples of policy types and arriving to the conclusion that combining different policies provides better results to the system under analysis. Poles [8] models remanufacturing under System Dynamics to evaluate strategies aimed at improving a production system. Zamudio-Ramirez [9] investigates the economic aspects related to automobile recycling in the United States of America using SD, the same country analyzed by Taylor [10], who employs the approach on the paper industry, including both forward and reverse flows, and discovers that sending more paper to recovery pathways does not guarantee an increase in paper reuse for new paper production. Moreover, Dong et al. [11] develop a model to comprehend the impacts generated by regulations focused on cleaner production in the context of the Chinese electroplating industry. At last, Georgiadis and Vlachos [12] use System Dynamics to assess decision making in the context of reverse logistics, and in Vlachos et al. [13] they adopt it for studying remanufacturing capacity planning in a closed-loop supply chain situation.

3   Rationale of the Work

The quantitative model is used to assess the effect of different scenarios of policy intervention on the adoption of recycled composites. The problem definition lies in the untapped value of disposed composites, while a circular alternative exists in the de-manufacturing of scrap and end-of-life (EoL) composites to produce recycled composites (rComposites). Perusing and promoting the circular route entails savings in raw materials and energy. To achieve a higher market share of rComposites, the SD model should capture the market dynamics of the composites sector while scenarios are built to showcase the long-term effects of different policies. In this section, the Causal Loop Diagram (CLD) is presented as a method of visualizing the interrelations between different variables in the system, embodying the mental model of causal relations. Next, the Stock and Flow map is presented and elements within are elaborated. The map depicts the dynamics of the technical system, while scenarios are simulated by manipulating proxy parameters that disturb the technical system in accordance with the CLD connections.

Causal Loop Diagram

The CLD, captured in Fig. 1, illustrates the causal relations between the technical system and the regulatory environment. CLDs are composed of variables that are interconnected by arrows with defined polarities that can be positive or negative. The arrows represent the causal links (cause-effect relations) between variables, while their polarity displays whether they vary in the same direction. Positive links mean the derivative of the dependent variable (effect) with respect to the independent (cause) is positive, or that the independent adds to the dependent. Negative links, on the other hand, means that the derivative is negative or that the independent variable subtracts from the dependent.

The technical system is composed of two subsystems: production and demand. Starting with the *demand subsystem* (Fig. 2), the expected demand of composites effects the production of virgin composites which satisfies the bulk of said demand. Yet, an increase in composites demand -an exogenous factor- also triggers the adoption of rComposites, which causes a positive effect on the rComposites demand and consequently reduces the production of virgin composites. The demand subsystem is not only connected to the production subsystem through the *production of virgin composites* variable, but also through connecting the demand of rCompositets to the de-manufacturing rate, creating the *demand feedback loop*. Feedback loops can either be reinforcing i.e., nurturing a certain behavior, or balancing i.e., halting a behavior. Demand feedback loop is a reinforcing one, meaning an increase in the de-manufacturing rate of rComposites will cause an increase of in-use rComposites thus higher expectations of rComposites demand which, after a time delay, leads to higher de-manufacturing rate.

Going into more details in the production system, depicted in Fig. 3, two more reinforcing feedback loops are noticed; namely: *composites circularity loop* and *scrap recycling loop*. The former captures the life cycle of recycled composites, since in-use composites after their useful life become EoL composites that in turn are collected and enter de-manufacturing process to become in-use composites once again. Collected composite materials have two pathways, either entering de-manufacturing processes or being controllably disposed. *Scrap recycling loop* tackles the other input source to de-manufacturing route, that is scrap from the production of both virgin and recycled composites. Collection of scrap composites reinforces collected composites which after processing and adoption become in-use composites. If the quality requirements are fulfilled, these reinforcing feedback loops sustain the input to de-manufacturing.

Production rates for both virgin and recycled composites are governed by resource constraints represented in lead times. While scrap rates are dependent on production

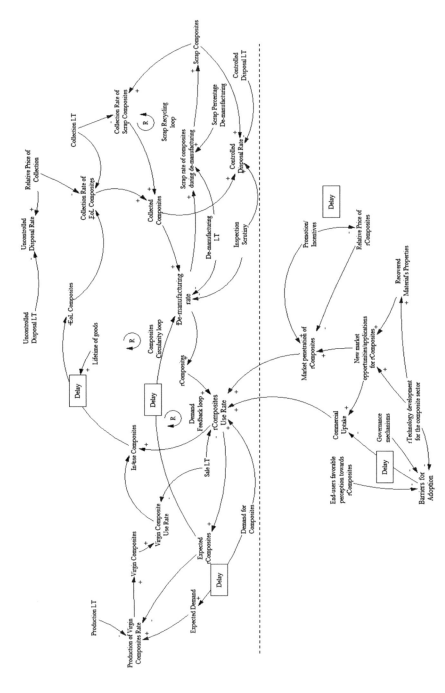

Fig. 1 Simplified causal loop diagram. Technical system upper half and regulatory environment bottom half. Developed using Vensim®

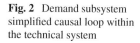

Fig. 2 Demand subsystem simplified causal loop within the technical system

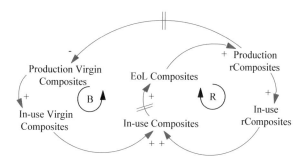

rates by considering the average rate of material scrapped during manufacturing and de-manufacturing activities represented by scrap percentage. Collection and disposal rates also respond to lead times, but as they present alternative routes, they depend also on a price comparison between the options captured *in relative price of collection*. Composite materials that are not collected are uncontrollably disposed, meaning they do not enter any circular route.

The second part of the CLD refers to the regulatory environment, which influences the technical system across different levels of analysis. Starting with the notable interlinks between policies and the technical system, these links serve two goals (1) setting the baseline for how the sector operates now, and (2) instigating disturbances to the technical system to create scenarios. Stand alone, the regulatory environment contains the subsystems *barriers for adoption, regulatory framework,* and *profitability of operations.* Barriers for adoption are affected by a set of different variables to paint a complete picture of the hurdles facing rComposite adoption, the effect is seen after a delay in the variable *commercial uptake.* This variable influences the technical system, particularly in the interlink with *rComposite use rate* in accordance with goal 2. While in the *regulatory framework* the effects are more related to goal 1, that is controlling the lead-times of different operations i.e., collection, disposal, and (de)manufacturing. For example, transportation regulations affect the collection lead times in the technical system, while EoL and extended producer responsibility (EPR) regulations influence the decision point where flow diverges into disposal and collection, i.e., *relative price of collection.* As for the profitability of operations, the subsystem captures the effect of technological advancement on the resource constraints of rComposite production, as well as connecting the different policies to *relative price of rComposites* variable.

Elaborating on the proxy variables rooting from the regulatory environment and influencing the technical system, Fig. 4 demonstrates the causes tree of *market penetration of rComposites* variable. The variable can be directly influenced by promotion activities, yet it is also subject to the available applications for rComposites, as well as the price difference between recycled and virgin composites. Each of these two causes have their own causes tree. For example, *relative price of rComposite* is influenced by a policy that affects the price of virgin composites, as well as another policy that affects the price of rComposites. While applications of rComposites are constrained by the quality of collected composites, and the recycling technology

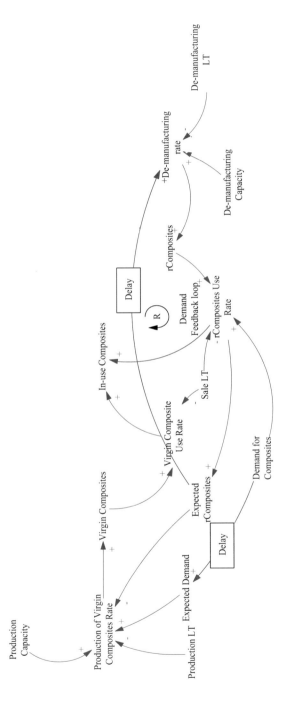

Fig. 3 Production subsystem causal loop in the technical system

used. Different causes that lead to changes in these proxy variables are formalized in the scenarios' definition section.

Stock and flow map

The stock and flow map provides the detailing of the model structure. In this subsection, elements are differentiated between stocks, flows, and other parameters. This map constitutes the quantitative model proposed. To capture policy effects, the time frame of the simulation needs to be higher than the lifecycle of composites. Since the composites use phase does not exceed 25 years, the model was assigned a time frame of 30 years. As previously mentioned, the logic behind this model is based on the existence of decision points which directs the flow of waste, along with proxy variables that mimic the effects of particular policies on the system. To capture realistic dynamics of the system, certain assumptions were necessary to narrow the focus of data collection and later on scenarios' definition. Particularly, the model tackles the carbon fiber market as a representative of FRP sector, and recycling as a representative of de-manufacturing activities. Detailing the assumptions will be discussed in the methodology section. Figure 5 captures the stock and flow model developed using AnyLogic software.

Systems are a network of stocks, flows and the information exchanged between them, through which stocks alter the flows. Stocks and flows can be represented by either diagramming or mathematics. Mathematically, stocks are the integrations over

Fig. 4 Market penetration causes tree

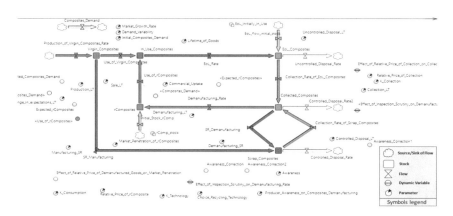

Fig. 5 Stock and flow map. Developed using AnyLogic®

time of their flows, consequently flows are represented as the derivative of stocks over time. In diagramming, convention dictates that stocks are described with rectangles, while arrows, or pipes, depict flows. Additionally, flows contain valves to highlight the presence of regulators that provide control over them. Table 1 details the stock and flow elements within the simulation environment. The distinction between stocks and flows can be noticed by checking the unit of measurement. Since stocks represent cumulation of flow, the measurement is done in quantities such as [tons]. While flows require additional information to convey their meaning, such as time. Thus, flows' measurement can be quantities per time unit such as [tons/week].

This model mostly relies on a state dependent system approach, which means changes in the state of stocks depends on inflows and outflows. However, other exogenous variables and constants can play a role in altering the interactions between stocks and flows. Including these additional parameters in the modelling environment is done by using the elements *dynamic variables* and *parameters*. *Dynamic variables* can be constant or dynamic, while *parameters* are strictly static. Table 2 lists these elements, along with one instance of a *table function* that serves as an input to the model by returning argument of composites demand values per year.

4   Methodology

Following the introduction to the model and explanation of elements within, this section focuses on the applied methodology. Since the output is a rank-order of scenarios, it is important to highlight the assumptions and mathematical formulation of the model. Scenarios are run as experiments with the response representing the market share of recycled composites. The metric *used_rcomposites* calculates the cumulation of in-use recycled composites circulating the system with a weekly instance. This is done by assigning a stock to the flow *use_of_rComposites*, the series is stored for each experiment. This approach is replicated for the *use_of_virgin_composites*. Market share is calculated by dividing the *used_rComposites* at the end of the simulation run by the total used composites. Equations (1–3) elaborate the formulation of the Market share of rComposites dependent variable.

$$\text{Market Share} = \frac{\text{Used rComposites}}{Used\ Virgin\ composites + \text{Used rComposites}} \tag{1}$$

$$Used\_Virgin\_Composites = \int_0^t Use\_of\_Virgin\_Composites \tag{2}$$

$$Used\_rComposites = \int_0^t Use\_of\_rComposites \tag{3}$$

Table 1 Stock and flow elements

Type of element	Name	Unit of measure	Description
Stock	Collected_Composites	[tons]	Represents the amount of waste composite material collected in the system
	EoL_Composites	[tons]	Represents the amount of composite material ending their use life and entering the end-of-life stage
	In_Use_Composites	[tons]	Represents the amount of composite material currently being used in applications by the market
	Scrap_Composites	[tons]	Represents the amount of composite material rejected either prior to or during processing
	Used_rComposites	[tons]	Represents the accumulated amount of recovered composites used by the system
	Used_Virgin_Composites	[tons]	Represents the accumulated amount of virgin composites used by the system
	Virgin_Composites	[tons]	Represents the amount of virgin composites available for the market
	rComposites	[tons]	Represents the quantity of recovered composites available for the market
Flow	Collection_Rate_of_EoL_Composites	[tons/week]	Represents the rate at which composites in the EoL phase are collected
	Collection_Rate_of_Scrap_Composites	[tons/week]	Represents the rate at which scrap composite materials are collected
	Composites_Demand	[tons/week]	Represents the market's demand of composite material
	Controlled_Disposal_Rate	[tons/week]	Represents the rate at which scrap composites are sent to disposal pathways

(continued)

Table 1 (continued)

Type of element	Name	Unit of measure	Description
	Controlled_Disposal_Rate2	[tons/week]	Represents the rateatwhich collected composites are sent to disposal pathways
	Demanufacturing_Rate	[tons/week]	Represents the processing rate of de-manufacturing activities
	EoL_flow_initial_stock	[tons/week]	Represents the rate of composite material known to be already in use and reaching EoL state
	EoL_Rate	[tons/week]	Represents the rate of composite material entering the market reaching the EoL state
	Production_of_Virgin_Composites_Rate	[tons/week]	Represents the rate of production of virgin composites
	SR_Demanufacturing	[tons/week]	Represents the rate of composite material scrapped duringde-manufacturing processes
	SR_Manufacturing	[tons/week]	Represents the rate of composite material scrapped during virgin manufacturing
	Uncontrolled_Disposal_Rate	[tons/week]	Represents the rate at which EoL composite material is discarded incorrectly
	Use_of_rComposites	[tons/week]	Represents therateatwhich recovered composites are employed by the market in applications
	Use_of_Virgin_Composites	[tons/week]	Represents the rate at which virgin composites are employed by the market in applications

Table 2 Dynamic variables and additional parameters

Type of element	Name	Unit of measure	Description
Dynamic variable	Effect_of_Relative_Price_of_Collection_on_Collection_Rate_of_EoL_Composites	N/A	Represents the impact of the relative price of collection on the rate of collection of EoL composites
	Effect_of_Inspection_Scrutiny_on_Demanufacturing_Rate	N/A	Represents the impact of the inspection scrutiny on the flow of collected composites into the de-manufacturing route
	Effect_of_Relative_Price_of_Demanufactured_Goods_on_Market_Penetration	N/A	Represents the impact of the relative price of the de-manufactured goods on their utilization by the market
	Expected_Composites_Demand	[tons/week]	Represents the volume of composites players expected to be demanded at a given moment
	Expected_rComposites	[tons/week]	Represents the volume of recovered composites expected to be entering the market at a given moment
Parameter	Awareness	N/A	Element that determines whether there is awareness aboutde-manufacturing pathways
	Change_in_expectations_LT	[weeks]	Represents the average delay for expectations to change in the face of new evidence

(continued)

Table 2 (continued)

Type of element	Name	Unit of measure	Description
	Collection_LT	[weeks]	Represents the average time taken to arrange and execute activities related to the collection of products
	Commercial_Uptake	N/A	Represents the share of the market willing to embrace de-manufactured products
	Controlled_Disposal_LT	[weeks]	Represents the average time taken to send materials to disposal pathways
	Demand_Variability	N/A	Represents the amplitude of random variations in demand
	Demanufacturing_LT	[weeks]	Represents the average time taken to perform the whole de-manufacturing process
	Demanufacturing_SR	N/A	Represents the share of material rejected by de-manufacturing processes
	Initial_Composites_Demand	[tons/year]	Represents the value of the yearly demand of composite materials at the simulation start time
	K_Collection	N/A	Measure of stakeholders' price sensitivity regarding collection activities

(continued)

Table 2 (continued)

Type of element	Name	Unit of measure	Description
	K_Consumption	N/A	Measure of stakeholders' price sensitivity on the consumption of FRP
	K_Technology	N/A	Represents the sensitivity to the input scrap to the recycling technology used
	Lifetime_of_Goods	[years]	Represents the average duration of a single use cycle of composite products
	Manufacturing_SR	N/A	Represents the share of input lost in the form of scrap by manufacturing processes
	Market_Growth_Rate	N/A	Represents the average yearly growth rate of the market's demand for composites during the simulation period
	Market_Penetration_of_rComposites	N/A	Represents the extent to which recovered composites can be employed in the applications of composites
	Accepted_EoL_Composites	N/A	Represents the percentage of collected composites that enter the de-manufacturing route

(continued)

Table 2 (continued)

Type of element	Name	Unit of measure	Description
	Production_LT	[weeks]	Represents the average time taken to perform the entire production process of composite materials
	Relative_Price_of_Collection	N/A	Represents the ratio between the cost of collecting and the cost of disposing EoL composites
	Relative_Price_of_rComposite	N/A	Represents the ratio between the price of a recovered and that of a virgin composite
	Sale_LT	[weeks]	Represents the average time taken to make and organize the activities related to the sale of products
	Uncontrolled_Disposal_LT	[weeks]	Represents the average time taken to get rid of EoL composite materials
Table function	EoL_Initially_in_Use	[tons/year]	Represents the yearly flow of EoL composites initially in use by the market at the start of the simulation

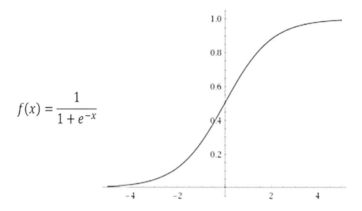

$$f(x) = \frac{1}{1 + e^{-x}}$$

Fig. 6 Sigmoid function and curve

As previously mentioned, the model depends on decision points that direct the flow of waste material either towards de-manufacturing activities or outside the system. The mathematical representation chosen for these decision rules is the sigmoid function, or S-shape curve, Fig. 6 shows the typical equation and corresponding sigmoid curve. The use of sigmoid function for decision rules is common in literature, Vlachos et al. [13] utilized it for reverse logistics decisions.

In this work, sigmoid functions are employed on 3 instances. (A) reverse logistics decision point of EoL composites collection, (B) reverse logistics decision point of waste disposal, and (C) market penetration decision point for rComposites. The regulation of flows out of the decision points in the stock and flow map is performed by adding an effect variable. This can be seen in the sigmoid formulation of dynamic variables:

A. *Effect_of_Relative_Price_of_Collection_on_Collection_Rate_of_EoL _Composites* represents the end user's decision whether to send the EoL composites to collection routes or just discard them by sending them to landfills. The model assumes consumers base their choice on economic factors, with a proxy of relative price of collection. The effect is adjusted by including a price sensitivity parameter (*K_Collection*), Eqs. (4) and (5) show this mathematical formulation.

$$\text{Effect\_of\_Relative\_Price\_of\_Collection\_on\_Collection\_Rate} \\ \_\text{of\_EoL\_Composites} = \frac{1}{(1 + e^{-K\_\text{Collection}*\text{Relative\_Price\_of\_Collection}})} \quad (4)$$

$$\text{Relative\_Price\_of\_Collection} = \frac{\text{Cost\_of\_Collection}}{\text{Cost\_of\_Disposal}} \quad (5)$$

B. *Effect_of_Inspection_Scrutiny_on_Demanufacturing_Rate* represents the de-
    manufacturers decision whether to accept the collected composites and allow
    them to enter de-manufacturing processing or reject them to be sent to appro-
    priate disposal. Like the previous variable, the effect is adjusted by a sensi-
    tivity parameter (*K_Technology*) that signifies the sensitivity of the recycling
    technology used to the input waste mix.
C. *Effect_of_Relative_Price_of_Demanufactured_Goods_on_Market_Penetration*
    presents composites exiting de-manufacturing processes with a range of market
    penetration according to their price in comparison to the price of newly manu-
    factured FRP. If these materials are cheaper than their virgin counterparts, there
    will be an incentive for their adoption. The effect is adjusted by multiplying a
    parameter of price sensitivity of users (*K_Consumption*) to the relative price of
    recycled composites (price of recycled composites/price of virgin composites).

Time delays are incorporated in this model as another source of dynamic behavior
in the system, along with the previously mentioned loops. In SD theory, delays
can either create oscillation and instability, or alternatively filter out noise. Time
delays are elements whose output laggardly trails the input, hence, inside every
delay there is an embedded stock, in which the difference between the output and
the input accumulates. There are 2 types of delays: material delays and information
delays. Material delays refer to delays to physical flow of materials, which entails
conservation of flow, i.e., elements leaving the delay stock can only do so if they
previously entered it. While information delays portray the progressive adjustment
of opinions and inferences based on the observation of current facts. In this case,
the stock is the belief itself, altered by the new information received, and there is no
conservation of flows involved. Equations (6) and (7) respectively show the general
formulation of material delay and the first order delay equation. While Eqs. (8) and
(9) showcase the formulation for information delay in a generalized form and first
order delay equation.

$$\text{Outflow}(t) = \text{Inflow}(t - \text{Average Residence Time}) \qquad (6)$$

$$\text{Outflow}(t) = \frac{\text{Accumulated inflow}}{\text{Average residence time}} \qquad (7)$$

$$\text{Reported Value}(t) = \text{Observed Value}(t - \text{Reporting Delay}) \qquad (8)$$

$$\text{Change in perceived value} = \frac{\text{Reported Value} - \text{Perceived Value}}{\text{Adjusted time}} \qquad (9)$$

The simulation baseline focuses on the market of carbon fiber reinforced polymers (CFRP), while the recycling technology used for de-manufacturing activities
is thermal recycling. Thermal recycling of CFRP is a viable economic solution for
industrial scale recycling. At the end of the simulation period (30 years) the market

share of rComposites is **5.8%**. Table 3 details the parameters' values for the baseline along with references when applicable.

Following the baseline run, selected parameters are manipulated to infer policy intervention. For each experiment, the parameters are assigned value ranges and step variations in accordance with the CLD and inputs from industry experts within the FiberEUse project. Table 4 illustrates the parameters' setting for different simulation scenarios. In *New Applications*, it can be noted that 2 scenarios are defined under it. The first refers to a scenario where the new applications for rComposites are coupled with an increase of accepted composites. While the second conservatively tackles just the increase in market penetration linked to broadening the use of rComposites in the CFRP sector. In *Technological Development*, the advancement of thermal recycling technologies -meaning more efficient operations- can lower the price of rComposites and increase the quality of rComposites which entails higher market penetration. In *Increase of perception*, awareness activities on the quality of rComposites directly affects the consumer uptake, while incentives to promote recycled components means intentionally lowering the price of rComposites. Whereas in *Trade Barriers*, the price of the virgin composites increases due to tariffs imposed, which also leads to an increase in production lead times. Finally in *Risk Aversion*, increasing the de-manufacturing rate by relieving the inherent risk in production of rComposites is related to an overall increase in rComposites stock.

5   Numerical Results

In this section, each scenario is analyzed and its corresponding effect to the market share of rComposites is reported. The scenarios are ordered in descending fashion relating to effect on the response parameter. Table 5 summarizes the scenarios, effect in increased market share, proxy parameters manipulated, and policy interpterion.

Elaborating on the numerical results, each scenario shall be presented in the following format: (i) tabulated presentation of parameters' step variations with their corresponding market share values at the end of simulation, and (ii) line chart for each setting with a weekly instance. Further, the trend of the results will be analyzed and tied to policy actions.

I.   **New Applications Scenarios**

Market penetration represents the extent to which recycled composites can be employed in different applications. The range of increase reported in Table 6 leads to the highest adoption effect, particularly if coupled with allowing more collected composites to enter the circular route. The effect of increasing the accepted input stream to de-manufacturing comes into play when coupled with a high market penetration. The takeaway here is that expanding the range of rComposites applications comes as a priority to unlock positive effects of improved collection rates and recoverable quality.

Table 3 Simulation baseline parameters values

Parameter	Value	Unit	Reference
Awareness	True	N/A	Assumption that composites sector is aware of recycling option
Change_in_expectations_LT	1	[weeks]	
Collection_LT	0,5	[weeks]	
Commercial_Uptake	60%	N/A	
Controlled_Disposal_LT	0.5	[weeks]	
Demand_Variability	5%	N/A	Low variation assumed based on the study of Vlachos, Georgiadis and Iakovou [13]
Demanufacturing_LT	1	[weeks]	Assumed to be equal to virgin production lead time
Demanufacturing_SR	15%	N/A	Assumed to be equal to virgin manufacturing scrap rate
Initial_Composites_Demand	48,488	[tons/year]	Source: FiberEUse
K_Collection	2.5	N/A	
K_Consumption	2.5	N/A	
K_Technology	10	N/A	
Lifetime_of_Goods	20	[years]	Source: Lefeuvre et al. [14]
Manufacturing_SR	15%	N/A	
Market_Growth_Rate	4%	N/A	
Market_Penetration_of_rComposites	15%	N/A	Source: FiberEUse
Producer_Awareness_on_Composites_Demanufacturing	100%	N/A	Assumption that players are aware of recycling option
Production_LT	1	[weeks]	Source: Vlachos, Georgiadis and Iakovou [13]
Relative_Price_of_Collection	32%	N/A	
Accepted_EoL_Composites	70%	N/A	

(continued)

Table 3 (continued)

Parameter	Value	Unit	Reference
Relative_Price_of_rComposite	60%	N/A	Source: Vo Dong et al. [15] The cost of the recycling process was used as a proxy for the cost of de-manufacturing
Sale_LT	0.5	[weeks]	
Uncontrolled_Disposal_LT	0.2	[weeks]	

Table 4 Parameter's settings for each scenario

Scenario	Parameter	Value range	Step variation
New applications I	Ratio of accepted EoL composite	70–90%	20%
	Ratio of market penetration	15–60%	15%
New applications II	Ratio of market penetration	15–60%	15%
Technological development	Relative price rComposites	60–40%	−10%
	Ratio of market penetration	15–35%	20%
Increase of perception	Commercial uptake	60–100%	20%
	Relative price rComposites	60–40%	−10%
Trade barriers	Production lead time virgin composites	1–5	4
	Relative price rComposites	60–20%	−20%
Risk aversion	De-manufacturing rate	Expected demand + 1 to 1.2 tons	0.1

In Fig. 7, the simulation baseline is represented by the yellow line. Further, the use of rComposites values with low market penetration coincide atop each other represented by the blue line. The effects of increasing market penetration while maintaining the baseline ratio of accepted composites is represented by the gray line. Finally, the double effects of increasing input stream and increasing market penetration is represented by the green line.

Table 5 Rank order of scenarios and policy summary

	Scenario name	Parameter(s)	Policy summary	Market share (%)
1	New applications	– Accepted EoL composites – Market penetration	– EoL regulation – Incentives for cross-sectorial collaboration – Compulsory use of recycled material – Green company policy/PR/image building	17.1
2	New applications	– Market penetration		15.8
3	Technological development	– Relative price of rComposites – Market penetration	– R&D subsidy – EPR regulation – Economic viability	12.8
4	Increase of perception	– Commercial uptake – Relative price of rComposites	– Consumer awareness activities – Sales subsidy	9.9
5	Trade barriers	– Relative price of rComposites – Production LT virgin composites	– Tariffs against import – Shortage in availability	6.9
6	Risk aversion	– De-manufacturing rate	– Production subsidy – Increase in number of agents	6.6
7	Baseline			5.8

Table 6 New applications scenarios results

Market penetration	Accepted composites	Market share rComposites (%)
0.15	0.7	5.9
0.45		14.1
0.6		**15.8**
0.15	0.9	5.9
0.45		14.1
0.6		**17.1**

II. Technological Development Scenario

Investing in the development of recycling technologies can directly affect the price of rComposites, since the recycling operations will be more efficient. Furthermore, new technologies can widen the scope of applications that rComposites can enter, since the quality of rComposites are expected to increase with newer technologies. The range of price decrease, reported in Table 7, can also be attributed to extended producer responsibility regulations. Since sorting activities of collected composites are responsible for a big portion of recycling cost. Thus, having a clean stream of waste would decrease the cost of de-manufacturing and consequently price of rComposites.

Fig. 7 New applications results trend for different settings

Table 7 Technological development scenario results	Relative price	Market penetration	Market share rComposites (%)
	0.6	0.15	5.87
	0.5		6.19
	0.4		6.47
	0.6	0.35	11.76
	0.5		12.31
	0.4		**12.78**

In Fig. 8, the market penetration changes lead to the creation of two groups. First the increase in market penetration coupled with decrease in rComposites price, represented by the green line. Then the gradual decrease of market share as the effect of price decrease is removed. The second group, represented by blue lines, show the case of just lowering the prices without an increase in market penetration.

III. **Increase in Perception Scenario**

Commercial uptake represents the willingness of FRP market to embrace recycled components in applications that rComposites can be utilized in. Increasing the commercial perception of recycled products has a direct effect on the rate of adoption and consequently the market share of rComposites. Awareness activities can play an important role in relieving the stigma on recycled options as inferior or lacking in quantities. Economic incentives are also essential to

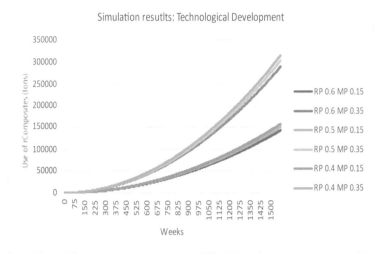

Fig. 8 Technological development results trend for different settings

encourage the sector to adopt recycled components, represented in Table 8 by relative price parameter.

In Fig. 9, the effect of increasing the commercial uptake parameter while decreasing the price of rComposites leads to higher adoption rate. It can be also noted that while decoupling the price parameter (green, dark blue and yellow lines) leads to comparatively lower market share, yet the effect of increasing uptake remains more significant.

IV. **Trade Barriers Scenario**

Trade barriers scenario tackles the volatility of raw material imports coming to the EU, these imports are crucial for the production of virgin composites. Imposing tariffs would lead to an increase of the price of virgin composites, represented in the parameter relative price. It can be noted that the range of decrease in this scenario is bigger to convey the price gap between virgin and recycled composites. Additionally, delays in shipment and other import hurdles

Table 8 Increase in perception scenario results

Commercial uptake	Relative price	Market share rComposites (%)
0.6	0.6	5.9
0.8		7.5
1		9.0
0.6	0.4	6.5
0.8		8.3
1		**9.9**

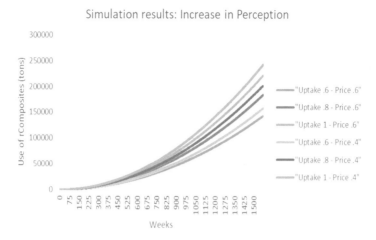

Fig. 9 Increase in perception results trend for different settings

would cause an increase in the production lead time for virgin composites manufacturing. Table 9 highlights the consequent increase in rComposites market share.

In Fig. 10, the effects of increased lead time and lower relative price are depicted. It is worth noting that on its own, the increase of production lead time did not yield significant change to the baseline; the parameter is mentioned for logical consistency. In order to reach significant increase in rComposites adoption, lead time increase should have been much higher which would not have conveyed a realistic market behavior. The effects of trade barriers are thus shown to be short-term, hence its low rank.

V. **Risk Aversion for De-manufacturers Scenario**

The production rate of de-manufacturing is responsible for the availability of rComposites in market, and consequently their adoption rate. This scenario envisions the effect of increasing the recycled composites inventory, as a sector, as a factor of de-manufacturers risk aversion. Table 10 shows how with small changes in production policies -stemming from higher risk tolerance -leads to high increase in market share of rComposites. From a regulatory perspective,

Table 9 Trade barriers scenario results

Production LT virgin composites	Relative price	Market share rComposites (%)
5	0.6	5.88
	0.4	6.49
	0.2	**6.91**

Fig. 10 Trade barriers results trend for different settings

Table 10 Risk aversion scenario results

De-manufacturers risk	Market share rComposites (%)
Base line	5.87
Expected demand + 1.1*t*	6.35
Expected demand + 1.2*t*	**6.58**

subsidizing recycling activities would encourage de-manufacturers to increase production leading to a boom in the rComposites sector. Another interpretation of aggressive production when compared to expected demand would be an increase in total number of players or de-manufacturers (Fig. 11).

6 Conclusion

In this Chapter, the impact of different policy actions on the deployment of the circular value-chain for composites has been evaluated through a mathematical approach. In particular, a model based on System Dynamics have been developed. The first step has been the creation of the Causal Loop Diagram to identify the interrelations among different actors in the composite sector. In the second step the dynamics of the system have been depicted through the use of Stock and Flow map. Finally, this methodology has been applied to different selected scenarios to identify the prioritization order of policies affected those specific scenarios. Table 11 summarizes policy actions against the level of impact on rComposites market share.

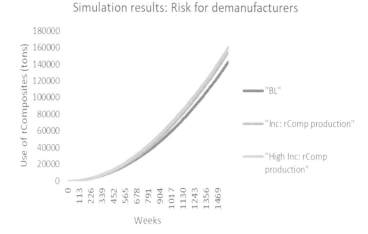

Fig. 11 Risk for de-manufacturers results trend for different settings

Table 11 Summary of policy actions

Policy actions	Level of impact
Creation of a unified legislation at EU level	High
Incentives for cross-sectorial collaboration	High
Force the reuse/recycling of composite materials (as in automotive sector)	High
Green trademarks and incentives for virtuous companies	High
R&D subsides and incentives	Medium
EPR regulation	Medium
Economic viability	Medium
Consumers' awareness activities	Medium
Sales subsidy for products embedding recycled materials	Medium
Tariffs against import of virgin materials	Low
Face shortage in availability	Low
Production subsidies	Low
Favor the increase in number of recyclers	Low
Tighten regulation on landfilling (till to prohibition)	Low
Facilitate cross-boundary transportation	Low

References

1. Forrester, J.W.: Industrial Dynamics. MIT Press, Cambridge (1961)
2. Sterman, J.D.: System dynamics: systems thinking and modeling for a complex world. In: ESD Internal Symposium (2002)
3. Nassehi, A., Colledani, M.: A multi-method simulation approach for evaluating the effect of the interaction of customer behaviour and enterprise strategy on economic viability of remanufacturing. CIRP Ann. Manuf. Technol. **67**, 33–36 (2018)
4. Scholz-Reiter, B., Freitag, M., De Beer, C., Jagalski, T.: Modelling dynamics of autonomous logistic processes: discrete-event versus continuous approaches. CIRP Ann. Manuf. Technol. **54**, 413–416 (2005)

5. Trailer, J.W., Garsson, K.: A system dynamics approach to assessing public policy impact on the sustainable growth rate of new ventures. New Engl. J. Entrepreneur. **8**(4) (2005)
6. Sterman, J.D.: Business Dynamics: Systems Thinking and Modeling for a Complex World. Irwin/McGraw-Hill, Boston (2000)
7. Wang, Y., Chang, X., Chen, Z., Zhong, Y., Fan, T.: Impact of subsidy policies on recycling and remanufacturing using system dynamics methodology: a case of auto parts in China. J. Clean. Prod. **74**, 161–171 (2014)
8. Poles, R.: System dynamics modelling of a production and inventory system for remanufacturing to evaluate system improvement strategies. Int. J. Prod. Econ. **144**, 189–199 (2013)
9. Zamudio-Ramirez, P.: Economics of automobile recycling. Massachusetts Institute of Technology (1996)
10. Taylor, H.F.: Modeling paper material flows and recycling in the US macroeconomy. Massachusetts Institute of Technology (1999)
11. Dong, X., Li, C., Li, K., Huang, W., Wang, J., Liao, R.: Application of a system dynamics approach for assessment of the impact of regulations on cleaner production in the electroplating industry in China. J. Clean. Prod. **20**, 72–81 (2012)
12. Georgiadis, P., Vlachos, D.: Decision making in reverse logistics using system dynamics. Yugoslav J. Oper. Res. **14**(2), 259–272 (2004)
13. Vlachos, D., Georgiadis, P., Iakovou, E.: A system dynamics model for dynamic capacity planning of remanufacturing in closed-loop supply chains. Comput. Oper. Res. **34**, 367–394 (2007)
14. Lefeuvre, A., Garnier, S., Jacquemin, L., Pillain, B., Sonnemann, G.: Anticipating in-use stocks of carbon fiber reinforced polymers and related waste flows generated by the commercial aeronautical sector until 2050. Resour. Conserv. Recycl. **125**, 264–272 (2017)
15. Vo Dong, P.A., Azarro-Pantel, C., Cadene, A.L.: Economic and environmental assessment of recovery and disposal pathways for CFRP waste management. Resour. Conserv. Recycl. **133**, 63–75 (2018)